AF864575

Introduction to Terahertz Electronics

Jae-Sung Rieh

Introduction to Terahertz Electronics

Springer

Jae-Sung Rieh
School of Electrical Engineering
Korea University
Seoul, Korea (Republic of)

ISBN 978-3-030-51841-7 ISBN 978-3-030-51842-4 (eBook)
https://doi.org/10.1007/978-3-030-51842-4

© Springer Nature Switzerland AG 2021
This work is subject to copyright. All rights are reserved by the Publisher, whether the whole or part of the material is concerned, specifically the rights of translation, reprinting, reuse of illustrations, recitation, broadcasting, reproduction on microfilms or in any other physical way, and transmission or information storage and retrieval, electronic adaptation, computer software, or by similar or dissimilar methodology now known or hereafter developed.
The use of general descriptive names, registered names, trademarks, service marks, etc. in this publication does not imply, even in the absence of a specific statement, that such names are exempt from the relevant protective laws and regulations and therefore free for general use.
The publisher, the authors, and the editors are safe to assume that the advice and information in this book are believed to be true and accurate at the date of publication. Neither the publisher nor the authors or the editors give a warranty, expressed or implied, with respect to the material contained herein or for any errors or omissions that may have been made. The publisher remains neutral with regard to jurisdictional claims in published maps and institutional affiliations.

This Springer imprint is published by the registered company Springer Nature Switzerland AG
The registered company address is: Gewerbestrasse 11, 6330 Cham, Switzerland

To my family: Sohee 소희*, Gyujin* 규진*, and Woojin* 우진

Preface

This book is based on the materials I prepared for a graduate-school-level course I have taught for several years in Korea University. The course was designed to provide the required knowledges for the students in the School of Electrical Engineering who are working on research topics around the THz electronics. For them, who are mostly involved in developing circuits and systems operating at frequencies beyond 100 GHz, quite a wide range of information and knowledge are on demand. They need high-frequency circuit design techniques combined with knowledges on microwave theories and device physics for their chip design. When chips are ready, they have to characterize and carry out experiments with them, definitely not a trivial task, for which the students are required to understand the propagation of THz beams in free space and waveguides. Often, they need to compare their results to or borrow ideas from THz vacuum electronic devices and optics-based THz components and systems. Also, they are expected to understand the basics of the application fields, such as spectroscopy, imaging, broadband communication, and radar, depending on their specific research topics. To prepare for the one-semester course that satisfies these needs, I had to collect information from many sources, such as technical papers, books, and, sometimes, from reliable online information, since there seemed no single textbook to serve this purpose. There are some textbooks intended to cover the general topic of THz, but I found them mostly written from the optics viewpoint. So, I decided to write one myself, which focuses on THz *electronics*.

As a glance at the table of content would suggest, this book is intended to provide the information as I listed above. After providing a brief overview of the THz band and the applications in Chap. 1, various THz signal sources and detectors are described in Chaps. 2 and 3. In Chap. 4, topics on THz propagation, such as Gaussian beam, antennas, and waveguides are introduced. As I believe that a basic knowledge on optics-based THz is still useful even for those working on THz electronics, the related basic concepts are highlighted in Chap. 5. Four selected THz applications are discussed in Chap. 6: THz spectroscopy, THz imaging, THz communication, and THz radar. As mentioned above, this book is mainly targeted at graduate students working on THz solid-state electronics. However, those professionals in the RF fields with electronics background, who are newly entering this emerging arena of THz electronics, will also find this book useful. I also hope this book will serve as a brief guide toward THz electronics for the researchers who are involved in THz vacuum electronic devices as well as those in the field of optics-based THz.

There was a question that hovered around me when I was about to embark on the task of writing, "Do we still need *books* as means to transfer knowledge?" We are living in a world where all the information is virtually a click away. We increasingly rely on online information with a higher level of graphic presentation, a tendency more obvious with young generations. So the question appeared legitimate from many accounts. Nevertheless, I managed to come up with justifications for me to move ahead. First, the value of information is greatly enhanced when they are *structured*. Structured information guides the readers through the relation between pieces of information. A book would serve this critical purpose around a focused topic. The readers may further develop a more detailed structure based on what is provided by the book. Second, the *interpretation* of information depends on who the author is. An analogy can be found in music. A piece of music can sound vastly different depending on who plays it, a consequence of different interpretations. The readers will perceive knowledges through the unique frame of interpretation provided by the author of the book. These couple of points encouraged me to drop the hesitation and move forward to writing.

It turned out that writing a book needed more devotion than I expected, requiring great patience and pain. The difficulty doubled when it was written in a language that is not my mother tongue. At the same time, however, the process provided me with lots of pleasure. It was a wonderful opportunity for me to accumulate a greater amount of knowledge and structure them in a systematic way. In a sense, I myself may be the first beneficiary of this book. With the book now released to public space, I wish it will benefit a wide range of students and researchers surrounding the topic of THz electronics.

Seoul, Korea Jae-Sung Rieh

Acknowledgments

This book would not have been completed in this form without the precious helps from so many intellectual and kind people around me. I first would like to thank the critical reviews on the early manuscripts provided by Prof. EunMi Choi (Ulsan National Institute of Science and Technology (UNIST)), Prof. Sanggeun Jeon (Korea University), Prof. Jinho Jeong (Sogang University), Prof. Byung-Sung Kim (Sungkyunkwan University), Prof. Moonil Kim (Korea University), Dr. Jung-Won Lee (Korea Astronomy and Space Science Institute (KASI)), Prof. Munkyo Seo (Sungkyunkwan University), Prof. Dongha Shim (Seoul National University of Science & Technology (SeoulTech)), Prof. Joo-Hiuk Son (University of Seoul), Prof. Ho-Jin Song (Pohang University of Science and Technology (POSTECH)), and Prof. Daekeun Yoon (National Chiao Tung University). The highly valuable comments by them definitely helped to improve the contents. I greatly appreciate my diligent students, Minje Cho, Doyoon Kim, Jungsoo Kim, Gihyun Lim, Heekang Son, and Junghwan Yoo, who provided great helps on the figures. I would also like to extend my gratitude to all my current and former students for extensive literature surveys reflected in the trend charts in the book, as well as the technical discussions in and out of the lab, which I enjoyed a lot. Further, I would like to take this opportunity to thank Prof. M. C. Frank Chang for having me stay at UCLA for my 1-year sabbatical, which allowed me a perfect environment to focus on writing, including an access to the engineering library located at the Boelter Hall. I remember that the peculiar scent emitted from the books sitting on the Boelter stacks somehow promoted the spirit of writing. About two-thirds of this book was written during my stay at the Westwood campus. Certainly, I appreciate Korea University for providing a comfortable yet highly stimulating environment for research. Finally, special thanks should go to my family: my wife Sohee and two kids, Gyujin and Woojin. They bore with me with great patience when I spent so much of my time at home writing, which otherwise would have been shared with them. Their understanding and encouragement have made this lengthy task successful in the end.

Seoul, Korea Jae-Sung Rieh

Contents

About the Author

Jae-Sung Rieh received the B.S. degree in electronics engineering from Seoul National University, Seoul, Korea, in 1991. He spent one and a half years in Korean Army for the mandatory military service before coming back to school for a graduate study in 1993. In 1995, he received the M.S. degree at Seoul National University, with a thesis entitled, "Fabrication of devices for HEMT MMIC's and design of low noise amplifiers." In the summer of 1995, he moved to the United States for a doctoral study, which resulted in his Ph.D. degree in electrical engineering from the University of Michigan, Ann Arbor, MI, USA, awarded in 1999. The dissertation title was "Development of SiGe/Si HBT's and their applications to microwave and optoelectronic monolithic integrated circuits."

In 1999, he joined IBM Microelectronics, Essex Junction, VT, USA, where he contributed to the development of 0.18 μm SiGe BiCMOS technology. In 2000, he moved down to IBM Semiconductor R&D Center, Hopewell Junction, NY, USA, where he was involved in the research and development activities for 0.13 μm SiGe BiCMOS technologies. In particular, he led the company efforts to develop the next-generation technology, which resulted in a 350-GHz SiGe HBT, the first Si-based transistor operating in the sub-millimeter wave band. It helped to push Si-based technologies toward THz electronics.

In 2004, he came back to Korea to pursue his long-desired career, teaching at university. He joined the School of Electrical Engineering, Korea University, located in Seoul, Korea, where he has been teaching and supervising students for more than 15 years. His major research interest lies in the mm-wave and terahertz devices and circuits. In 2012, he spent his first sabbatical at Submillimeter Wave Advanced Technology (SWAT) team in JPL, Pasadena, USA, which greatly promoted his interest in the THz research. For another sabbatical year spread over the 2018–2019 period, he was with High Speed Electronics Lab (HSEL) in UCLA, Los Angeles, USA, during which he started to work on his book *Introduction to THz Electronics*.

Prof. Rieh was a recipient of IBM Faculty Award (2004) and two-time co-recipients of IEEE EDS George E. Smith Awards (2002 and 2006). He was also co-awarded IEEE Microwave and Wireless Component Letters Tatsuo Itoh Best

Paper Award (2013). He has served as an Associate Editor of the IEEE Microwave and Wireless Components Letters (2006–2009) and the IEEE Transactions on Microwave Theory and Techniques (2010–2013). He is a Fellow of IEEE.

Chapter 1
Introduction

In this opening chapter, the background and basic concepts of terahertz electronics will be presented along with a brief overview of the topics to be covered throughout this book. A discussion on the definition of the terahertz (THz) band is first provided, followed by a review of various unique properties of the THz band that have attracted many researchers for scientific interests and practical applications. In the next section, a preview on the technical approaches to the generation and detection of THz waves is presented, which will serve as a quick reference to the readers before detailed descriptions are presented in the following chapters. In addition, a brief overview of the application areas of the THz band will be provided, which will be more detailed in Chap. 6 of this book.

1.1 Terahertz Band

A proper starting point of the discussion on the terahertz electronics would be the review of the definition of the "terahertz (THz) band." A clear definition of the THz band, however, does not exist as of today, and it varies over different authors and literatures. The lower boundary of the band falls on either 100 or 300 GHz, while the upper boundary hovers around 3 THz, 10 THz, and sometimes 30 THz. Although it is supposed to be a technical term, a proper definition may follow the rule of general languages in accepting terms, say, the popular usage by the users. A scan over the existing literature with a standard search engine reveals that "100 GHz–10 THz" is the most widely used definition with a reasonably large margin, followed by "100 GHz–3 THz" and "300 GHz–10 THz." With this usage pattern identified, "100 GHz–10 THz" may serve as a reasonable definition of the THz band, unless there arises a strong shift in the usage preference in the future. Hence, this definition will be used in this book. Another related term occasionally found in the literature is the sub-terahertz band. Its definition is as vague as that of the THz band, popular two versions being "100 GHz–1 THz" and "100 GHz–300 GHz." Again, based on the

© Springer Nature Switzerland AG 2021

J.-S. Rieh, *Introduction to Terahertz Electronics*,
https://doi.org/10.1007/978-3-030-51842-4_1

usage counts in the literature, the former can be accepted as the definition of the sub-terahertz band, literally indicating the band below 1 THz.

The discussion on the band definition naturally brings us to the definitions of other frequency bands surrounding the THz band. An overview of the bands outside the THz band may provide an informative backdrop, against which the relative position of the THz band can be better pictured from a macroscopic point of view. There are various frequency band notations used by radio frequency (RF) community. The three most prevailing ones are the ITU (International Telecommunication Union) frequency bands, IEEE (Institute of Electrical and Electronics Engineers) radar bands, and the waveguide bands.

The ITU band notation, shown in Table 1.1, divides the spectrum from 0.03 Hz up to 3 THz by a decade span and assigns band numbers and symbols [1]. Some of the symbols, such as VHF (very high frequency) and UHF (ultra high frequency), are familiar to the general public as those band names were adopted for the carriers of

Table 1.1 ITU band notation

Band number	Symbol	Frequency range	Wavelength range	Metric subdivision
−1		0.03–0.3 Hz	1,000,000–10,000,000 km	Gigametric waves
0		0.3–3 Hz	100,000–1,000,000 km	Hectomegametric waves
1		3–30 Hz	10,000–100,000 km	Decamegametric waves
2		30–300 Hz	1000–10,000 km	Megametric waves
3	ULF	300–3000 Hz	100–1000 km	Hectokilometric waves
4	VLF	3–30 kHz	10–100 km	Myriametric waves
5	LF	30–300 kHz	1–10 km	Kilometric waves
6	MF	300–3000 kHz	100–1000 m	Hectometric waves
7	HF	3–30 MHz	10–100 m	Decametric waves
8	VHF	30–300 MHz	1–10 m	Metric waves
9	UHF	300–3000 MHz	10–100 cm	Decimimetric waves
10	SHF	3–30 GHz	1–10 cm	Centimetric waves
11	EHF	30–300 GHz	1–10 mm	Millimeteric waves
12		300–3000 GHz	0.1–1 mm	Decimillimetric waves
13		3–30 THz	10–100 μm	Centimillimetric waves
14		30–300 THz	1–10 μm	Micrometric waves
15		300–3000 THz	0.1–1 μm	Decimicrometric waves

1. *Abbreviations*: *ULF* ultra low frequency, *VLF* very low frequency, *LF* low frequency, *MF* medium frequency, *HF* high frequency, *VHF* very high frequency, *UHF* ultra high frequency, *SHF* super high frequency, *EHF* extremely high frequency

2. Bands 3–15 are from ITU Radio Regulations [1]. Bands −1 to 2 are the extension proposed by URSI (International Union of Radio Science).

3. Band number N extends from 0.3×10^N to 3×10^N Hz.

4. Frequency lower limit is exclusive; upper limit is inclusive.

television broadcasts. Although some may feel that the symbol expressions are too generous (e.g., "ultra high" frequency represents a band of merely 300 MHz to 3 GHz), the nomenclature structure is highly systematic and covers the entire frequency range that would be of practical use from electronics point of view. It is also noteworthy that the origins of the widely adopted band definitions, millimeter-wave (30–300 GHz) and sub-millimeter-wave (300 GHz–3 THz), can be traced back to this ITU notation, corresponding to "millimetric waves" (Band 11) and "decimillimetric waves" (Band 12), with a slight modification in the names. Related to this, it can be pointed out that the definition of the THz band as discussed above ranges across the millimeter-wave and the sub-millimeter-wave bands.

IEEE band nomenclature, officially called "IEEE Standard Letter Designations for Radar-Frequency Bands" [2], is shown in Table 1.2. The band designations for frequencies lower than 1 GHz in this nomenclature were apparently borrowed from the ITU notation with a small change (the range for UHF band is truncated), while those for beyond 1 GHz inherit the notations from the radar community. The letter assignments for the bands, such as L, S, C, X, Ka, K, Ku, V, and W, do not appear to follow any obvious rule or pattern and may look arbitrary at first glance. As such, getting familiar with them has been a kind of ritual process that new microwave and radar engineers have to go through. In fact, however, they are not totally arbitrary, and a brief reference to their historical background would help readers to understand (and remember) the designation as introduced in [3]: The wavelength used for the early radars was 23 cm, while a shorter wavelength of 10 cm was later adopted, each leading to L (long) and S (short) symbols for the corresponding frequency bands. A fire control radar was later introduced, for which letter X indicates a spot marking, leading to X-band. C is for an intermediate band located in between, thus 'compromising' the advantages of S-band and X-band. K-band took the initial from a German word 'kurtz' (meaning 'short') indicating a higher frequency, while Ku and Ka represent K-under and K-above, respectively. V indicates "very"

Table 1.2 IEEE standard letter designations for radar frequency bands

Band designation	Frequency range	Comments
HF	3–30 MHz	
VHF	30–300 MHz	
UHF	300–1000 MHz	
L	1–2 GHz	L for "long" wave
S	2–4 GHz	S for "short" wave
C	4–8 GHz	C for "compromise" between S and X band
X	8–12 GHz	X for "cross" (or spot markings)
Ku	12–18 GHz	Ku for "kurz-under"
K	18–27 GHz	German "kurz" means short
Ka	27–40 GHz	Ka for "kurz-above"
V	40–75 GHz	V for "very" high frequency band
W	75–110 GHz	W follows V in the alphabet
mm	110–300 GHz	

high frequency and W is a letter that simply follows V, both of which were newly added in the revision of the IEEE standard made in 1984. With the current version of the IEEE standard, revised on 2002 [2], the frequency band beyond 110 GHz is designated simply as "mm," indicating the mm-wave. Hopefully, new designations may appear in the next revision and assign additional letter designations to expand the frequency range covered by this standard.

There are other letter-based band definitions widely used among microwave engineers, including Q-band, U-band, E-band, F-band, D-band, and G-band. They are not part of IEEE band standard described above. Instead, they are related to the notation system based on the rectangular waveguides, with one-to-one mapping on to various bands defined by the waveguide standard as shown in Table 1.3. The waveguide standard notation starts with "WR" (Waveguide Rectangular), followed by a number that indicates the broad-side width of the inner dimension of a rectangular waveguide, expressed in terms of hundredths of an inch (=10 mils). As will be described in detail in Sect. 4.3.1, the inner dimension of a metallic rectangular waveguide determines the operation frequency range of the waveguides as shown in Table 1.3 and thus appears in the band notation to indicate the range. One may notice that V-band and W-band, a part of IEEE band standard, are also included in this list, although the range for V-band in this notation (50–75 GHz) slightly differs from the V-band defined in the IEEE standard (40–75 GHz). There are a few other band designations beyond 220 GHz sparsely used by some such as H-, Y-, or J-bands, although their usage has been limited and not fully accepted yet.

Other definitions have also been used. Frequency bands that are frequently found in relation to AM broadcasting or amateur radio communication are short-wave (SW), medium-wave (MW), and long-wave (LW), as described in Table 1.4. These bands are roughly distributed over HF, MF, and LF of the ITU band notation. Note that most of the AM broadcasting systems are based on medium-wave (MW) from

Table 1.3 Band notation based on the rectangular waveguide standard

Band designation	Frequency range (GHz)	Waveguide standard
Q band	33–50	WR22
U band	40–60	WR19
V band	50–75	WR15
E band	60–90	WR12
W band	75–110	WR10
F band	90–140	WR8.0
D band	110–170	WR6.5
G band	140–220	WR5.1

Table 1.4 Frequency bands based on wavelength

Band designation	Frequency range
Long-wave (LW)	Roughly below 300 kHz
Medium-wave (MW)	526.5–1606.5 kHz (EU) 525–1705 kHz (USA)
Short-wave (SW)	1700 kHz–30 MHz

this definition, while some AM broadcasts are based on short-wave and long-wave. The term "microwave" is widely used as a general term indicating higher parts of radio frequencies, but the official definition of microwave band is the frequency range of 300 MHz–300 GHz [4]. It comprises UHF, SHF, and EHF of the ITU notation.

In the above, the frequency band definitions have been reviewed for radio frequencies. If we move our attention to the other side of the THz band, there are band definitions in the optical spectrum. One major difference encountered as we enter the domain of optics is that the electromagnetic bands are now described in terms of wavelength instead of frequency as was the case for radio frequencies. This probably arises from the fact that most optical phenomena are better described in terms of wavelength than frequency. In the discussion below, we will present the optical bands mainly in terms of wavelength to follow the convention, but the corresponding frequency ranges will also be provided as an effort to view the optics from radio frequency perspective. The optical band with the lowest frequency (longest wavelength) is the infrared band, which is sequentially followed by visible light, ultraviolet, X-ray, and γ-ray bands. The boundaries of each optical band, however, are not clearly defined and vary over the sources. You will notice that even for visible light band, the boundaries slightly vary depending on where you find the information. Table 1.5 lists the boundaries of the optical bands in terms of the frequency and wavelength following a standard college textbook [5]. The infrared band can be divided into smaller sub-bands, typically into the far-infrared (FIR), mid-infrared (MIR), and near-infrared (NIR) bands, although there are other ways of dividing the band. FIR is the region where optics and the traditional radio engineering encounter each other. It is often defined as 15 μm–1 mm (300 GHz–20 THz), while the upper boundary (15 μm) is sometimes set differently. By most of the definitions, FIR contains the sub-millimeter-wave band (300 GHz–3 THz) and in great part overlaps with the THz band (100 GHz–10 THz) as discussed earlier.

The band definitions introduced so far are based on either *frequency* or *wavelength*. However, electromagnetic waves can be described by other physical quantities as well. In some occasions, *wavenumber*, defined as an inverse of the wavelength, is used. The readers will notice that most of the spectroscopy data are presented in terms of wavenumber. *Photon energy* is another way to represent electromagnetic waves, which assumes that the energy of electromagnetic waves is quantized into photons. Photon energy $h\nu$ can be converted to *temperature* with a relation of $kT = h\nu$, where k and h are Boltzmann constant and Plank constant, respectively. In this case, the temperature T roughly corresponds to the temperature

Table 1.5 Optical frequency bands

Band designation	Frequency range	Wavelength range
Infrared	300 GHz–385 THz	1 mm–780 nm
Visible light	385 THz–770 THz	780–390 nm
Ultraviolet	770 THz–34 PHz	390–9 nm
X-ray	24 PHz–50 EHz	12.5–0.006 nm
γ-ray	>2.4 EHz	<0.125 nm

Table 1.6 Conversion between frequency and various physical parameters: (a) Shown in the decades of frequency, (b) Shown in the decades of wavelength

Frequency	Time period	Wavelength	Wavenumber	Photon energy	Temperature
(a)					
1 kHz	1 ms	300 km	0.0033 km^{-1}	4.14 peV	48 nK
10 kHz	100 μs	30 km	0.033 km^{-1}	41.4 peV	480 nK
100 kHz	10 μs	3 km	0.33 km^{-1}	414 peV	4.8 μK
1 MHz	1 μs	300 m	0.00003 cm^{-1}	4.14 neV	48 μK
10 MHz	100 ns	30 m	0.0003 cm^{-1}	41.4 neV	480 μK
100 MHz	10 ns	3 m	0.003 cm^{-1}	414 neV	4.8 mK
1 GHz	1 ns	30 cm	0.03 cm^{-1}	4.14 μeV	48 mK
10 GHz	100 ps	3 cm	0.33 cm^{-1}	41.4 μeV	480 mK
100 GHz	10 ps	3 mm	3.33 cm^{-1}	414 μeV	4.8 K
1 THz	1 ps	300 μm	33.3 cm^{-1}	4.14 meV	48 K
10 THz	100 fs	30 μm	333 cm^{-1}	41.4 meV	480 K
100 THz	10 fs	3 μm	3333 cm^{-1}	414 meV	4800 K
1 PHz	1 fs	300 nm	33,333 cm^{-1}	4.14 eV	48,000 K
(b)					
3 kHz	333 μs	100 km	0.01 km^{-1}	12.4 peV	144 nK
30 kHz	33.3 μs	10 km	0.1 km^{-1}	124 peV	1.44 μK
300 kHz	3.33 μs	1 km	1 km^{-1}	1.24 neV	14.4 μK
3 MHz	333 ns	100 m	0.0001 cm^{-1}	12.4 neV	144 μK
30 MHz	33.3 ns	10 m	0.001 cm^{-1}	124 neV	1.44 mK
300 MHz	3.33 ns	1 m	0.01 cm^{-1}	1.24 μeV	14.4 mK
3 GHz	333 ps	10 cm	0.1 cm^{-1}	12.4 μeV	144 mK
30 GHz	33.3 ps	1 cm	1 cm^{-1}	124 μeV	1.44 K
300 GHz	3.33 ps	1 mm	10 cm^{-1}	1.24 meV	14.4 K
3 THz	333 fs	100 μm	100 cm^{-1}	12.4 meV	144 K
30 THz	33.3 fs	10 μm	1000 cm^{-1}	124 meV	1440 K
300 THz	3.33 fs	1 μm	10,000 cm^{-1}	1.24 eV	14,400 K

of a blackbody that would emit photons with an energy around $h\nu$ (to be more precise, the most probable photon energy from a blackbody with a temperature of T is 2.8 kT rather than kT). The conversions between these parameters are summarized in Table 1.6, in terms of the decades of frequency and wavelength separately. For a quick reference, 1 THz corresponds to a wavelength of 300 μm, wavenumber of 33.3 cm^{-1}, photon energy of 4.14 meV, and temperature of 48 K.

1.2 Properties of THz Waves

One of the well-known properties of THz waves is the high absorption in water. The absorption spectra and the causes of the absorption vary over the state of water. Hence, the discussion needs to be made separately for three phases of water: water vapor, liquid water, and ice. For water vapor, the absorption results mostly from the transitions related to the rotational modes, the vibrational modes, and the electron states of the water molecules. The energy associated with the transitions between rotational modes is relatively small and corresponds to the THz band, leading to the absorption of THz waves, whereas vibration-mode and electron-state transitions affect shorter wavelength (higher frequency) regions. For liquid water, the hydrogen bonding between water molecules, which results from the shorter distances between water molecules than in vapor state, suppresses the rotations of the water molecules. Instead, the vibration modes are affected by the hydrogen bonding, and the relevant energy is shifted into the THz region, dominating the absorption in the THz band. For ice, the situation is similar to the liquid water case, with a stronger inter-molecular interaction by the hydrogen bonding. The water absorption spectra for all three water phases are shown in Fig. 1.1 [6]. It is interesting to note that the absorption profile for liquid water (and ice) does not show sharp peaks as shown for water vapor absorption, which is a result of the broadened peaks due to inter-molecular interaction in condensed state. Overall, it is obvious from the figure that the water absorption in the THz band by water, regardless its phase, is much stronger than in the adjacent bands, such as visible light and lower-frequency radio bands. Internal to the THz band, the absorption tends to increase with frequency. For

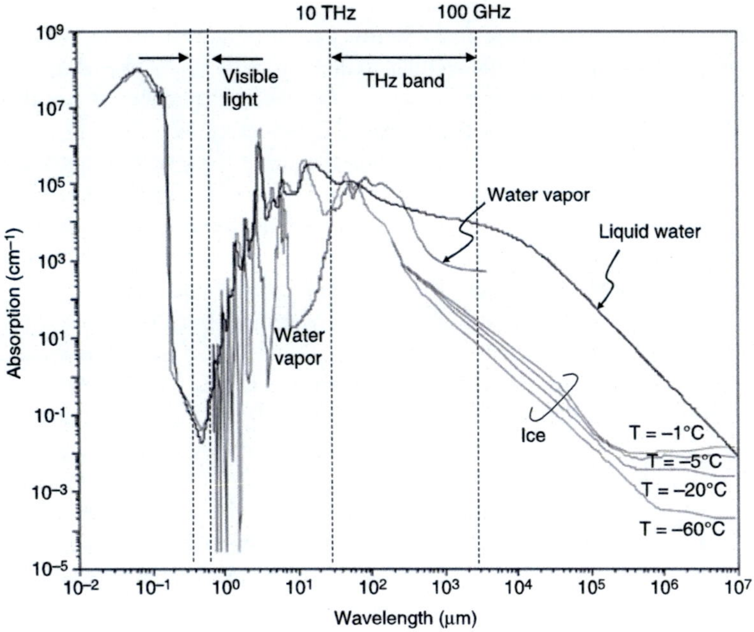

Fig. 1.1 Water absorption spectra for the three phases of water: water vapor, liquid water, and ice. (Adapted by permission from Springer Nature: [6] © 2007)

example, the absorption coefficient for liquid water at 10 THz is around 1000 cm^{-1}, while it is below 100 cm^{-1} at 100 GHz. The absorption levels by water vapor and ice inside the THz band show even a shaper increase with frequency as indicated by the profiles.

Another notable property is that various optically opaque dielectric materials, such as paper, clothe, leather, wood, and most of plastics and ceramics, are transparent to THz waves. One may be tempted to see what is behind such transparency observed with the listed materials. Obviously, the transparency of dielectric materials is closely linked to the absorption level of electromagnetic waves in the given material as represented by the absorption coefficient, although it is also partly related to the reflection and scattering. The frequency dependence of the absorption coefficient is related to so many macroscopic and microscopic properties of the material and cannot be discussed in a general way. The readers may recall the basic electromagnetic theory introduced in college classes, in which the absorption coefficient (α) of electromagnetic waves in a dielectric material with a finite loss is given as:

$$\alpha = \omega\sqrt{\frac{\mu\varepsilon}{2}\left[\sqrt{1+\left[\frac{\sigma}{\omega\varepsilon}\right]^2}-1\right]}, \tag{1.1}$$

where μ, ε, and σ are the permeability, permittivity, and conductivity of the material, respectively, and ω is the angular frequency. It shows a general tendency for α that increases with increasing frequency. However, the reality is not that simple and affected by lots of other factors as well. For example, μ, ε, and σ in the equation, which are rather macroscopic parameters, have their own frequency dependences, which will also affect the trend. More importantly, in the microscopic level, the transitions between energy states inside atoms, molecules, and lattices constituting the material would lead to a great number of absorption lines that are specific to the incident photon energy or electromagnetic wave frequency. This will also strongly affect the frequency dependence of the absorption. In fact, the frequency dependence of the water absorption shown in Fig. 1.1 is mostly dominated by this factor. With all these observations given, there seems no single theory that explains the behavior of THz waves in dielectric materials. Instead, all the factors mentioned above rather collectively affect the transparency observed with the materials listed above. As will be discussed in further detail later, the transparency of those materials is surely a highly favored property for various practical applications including imaging. It is fair to note, though, that such transparency may apply to the lower radio frequency bands as well, even with less absorption. However, the THz waves additionally benefit from the smaller wavelength, which is favored for higher resolution desired in imaging applications.

Reflection on metallic surfaces is also often cited as a major property of THz waves. The high reflectivity is useful to obtain a high contrast between metallic and non-metallic surfaces in imaging applications, among other advantages. As a matter of fact, the reflection on metallic surfaces, which results from the high free-electron density in metals, is a fundamental property of electromagnetic waves in general, not limited to the THz band. At this point, however, it would be worthwhile to briefly review the frequency-dependent behavior of the reflection near the metallic surface. The reflection of electromagnetic waves on metallic surfaces is, in fact, not a total reflection unless the conductivity of the metal is infinite. With a very large conductivity, α in Eq. (1.1) is reduced to $\sqrt{\omega\mu\sigma/2}$, which is basically the absorption coefficient of a conducting material. The inverse of this parameter, or $\sqrt{2/\omega\mu\sigma}$, has a dimension of length and is called the skin depth, which indicates the average depth of electromagnetic wave penetration below the metallic surface. To be more precise, it is the depth at which the field strength is reduced down to 36.8% (1/*e*) of the original value at the surface. It shows that the extent of the finite penetration of electromagnetic waves into metal depends on the frequency, which decreases with increasing frequency. Hence, it can be said that a more ideal reflection on metallic surfaces can be obtained for THz waves when compared with lower-frequency waves. We will come back to the issue of the skin depth in Sect. 4.3.2 with more practical situations.

An outstanding advantage of THz waves for imaging applications, particularly against the prevailing X-ray imaging, is that THz radiation is non-ionizing radiation, imposing little hazardous effects on human body when exposed. Ionizing radiation is a kind of radiation that carries photon energy large enough to detach electrons from atoms or molecules, leading to their ionization. An exposure of a human body to a high dose of ionizing radiation may cause damage to DNA, leading to increased possibility of developing cancer afterwards. When the energy is larger than ~10 eV, we say the radiation is ionizing. Typical X-ray imaging systems for medical purpose involves photon energies close to 100 keV. Hence, accumulated exposures to a high-dose X-ray radiation may be harmful for humans. In contrast, the energy corresponding to THz waves falls on the order of tens of meV at most, remote from the energy range of concerns. This clearly indicates that THz imaging is a safe imaging modality, especially when compared with X-ray imaging.

1.3 Approaches to THz Systems

The THz band is situated between the spectra covered by the traditional electronics and optics, both of which have been heavily explored for various science and engineering purposes. In contrast, the THz spectrum has been relatively less exploited so far. For this reason, the THz band is often referred to as the "THz gap." The relatively slow development of the THz band for applications can be attributed to at least a couple of major challenges. First, the attenuation level of THz waves in the Earth's atmosphere is high. As shown in Fig. 1.2 [7], the largest attenuation occurs for a range of 1–10 THz, the core portion of the THz band, which arises mostly from the strong absorption by water molecules in the atmosphere. The high attenuation will limit the long-distance propagation of the THz waves in the air. Second, we lack devices to properly manipulate THz waves, including generation and detection. This is symbolically revealed by Fig. 1.3 [8],

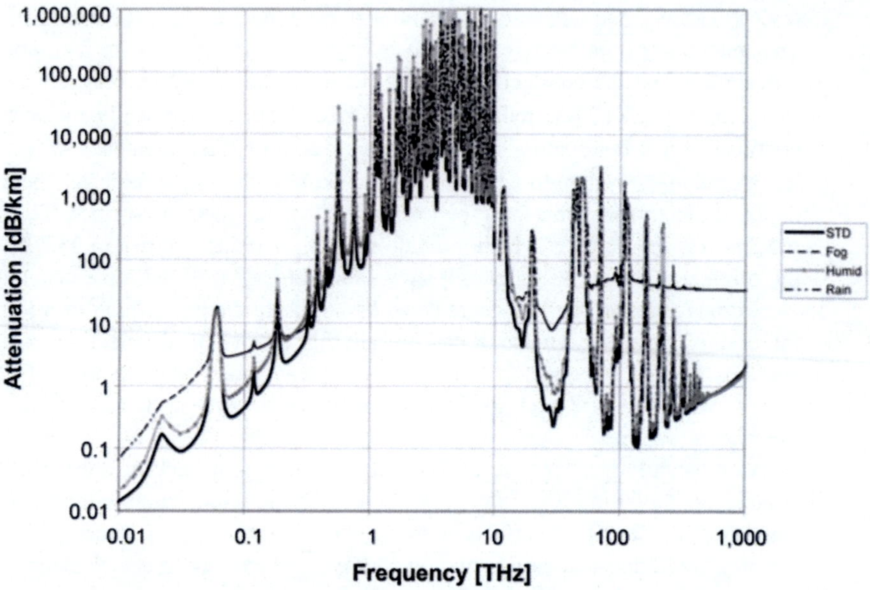

Fig. 1.2 Attenuation of electromagnetic waves in the atmosphere of Earth in various conditions. The largest attenuation occurs for 1–10 THz frequency range. (© 2007 IEEE [7])

Fig. 1.3 Output power of various solid-state signal sources, showing that the lowest output power occurs near 1 THz. (© 2005 IEEE [8])

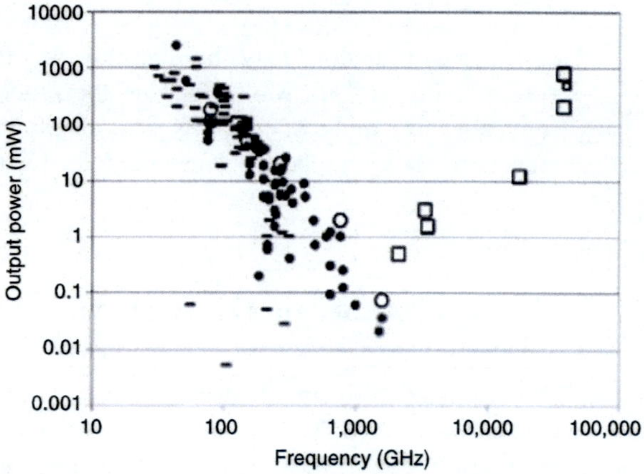

which shows the output power from solid-state signal sources as a function of frequency. When approached from low frequencies based on electronic devices, the output power decreases with increasing frequency, while the power decreases with decreasing frequency when approached from high frequencies with optical devices. As a result, there exists a region where neither approach provides a sufficient output, which is around the THz band as obviously indicated in the figure. While the first challenge is not something we can engineer, the second one can be overcome with technological advances. In fact, a great deal of advances are being

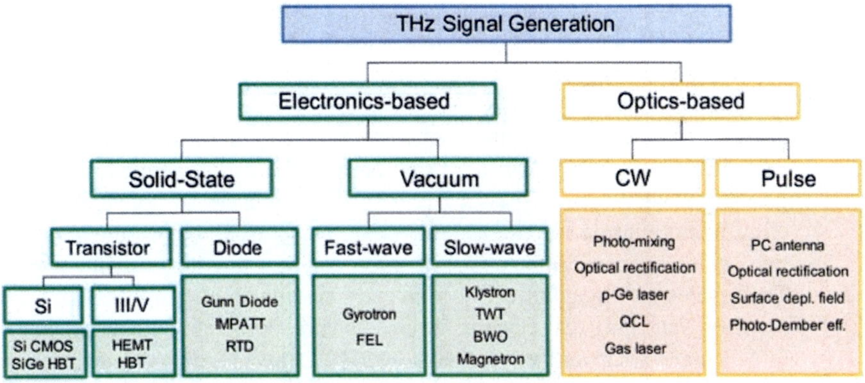

(a)

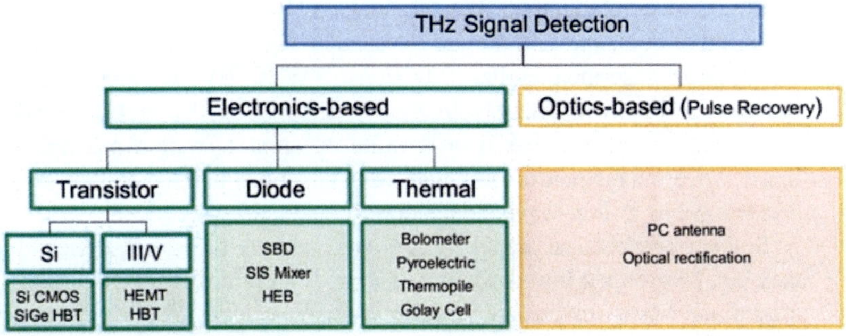

(b)

Fig. 1.4 Various techniques for THz systems. (**a**) Signal generation (**b**) Signal detection

made in device technologies, both electrical and optical, to narrow down the THz gap. Although the details will be discussed in Chaps. 2 and 3 (electrical methods) as well as in Chap. 5 (optical methods), it would be helpful to take an overview of the approaches to build up THz generation and detection systems. This section will guide the readers through various techniques to generate and detect THz signals, which are highlighted in Fig. 1.4.

1.3.1 Approaches to THz Signal Generation

The signal generation has long been a major bottleneck for exploiting the THz band, and, as such, extensive efforts have been made to develop THz sources with sufficient output power. Those efforts have been recently rewarded by some promising results, from both electrical and optical fronts, as will be briefly reviewed below.

Let us first start with the devices with the electrical approach. There are two distinct electrical approaches to generate THz signals, one based on vacuum devices and the other based on solid-state devices. THz signal generation with vacuum devices is basically an extension of the efforts to generate microwave signals with vacuum devices, which can be traced back to the early twentieth century. Signal generation with vacuum devices is basically a process of converting modulated electron current into electromagnetic wave radiation. The electron current modulation, which happens inside a vacuum tube, can be either longitudinal or transverse. The devices with longitudinal electron current modulation, a process that necessarily involves bunching of electron beam, include klystrons, traveling-wave tubes (TWTs), and backward wave oscillators (BWOs). Categorized as the devices with transverse modulation are gyrotrons and free electron lasers (FELs), the former with helical motion and the latter with wiggling motion of electrons. Inside these devices with transverse modulation, the phase velocity of the electromagnetic wave exceeds the speed of light, and thus they are also called fast-wave devices. On the other hand, the devices described earlier (klystrons, TWTs, BWOs) are called slow-wave devices as the wave phase velocity is smaller than the speed of light. Another vacuum device that is based on a quite different operation principle from those listed above is a magnetron, in which electrons travel in the radial direction. It can be categorized as a slow-wave device, too.

Solid-state electrical signal sources are typically based on semiconductors, and they can be grouped into diode-based sources and transistor-based sources. Widely used diode-based sources are resonant tunneling diodes (RTDs), Gunn diodes, and IMPATT (IMPAct ionization Transit Time) diodes. They are all based on different physical mechanisms as will be described in Chap. 2 in detail. There are sources based on more conventional diodes, such as *p-n* junction diodes and Schottky diodes, which are in fact oscillator circuits employing a resonant structure realized with the diodes. A more preferred approach for oscillator circuits, though, is to employ transistors as the core circuit components. A variety of circuit topologies including transistors can create oscillating current, which can be radiated through an antenna or can be delivered to neighboring circuit components. Realizing transistor-based oscillators operating at the THz band is challenging and has been limited due to the transistor performance that is not sufficient for such a high-frequency band. However, the barrier is getting lowered with the recent progress in semiconductor process technologies. Figure 1.5 shows the transistor operation speeds reported by various technologies based on f_T (cutoff frequency, where the current gain becomes unity) and f_{max} (maximum oscillation frequency, where the power gain becomes unity), which indicates that transistors are now capable of operating at several-hundred GHz or even beyond 1 THz depending on technologies. With the help of this advancement in semiconductor technologies, a growing number of transistor-based signal sources are now being reported based on various technologies, as will be discussed in Chap. 2.

Optical methods are a more traditional mode for THz signal generation, which is a major reason for the dominating position of the optics-based approach for THz research till today. While the signal generation with electronic devices is mostly continuous wave (CW), a more prevailing form of THz signals generated with

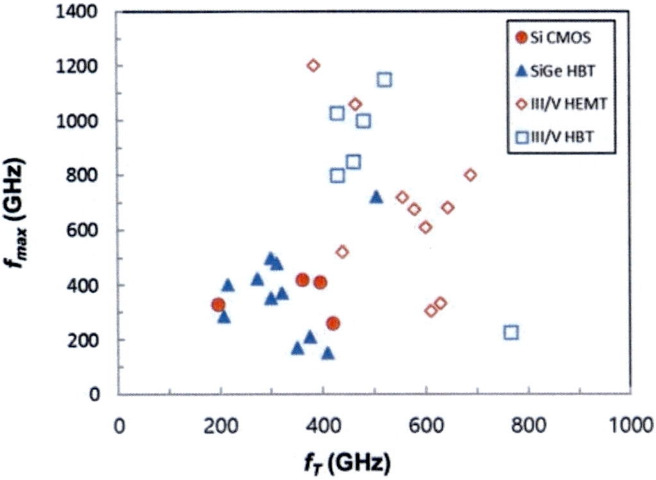

Fig. 1.5 Transistor operation speeds reported by various semiconductor process technologies represented in terms of f_T (cutoff frequency) and f_{max} (maximum oscillation frequency). (From various sources)

optical methods is a pulse. Ultrashort THz pulse generation is based on either photoconductive (PC) antenna or optical rectification method with electro-optic (EO) materials. While there are various options to generate THz pulses with the PC approach, the basic principle is to cause a THz pulse radiation with a very fast change in the electric current in a photoconductive material excited by an incident femtosecond laser beam. As for the EO approach, a pulse is generated as a result of the nonlinear response of electro-optic materials to femtosecond laser beams incident with a strong optical field. CW THz waves can also be generated with optical methods, the most popular mechanism being photomixing. When two CW laser beams with a slight frequency difference ($\Delta\omega = \omega_1 - \omega_2$) are combined, a beating CW signal with a frequency corresponding to the frequency difference $\Delta\omega$ is generated. When the frequencies of the two lasers are tuned as desired, a beating signal may fall on the THz band. Another optical approach to obtain CW signals is to develop a laser with a very long wavelength that enters the regime of the THz band, for which transitions between very closely located energy levels need to be exploited. The optical methods for THz generation will be treated in Chap. 5, separately from those based on electrical methods.

1.3.2 Approaches to THz Signal Detection

THz detectors are another key component to complete THz systems together with THz sources. There have been various types of THz detectors developed, for which high responsivity and low noise level are desired. They can be categorized into thermal detectors, square-law detectors, and heterodyne detectors. Also, THz pulses can be detected with optical methods.

For thermal detectors, the temperature rise due to THz wave absorption is exploited to estimate the incident power level, for which various techniques can be employed. Bolometers convert the temperature rise into resistance change of the constituting material, which is in turn converted into a measurable electrical quantity such as current or voltage. Golay cell is another popular THz detector, in which the temperature rise of the gas contained in a cell due to THz wave illumination causes inflation of the gas cell, which alters the reflection path of a light incident on the cell surface because of the deformed surface by the inflation. In pyroelectric detectors, the temperature change due to illumination causes change in the polarization property of pyroelectric materials, which can be converted to voltage change across a slab of such materials and then measured. Thermopile makes use of the temperature difference established across a material by illuminating THz waves on one side of the material, which is converted into a voltage difference by Seebeck effect. In a sense, these thermal detectors can be categorized as detectors based on electrical approach, although the details vary over the device types.

Square law detectors exploit the nonlinear response property of devices or circuits, which converts incoming CW sinusoidal signal into DC signal (as well as other harmonic frequency components). By detecting the amplitude of the generated DC signal, the strength of the incident signal can be determined. Since the measured DC voltage amplitude is proportional to the square of the input signal voltage, they are called square law detectors. In fact, because the input power is proportional to the square of the input signal voltage for a given load, the measured DC voltage is proportional to the input power for normal operation. For square law detectors, in principle, any nonlinear device can be used, while diodes and transistors are widely adopted for compatibility with other electronic parts.

For heterodyne detectors, the input signal is first frequency down-converted to a lower frequency then detected at intermediate frequency (IF). For the frequency down-conversion, a mixer is a key component, and thus those mixers operating at the THz band have long been considered as effective THz detectors. Traditional THz mixers are based on diode structures, which include Hot Electron Bolometers (HEBs), Semiconductor-Insulator-Semiconductor (SIS) mixers, and Schottky Barrier Diode (SBD) mixers. With the advent of various high-frequency semiconductor technologies, heterodyne receivers based on transistor circuits have also reached sufficiently high operation frequencies to be used in the THz band. With this approach, a high-level integration is possible, which would comprise a mixer, local oscillator, IF components, and sometimes a low-noise amplifier at the front, all on a single chip. The detection based on the electrical methods will be detailed in Chap. 3.

Optical methods can also be used for THz detection, mostly for the recovery of incoming THz pulses. THz signals applied on photoconductive (PC) material can be detected when an ultrashort optical pulse from a femtosecond laser is simultaneously applied, as the pulse triggers a current with photo-generated carriers. With the sampling process (also called gating process) based on the controlled time delay of the optical pulses arriving at the detector in synchronization with THz pulses, the profile of the incoming THz pulses can be recovered. Electro-optic (EO) material can also be used to recover THz pulses. Linearly polarized optical beam incident on an electro-optic material will go through a polarization-dependent phase delay with the presence of THz field applied on the material. The phase difference between the two polarized beams emerging from the EO material will determine the strength of the incoming THz signal. In combination with a proper sampling process, it will enable the recovery of the THz pulse shapes, similar to the case with PC material. Details of the optical detection will be described in Chap. 5.

1.4 THz Applications

It should be fair to say that the interests on the THz band have been so far more inclined toward the research purposes than the practical ones, especially when compared with the neighboring spectra heavily explored by electronics and optics for many useful practical applications. However, there are limited but important application fields that have long relied on the THz spectrum. Also, on a more recent but promising side, there are rapidly emerging practical THz applications that are expected to be more closely linked to our daily lives. Some of these THz application areas are introduced below, which will be described in more detail in Chap. 6.

Spectroscopy is the analysis of the interaction between matter and electromagnetic waves as a function of frequency or wavelength, usually based on the emission or absorption properties of the matter. The response of a material to electromagnetic waves provides a great amount of physical and chemical information of the material, and the spectroscopy is a useful tool to extract such information from the given material. The readers should be aware that the key clues for understanding the internal structure of atoms came from the spectral lines emitted by the atoms, as spectroscopic methods were successfully applied to the atomic study in the late nineteenth century. Different ranges of spectrum carry different information. THz spectrum, for example, contains critical information on the properties of gas molecules, since the photon energy of the THz band corresponds to the transition energies in the rotational and vibrational modes of the molecules as mentioned earlier. This makes THz spectroscopy a precious tool for the study of the behavior of gas molecules. Traditionally, THz spectroscopy was mainly based on the frequency-dependent reflection behavior of gratings, but Fourier-transform spectrometer that emerged later has extended the capability of THz spectroscopy. This was followed by the THz time-domain spectroscopy (TDS), which became a mainstream of modern THz spectroscopy till today. See Sect. 6.1 for more details of these methods.

Imaging refers to a process of mapping the locational property of an object to another space, mostly 2D but sometimes 3D, based on the spatial distribution of the radiation from the object that comes from reflection, transmission, or self-generation. When the radiated electromagnetic wave belongs to the THz band, we call it THz imaging. While there are lots of imaging modalities including the most conventional one based on visible light, THz imaging has its own attractive features. The major unique properties of THz waves described in Sect. 1.2 can be mostly converted to the advantages of THz imaging. The high water absorption level of THz waves results in high-contrast images with materials with a high water content. The transparency of optically opaque materials to THz waves offers opportunities for security imaging, especially useful when it is applied to humans owing to the non-ionizing characteristics of THz waves. When compared with microwave imaging, THz imaging provides an enhanced resolution with the shorter wavelength. As for the actual methods to practice THz imaging, there are many different options. There are active imaging and passive imaging, while active imaging can be obtained with either transmission or reflection mode. It may be based on THz pulse or CW THz wave. THz imaging can also be obtained with either single-pixel detectors or array detectors. The details of these various modalities are described in Sect. 6.2.

Wireless communication defines our modern lives in various ways in the form of cellular networks as well as local connectivity, such as WiFi, Bluetooth, zigbee, and so forth. We are always hungry for higher data rate for quick downloads and uploads with these networks. One of the most efficient ways to achieve such a high data rate is to increase the bandwidth, which in turn can be realized with a higher carrier frequency. From this perspective, the THz band is a highly attractive and preferred candidate when compared with the currently prevailing frequency band of 1–2 GHz band as well as to the emerging mm-wave band. Of course, there are great challenges for the THz waves to be successfully applied to the wireless communication, which arise from the transmitter/receiver performance limit as well as the attenuation due to Earth's atmosphere and high free-space path loss (FSPL). However, there are intensive efforts growing recently to overcome these challenges and pave the way for THz communication systems in various application scenes, which will be more detailed in Sect. 6.3.

Radar (RAdio Detection And Ranging) is a system that determines the range (distance) and velocity of a target object based on the radio wave detection. While its application has long been driven by military needs, its commercial application is attracting growing interests recently, one good example being the car radars based on 77–79 GHz frequency range. While the application of the THz band to radars is still on its infant stage, there are obvious advantages expected for THz-based radars. The shorter wavelength of THz waves than usually used for current radars will provide an improved resolution for the ranging data acquired. Also, for imaging radars, which are gaining increasing attention, the image resolution will be improved, too, with enhanced angular accuracy. Although limited by the same set of challenges that affect the other applications, the expected advantages will push the development of THz radars for various practical applications. The details will be discussed in Sect. 6.4.

References

1. ITU, *Radio Regulations* (ITU, Geneva, 2016)
2. IEEE, *IEEE Standard Letter Designations for Radar-Frequency Bands* (IEEE, New York, 2002)
3. Microwaves101, https://www.microwaves101.com/encyclopedias/frequency-letter-bands. Accessed 2020
4. D.M. Pozar, *Microwave Engineering* (John Wiley & Sons, New York, 2009)
5. E. Hecht, *Optics* (Addison Wesley, Reading, MA, 1998)
6. B. Wozniak, J. Dera, *Light Absorption in Sea Water* (Springer, New York, 2007)
7. R. Appleby, H.B. Wallace, Standoff detection of weapons and contraband in the 100 GHz to 1 THz region. IEEE Trans. Antennas Propag. **55**(11), 2944–2956 (2007). https://doi.org/10.1109/TAP.2007.908543
8. T.W. Crowe, W.L. Bishop, D.W. Porterfield, J.L. Hesler, R.M. Weikle II, Opening the terahertz window with integrated diode circuits. IEEE J. Solid State Circuits **40**(10), 2104 (2005)

Chapter 2
THz Sources and Related Topics

In this chapter, a review on various electrical methods to generate THz signals and related techniques will be presented. Traditional THz sources based on vacuum devices and diodes will be first discussed, followed by oscillator circuits with transistor technology operating at the THz band. Discussed next will be two effective ways to boost the oscillation frequency of THz signal sources, the *n*-push technique and frequency multiplication. This will be followed by the description of two widely used power-boosting techniques, amplification and power combining. Lastly, beamforming, which is a special case of power-combining technique that additionally enables the steering of beams with improved directivity, will be treated as a separate topic.

2.1 Vacuum Device Sources

THz signal generation with vacuum devices is an extension of the efforts to generate microwave signals based on vacuum devices, which can be traced back to the early twentieth century. Signal generation with vacuum devices is basically a process of converting a modulated electron current into electromagnetic wave radiation. The signal output power obtained from vacuum sources is typically much larger than the output power from solid-state devices. This is mainly because the electron current in vacuum devices, from which the radiation is extracted, is a *convection* current through the vacuum, whereas the electron current in solid-state devices is basically a *conduction* current that travels through a medium that necessarily involves carrier scattering. The absence of scattering inside the vacuum devices leads to a more efficient transport as well as low heating effect, a big advantage of vacuum device-based signal generation.

Inside a vacuum device, or tube, electron current modulation is applied for signal generation. The electron current modulation can be either longitudinal or transverse. The devices with *longitudinal* electron current modulation, a process that typically

© Springer Nature Switzerland AG 2021

J.-S. Rieh, *Introduction to Terahertz Electronics*,
https://doi.org/10.1007/978-3-030-51842-4_2

involves bunching of the electron beam along the longitudinal direction, include klystrons, traveling-wave tubes (TWTs), and backward wave oscillators (BWOs). Categorized as the devices with *transverse* modulation are gyrotrons and free electron lasers (FELs), the former with helical motion and the latter with wiggling motion of electrons. Internal to this type of devices, the phase velocity of the generated electromagnetic wave is larger than the speed of light, so they are often called fast-wave devices. In contrast, the devices with longitudinal modulation, as mentioned earlier, are called slow-wave devices, with the phase velocity smaller than the speed of light. Yet another vacuum device, which is based on a quite different operation principle than those mentioned above, is the magnetron. It does not involve electron modulation but still can be categorized as a slow-wave device if the aforementioned definition is applied.

In this section, a brief snapshot of some selected vacuum devices is provided. For a more detailed description, the authors are encouraged to refer to other sources focusing on vacuum devices [1–4].

2.1.1 *Klystrons*

The klystron is essentially an amplifying device based on a vacuum tube, which can be converted to an oscillator for signal generation when needed. It was invented by the Varian brothers at Stanford back in 1939 [5]. It modulates the electron beam emitted by an electron gun inside a cavity resonator as triggered by the input signal, generating a stream of electron bunches along the tube. The bunched electrons excite electromagnetic waves at another cavity resonator located on the opposite side of the tube, and the generated electromagnetic waves are extracted as the device output signal. The power level of the output signal generated in this way is much larger than that of the input signal. Hence, amplification is attained with the device.

The operation can be described in more detail with the help of Fig. 2.1a, which shows the typical structure of a klystron. Electrons are emitted from the cathode that is heated for thermal emission and then accelerated by the electric field established by the anode at a more positive potential. The unit composed of the cathode and the anode is called the electron gun. The electron beam emerging from the gun maintains a steady flow with a uniform density along the traveling (longitudinal) axis until it reaches the first cavity resonator, which is called the "buncher" cavity. This cavity with a circular cross section serves as a cylindrical resonant cavity. That is, when it is excited by the AC input signal supplied through a coaxial cable or waveguide, an alternating electric field is induced across a pair of grids that are located at the input and output boundaries of the cavity. The alternating electric field, which is of the same frequency as the input signal, periodically accelerates and decelerates the electrons that pass through the cavity. In consequence, as the modulated electrons stream out of the cavity and travel through the space along the tube between the two cavities (called the drift space), *accelerated* electrons tend to catch up the

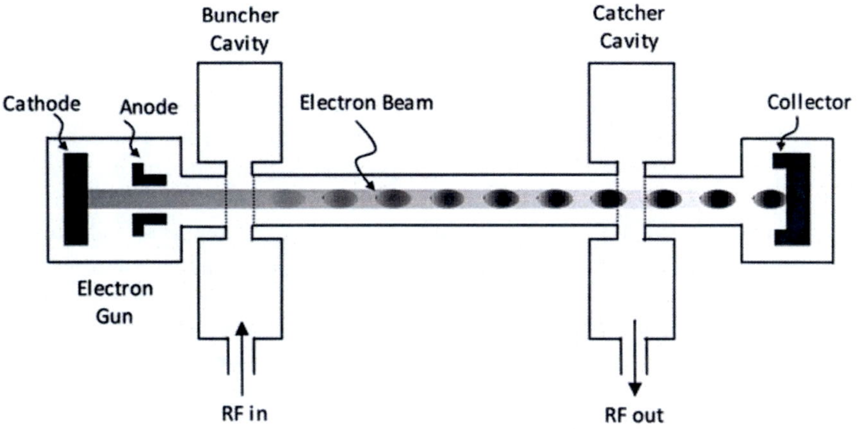

(a)

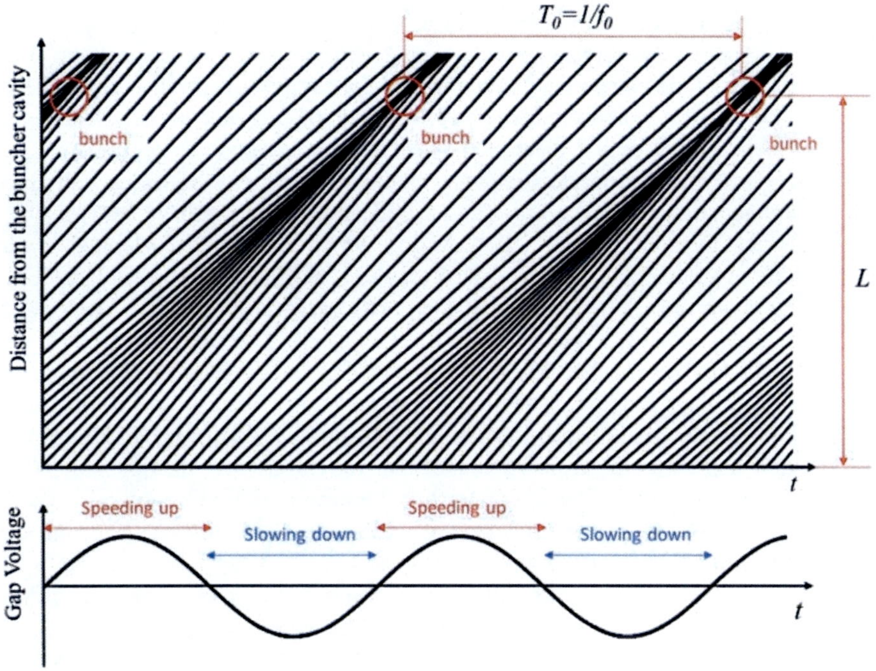

(b)

Fig. 2.1 (**a**) Schematic of a typical klystron. (**b**) Space–time diagram of electron beam showing the formation of electron bunches *decelerated* electrons that move ahead. As a result, a series of electron bunches are formed along the axis by the time they arrive at the second cavity.

This bunch-forming process can be better understood with Fig. 2.1b, which plots the distances traveled by the velocity-modulated electrons as a function of time [6]. Electron bunching induced by the periodic modulation is the most pronounced at location *L*. This is the location where the second cavity is located. Note that this cavity, called the "catcher" cavity, has the same dimension and thus the same resonant frequency as the buncher cavity. The electrons passing through the catcher cavity excites electromagnetic waves inside the cavity with the same frequency as the input signal. The generated electromagnetic waves are extracted as the output signal at the output port. It shows a much larger power level than the input signal power, as it has gained energy from the electron beam passing through the cavity. To be more precise, the induced electrical field across a pair of grids at the catcher cavity acts to decelerate the electrons and reduce their kinetic energy, and the energy "stolen" from the electrons is transferred to electromagnetic waves being generated. In fact, the induced electric field at the catcher cavity, triggered by the electron bunches, is an alternating field, and it will not only decelerate but also partly accelerate the electrons passing through the cavity as was the case for the buncher cavity. There is a difference, though. The alternating field at the catcher cavity acts on bunched electrons, whereas the field at the buncher cavity acts on a uniform flow of electrons. The deceleration cycle at the catcher cavity coincides with the peak of the bunches, while the acceleration cycle falls on the valleys between the peaks. Hence, there are far more electrons decelerated than those accelerated, leading to a net reduction in the energy of the electron beam. The electrons that have lost kinetic energy to the catcher cavity will eventually arrive at the collector located at the end of the tube and exit the device.

So far, the standard klystron that contains *two* cavities has been considered to explain the operation. However, additional cavities can be added to the device to improve the performance, such as gain or bandwidth. This type of klystron is called the multi-cavity klystron. There are other variants as well: the reflex klystron with a single cavity, the extended interaction klystron (EIK) with multi-gap cavities, the optical klystron operating at optical frequencies, and so forth. Although the basic function of klystrons is the amplification as described above, it can always be modified to work as an oscillator by placing a proper feedback path between the buncher and catcher cavities, as suggested by the original paper by Varian brothers [5]. In this case, any electrical noise in the system will be amplified repeatedly around the feedback loop to establish a sustaining oscillation with a sufficient output power, in which the oscillation frequency will be the same as the resonant frequency of the cavities. Also, klystrons may be operated in the pulsed mode instead of the standard continuous wave (CW) mode when needed. Modern high-frequency klystrons are known to operate up to a few hundreds of GHz and the typical output power can reach beyond 10 W (for CW operation) near 100 GHz [7, 8]. The main challenge for THz operation of klystron is to reduce the electron beam diameter to a level much smaller than the operating wavelength. The heat management and the tight assembly accuracy, which become more challenging with increasing frequency, also need to be addressed for THz applications [9].

2.1.2 *Traveling-Wave Tubes (TWTs)*

The traveling-wave tube (TWT) is another type of vacuum device used for the amplification of electromagnetic waves applied as a signal. Although many researchers have contributed to developing the concept of the early TWT, Kompfner is widely credited for the invention of TWT in the early 1940s [10]. The overall operation principle of TWTs [11] is not much different from that of klystrons. Electrons are emitted from an electron gun, and electron bunches are formed as they travel down to the collector along the tube. The electron bunches, created by the input signal, will end up generating the output signal with a much larger power than the input signal by transferring a part of their kinetic energy to electromagnetic waves. The main difference, however, lies in the manner the electron bunches are created and its energy is delivered to the output signal. This will be described below with its conceptual diagram provided in Fig. 2.3a.

Remember the electron bunches in klystrons are created by the electric field of the standing wave *localized* inside a cavity. On the other hand, the electron bunches in TWTs are created by the electric field of the wave *traveling* along the tube, which is the origin of the device name. For the electrons emitted from the electron gun to effectively interact with the traveling wave injected at the input port, the electron movement needs to be synchronized with the phase of the electromagnetic wave. That is, the electron velocity should be same as the phase velocity of the electromagnetic wave. In this way, the electrons can remain stationary against the electric field profile of the traveling wave, when viewed from a reference frame moving at the same speed as the electrons and the wave. This is a preferred situation for a consistent mutual interaction between the electrons and the wave. One issue in this process is that the electron velocity established by the electron gun is typically much smaller than the wave velocity, which is in fact the speed of light in vacuum. Hence, a mechanism to slow down the traveling wave is required, which is typically achieved by employing a slow-wave structure. There are various forms of slow-wave structures. A traditional one that has been widely used for TWTs (and also for backward wave oscillators to be described later) is the helix, or a coiled wire, as is assumed in the device depicted in Fig. 2.2a. In this case, the phase velocity of electromagnetic waves guided by the helix will be reduced by a factor of $\sin\psi$ when the wave travels along the longitudinal axis, where ψ is the angle between the tilted coil and the vertical (transverse) axis (see Fig. 2.2a).

With the slow-wave structure in act and the velocities in synchronization, some electrons will experience the accelerating phase of the field while some will experience the retarding phase of the field as they travel along the tube, depending on their relative location against the periodic field profile. As a result, electrons streaming down the path will be squeezed into a series of bunches as is described in

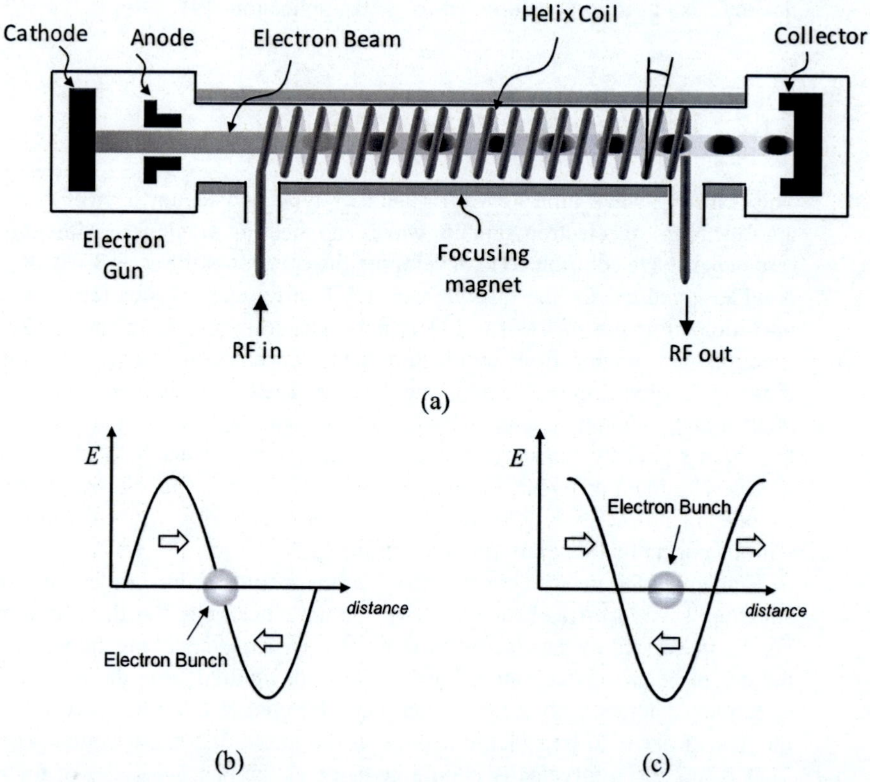

Fig. 2.2 (**a**) Schematic of a typical TWT. (**b**) Relative position of the longitudinal electric field and electrons to generate electron bunches. (**c**) Relative position of the longitudinal electric field and electrons to transfer electron kinetic energy to electromagnetic wave by electron deceleration. (The direction of the electric field is shown with reference to electrons for convenience. The actual field is in the opposite direction)

Fig. 2.2b. After the bunch formation, the next step is for the electron bunches to lose some of their kinetic energy, so that it can be used to build up the output signal amplitude. For this to happen, the electron bunches need to be decelerated as they travel toward the collector. With the configuration given in Fig. 2.2b, however, no deceleration of electron bunches will be obtained as the electrons are located at zero-field positions. But if we assume the electrons were initially moving slightly faster than the phase of the field, electrons would gradually catch up the regarding phase of the field, leading to the configuration depicted in Fig. 2.2c. In this situation, the electrons will begin to lose the kinetic energy, which will be transferred to the output signal that grows until the electron reach the collector, completing the amplification action. This energy transfer mechanism is sometimes interpreted as a *Cerenkov (Cherenkov) radiation*, in which any charged particle traveling at a speed larger than the phase velocity of electromagnetic waves traveling in that medium will radiate electromagnetic waves. This can be compared to another type of mechanism,

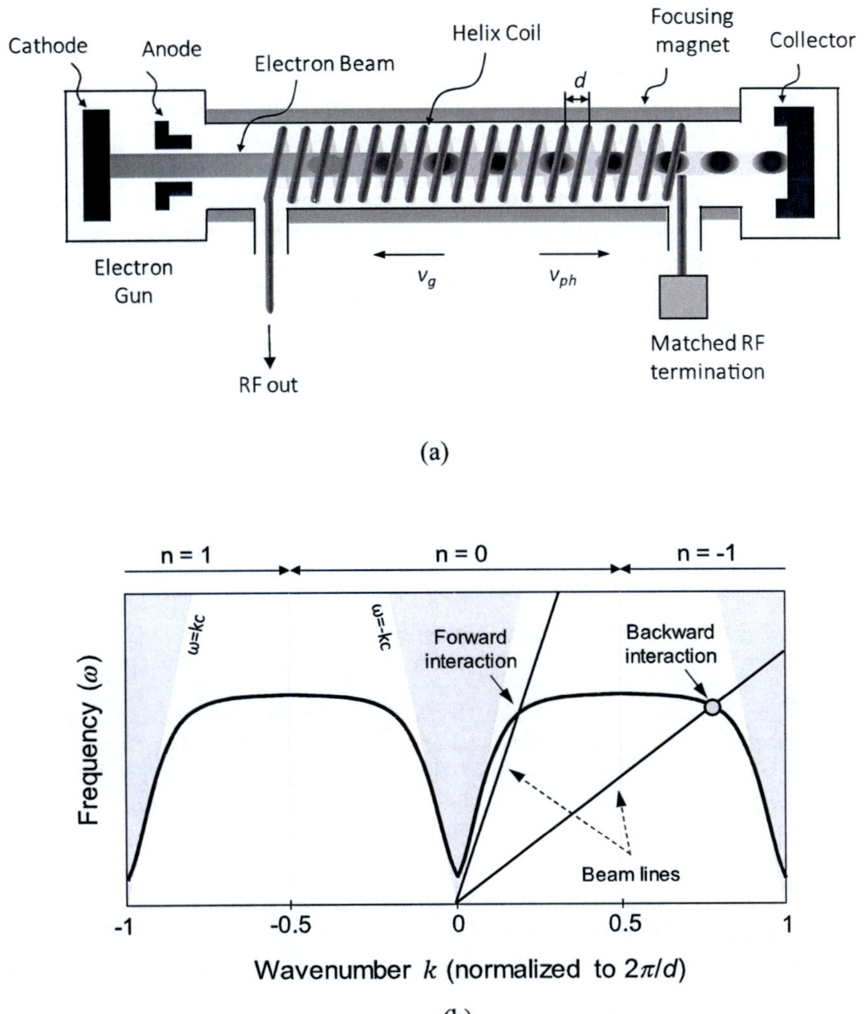

Fig. 2.3 (**a**) Schematic of a typical BWO. (**b**) Dispersion relation of electromagnetic wave inside a BWO showing the group velocity and phase velocity. (© 2011 IEEE [7])

which is called the *transition radiation*. In this latter case, electrons will radiate when they pass through a boundary between two media with different refraction indices or, in a more general term, travel in a medium with non-homogeneities. In fact, the energy transfer at the catcher cavity of klystrons as described earlier can be viewed as an example of the transition radiation. There is yet another type of transition, *bremsstrahlung radiation*, which will be discussed shortly.

In the operation of TWTs described above, a helix was assumed to be employed as the slow-wave structure. As the dispersion (speed distribution over frequency) with a helix is known to be very small, TWTs with a helix typically show excellent broadband characteristics. Sometimes, however, other slow-wave structures are employed, such as the cavity-coupled structures that exhibit narrowband but high output power. It is also noted that typical TWTs employ a focusing magnet to tightly confine the electron beam around the center with a magnetic field formed along the longitudinal axis. For this purpose, either electromagnet with a solenoid or permanent magnets arranged in a periodic alternating configuration can be used.

The operation frequency of TWTs can reach up to several hundreds of GHz, as indicated by a recent report that demonstrated a TWT amplifier working beyond 600 GHz [12]. Typical average output power can reach well beyond 100 W near 100 GHz [13]. Although the operation frequency of TWTs has already reached deep into the THz band, there are still challenges to further increase the operation frequency. It is known that the main issues mostly involve the circuit size reduction for increased frequency, which also requires a precise beam alignment. With these issues addressed, the operation will be extended to cover higher frequency bands around 1 THz.

2.1.3 *Backward Wave Oscillators (BWOs)*

The backward wave oscillator (BWO) is an oscillator based on vacuum tubes, the invention of which is traced back to the early 1950s [14, 15]. As opposed to klystrons and TWTs described above, which are basically amplifiers, BWO is intrinsically an oscillator. It can be regarded as a TWT modified for oscillator function. Hence, BWOs share the same main structure with TWTs such as the electron gun and the slow-wave structure. They also share the same mechanism for electron bunch formation. Instead of applying an external feedback loop to achieve oscillation, which is a typical method of converting an amplifier to an oscillator as was the case for the klystron oscillator described above, BWOs are elegantly designed to excite electromagnetic waves with a *negative* group velocity to meet the oscillation condition inside the device. As a result, the electromagnetic waves established in the tube show a group velocity and a phase velocity with the opposite directions, the former traveling backward while the latter traveling forward. This is how the device name bears "backward wave." Operation principle is described below with the help of Fig. 2.3a that depicts a typical BWO with a helix as the slow-wave structure.

In order to understand the opposite direction (sign) for the group velocity and the phase velocity obtained in BWOs, the readers are referred to the definition of the two velocities, which are shown below for the sake of the readers' convenience:

$$v_{\mathrm{g}} = \frac{\partial \omega}{\partial k}, \quad v_{\mathrm{ph}} = \frac{\omega}{k}, \tag{2.1}$$

where v_g and v_{ph} are the group velocity and phase velocity of the wave, respectively, ω is the angular frequency, and k is the wave vector. As in the dispersion relation for the electromagnetic waves inside a BWO as shown in Fig. 2.3b, there exist regions where the group velocity and the phase velocity become opposite. The example point indicated in the plot belongs to the first backward spatial harmonic, for which the group velocity is negative and the phase velocity is positive. Note that, in principle, there may exist a huge number of harmonics due to the periodicity of the slow-wave structure. In BWO operation, the phase velocity is aligned to the electron beam velocity for bunch formation, in the forward direction toward the collector, while the group velocity is formed backward toward the electron gun. As the wave energy is carried by the group velocity, the kinetic energy taken by the electromagnetic waves from the electron beam is transported backward, and thus, the generated electromagnetic waves grow along the backward direction. Hence, when a wave completes a round trip along the signal path inside the tube, its magnitude increases, and when it repeats, oscillation is achieved. The output signal is extracted from the output port, typically located near the electron gun side of the tube.

As was the case for TWTs, various types of slow-wave structures may be applied although a helix was assumed in Fig. 2.3a. Also, BWOs can be categorized into either O-type or M-type (true for TWTs as well). The types are distinguished by the direction of the static magnetic field introduced in the device. For O-type devices, the direction is longitudinal, or along the direction of electron flow, whereas for M-type devices, the direction is transverse, or perpendicular to the direction of electron flow. Figure 2.3a describes an O-type BWO structure. The operation frequency of BWOs of today has exceeded 1 THz [16]. The maximum output power obtained with BWOs near 100 GHz is around 1 W. Variants of BWO have also been developed for improved output power, which include clinotron [17] and orotron [18], adopting a tilted tube and Fabry–Pérot cavity for structural variation, respectively. Around 100 GHz, for example, they show improved power by a factor of up to 10 compared to conventional BWOs [7].

2.1.4 Gyrotrons

The gyrotron is a high-power oscillator based on a vacuum tube. Its first demonstration was made in 1964 by a group of Russian researchers [19]. In a sense, gyrotrons are similar to other vacuum devices described above such as klystrons, TWTs, and BWOs, in that they are based on a tube along which electron beam travels and generates electromagnetic waves. However, gyrotrons contrast with those devices in at least three main aspects. First, they do not employ slow-wave structures that are often used in other devices to slow down the phase velocity of electromagnetic waves along the longitudinal direction. Second, electrons travel at a very high speed, close to the speed of light, so that relativistic effects need to be considered. Third, the electrons travel with a helical motion, or they gyrate as they travel along the tube. These differences end up with a significantly higher output

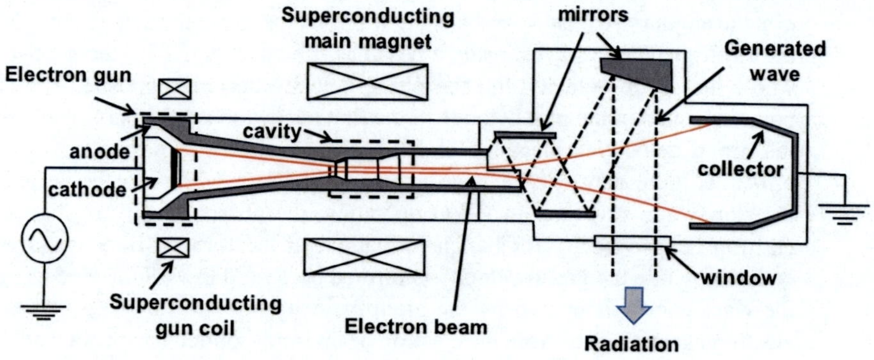

Fig. 2.4 Schematic of a typical gyrotron. (© 2011 IEEE [7])

power compared to those generated by other devices described above. A typical structure of a gyrotron is depicted in Fig. 2.4 (Adapted from [7]).

In gyrotrons, electrons are emitted by the electron gun, as they emerge from the cathode and get accelerated by the electric field established by the anode at a more positive potential. This emission process is similar to that of the devices with a slow-wave structure (so-called the slow-wave devices) described above, but the applied potential at the anode is much larger in the case of gyrotrons, resulting in electrons with a higher exit speed. Also, the electrons are generated from the cathode surface area surrounding the cathode tip, not from the tip end, so the electron beam forms a hollow tube, its cross section being annular-shaped. As the emitted electron beam enter the cavity with a strong axial magnetic field along the tube, which is established by an external magnet, typically a superconducting solenoid coil, the electrons begin to gyrate, which will create an azimuthal (rotational) velocity component in addition to the existing longitudinal and transverse velocity components. The gyrating electrons radiate, generating electromagnetic waves. This is an example of bremsstrahlung radiation, which occurs when an electron (or any charged particle) is decelerated due to external electric or magnetic fields [1]. Bremsstrahlung radiation and the two other types of radiations mentioned above, Cherenkov radiation and transition radiation, are known to be the three basic types of electromagnetic radiation generated by charged particles.

A more detailed look on the gyrotron operation can be made with the gyrating angular frequency of electrons:

$$\omega = \frac{qB}{m}, \tag{2.2}$$

where B is the magnitude of the longitudinal magnetic field along the tube, q is the elementary charge, m is the electron mass (distinguished from the rest mass of electron m_0), which increases with the electron speed (v_e) by Lorentz factor $\gamma = 1/\sqrt{1-(v_e/c)^2}$. The increase in m is negligible at lower speed but becomes significant for electrons traveling at a relativistic speed, which is the case for electrons inside gyrotrons. The electrons in a helical motion are in random phase when they are first emitted and the radiated electromagnetic waves are yet incoherent. However, as they travel into the resonator cavity, the electrons that lose kinetic energy to the electromagnetic waves are slowed down and become lighter, leading to increased angular frequency and thus attain a phase lead. On the other hand, electrons that gain kinetic energy speed up and become heavier, leading to a decreased angular frequency and a phase lag. As a result, phase-bunching is achieved for the electrons with aligned phase and the radiated electromagnetic waves becomes coherent. The electrons that have completed the travel across the resonant cavity arrive at the collector and exit the device. The generated electromagnetic waves are extracted through a window located on the side (as is the case for the structure shown in Fig. 2.4) or on the end of the cavity.

In fact, fast-wave vacuum devices generating coherent radiation with gyrating motion of electrons, or *gyro-devices*, can be realized based on other types of devices described earlier such as klystrons, TWTs, and BWOs by modifying the basic structures. In those cases, they are called gyro-klystrons, gyro-TWTs, and gyro-BWOs, respectively, operating as either oscillator or amplifier according to the function they serve. To distinguish from these other types of gyro-devices, the device described in Fig. 2.4 is sometimes called the *gyro-monotron*, a name derived from the fact that it is based on a single-cavity structure. In this case, the term *gyrotron* can be reserved for a more general use, referring to all the types of gyro-devices, also known as electron cyclotron masers (ECMs) or cyclotron resonance masers (CRMs).

One advantage of gyrotrons for THz operation is that increasing oscillation frequency can be achieved with an increased magnetic field as indicated by Eq. (2.2), not necessarily requiring a scaling of the device structure dimension as was the case for the slow-wave devices described above. For oscillation frequencies higher than readily available with a reasonable strength of the magnetic field, harmonics are typically utilized. The operation frequency of gyrotrons has continuously increased since its invention, leading to oscillation frequencies now exceeding 1 THz [20]. Around 100 GHz, output powers higher than 1 MW have been reported for CW operation, while pulse-mode operation can be employed for even higher achievable peak powers [21].

2.1.5 Free Electron Lasers (FELs)

The free electron laser (FEL) is a device that generates coherent radiation on a very wide range of spectrum, from microwave to X-ray. Oscillation with an FEL was first demonstrated in 1977 [22], although its basic concept can be traced back to as early as 1951 [23]. Being a fast-wave device, it shares more common features with gyrotrons than other devices, mostly on the basic fact that electromagnetic waves are generated with bremsstrahlung radiation by electrons traveling at a relativistic

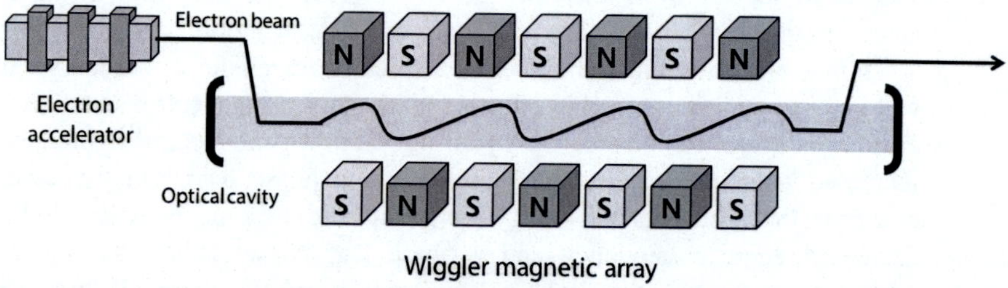

Fig. 2.5 Schematic of a typical FEL. (Adapted by permission from Springer Nature: [24] © 2009)

speed inside a cavity. However, there are major differences that characterize the FEL as follows: electron speed is heavily relativistic (very close to the speed of light) in FELs while it is only moderately relativistic (comparable to the speed of light) in gyrotrons; electrons travel in a wiggling motion in most FELs while they travel in a helical motion in gyrotrons; there are mirrors located on both ends of the cavity in FELs, which are absent in typical gyrotrons. A sketch of a typical FEL is presented in Fig. 2.5 [24].

Since electrons moving at very high speed are desired in FELs, the electron-emitting element is a bit different from other vacuum devices. Rather than a conventional electron gun composed of a cathode and an anode, more sophisticated and powerful *electron accelerators* are often adopted for electron emission. The emitted electrons travel through a path flanked on both sides by an array of magnets with alternating polarity placed along the longitudinal direction, which is called the undulator. Due to the alternating periodic lateral magnetic field, the electron beam follows a zigzag or wiggling path. With a periodic change of the velocity direction, the electrons radiate, which is initially incoherent. However, as the generated electromagnetic waves are reflected back and forth by the pair of mirrors at the cavity ends and spatially confined within the cavity with proper modes established, electron bunches are formed, with its period aligned to the wavelength of the electromagnetic waves. Accordingly, the radiation from the bunched electrons becomes coherent. With this process, a high-power coherent radiation is obtained, which is extracted at the output port through a coupler.

In fact, FELs are not exactly a laser by a strict definition. As indicated by the full name, Light Amplification by Stimulated Emission of Radiation, a laser involves a *stimulated* emission. FELs are based on *spontaneous* emission from the accelerated electrons (see Sect. 5.2.3 for a brief description of emission types and laser operation). Nevertheless, FELs share a key feature with the conventional lasers in that they both employ a cavity bounded by a pair of mirrors that help to confine the radiation for high density and generate coherent radiation. This is how the original term is still maintained, which was first suggested in 1973 [25].

The operation frequency of FELs is mainly determined by the spatial period of the magnet arrays and Lorentz factor, with an additional affect by the magnetic field strength. Hence, FELs can cover a wide range of spectrum with a proper undulator design and control over the electron acceleration, including the THz band. It is noted that a typical FEL is not a type of device that can be hand-carried or installed on a laboratory table top, but rather a bulky facility, shared by researchers from a wide range of locations. The operation frequency range and achievable output power vary over different facilities. The Novosibirsk FEL facility in Russia, for example, operates over a frequency range of 1–3 THz with an average power and a peak power up to 0.5 kW and 1 MW, respectively [26].

2.1.6 Magnetrons

The magnetron is another type of vacuum tube-based oscillators, which has found its way in a wide range of applications. While it is best known for its popular adoption for microwave ovens ubiquitous over the world, it was a key component in military applications in its early days. The magnetron was developed in the first half of the twentieth century based on cumulative contributions by a number of investigators around the world. A pioneering work by Hull in 1921 [27] that detailed the key operation principles of magnetrons triggered some earlier versions of magnetron development, while more practical versions with a higher output power based on a structure similar to modern cavity magnetrons were demonstrated in 1940 by Randall and Boot [28]. The structure and operation of magnetrons are substantially different from the other vacuum devices described so far, mainly in that the electrons in magnetrons travel in the transverse direction in perpendicular to the magnetic field, a key feature of M-type devices.

The basic structure of a typical magnetron is shown in Fig. 2.6. As can be seen from the cross section of the magnetron, the cathode, a cylindrical rod, is located at

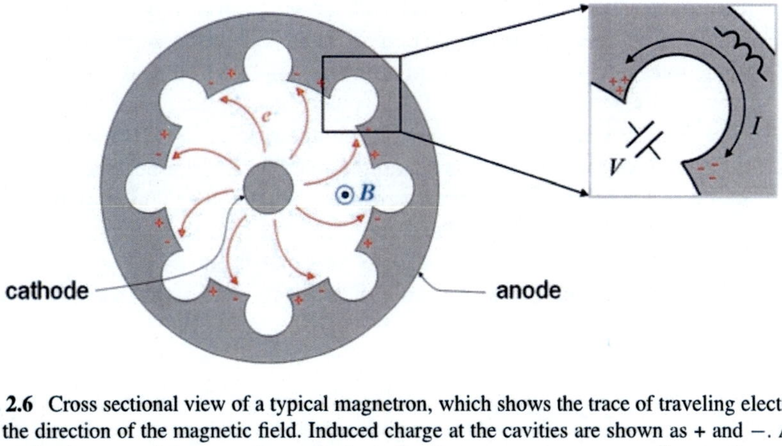

Fig. 2.6 Cross sectional view of a typical magnetron, which shows the trace of traveling electrons and the direction of the magnetic field. Induced charge at the cavities are shown as + and −. Also shown in the zoomed box are the equivalent circuit components corresponding to the cavity resonator

the center along the tube axis and is concentrically surrounded by the anode, which contains multiple circular cavity resonators along the perimeter. A magnetic field is applied by an external magnet along the longitudinal (axial) direction, while a transverse (radial) electric field is established between the cathode and the anode. Electrons emitted from the cathode travel initially toward the anode in the radial direction as affected by the electric field. Soon after the emission, the electrons experience a force in the azimuthal (rotational) direction due the magnetic field, which tends to prevent electrons from landing on the anode surface, causing circulation of the electrons through the space between the anode and the cathode. When they eventually arrive at the anode, the electrons are localized near the opening gap of the cavities on one side, inducing an electrical field across the gap of each cavity. This opening gap acts as a capacitor while the path around the cavity in the anode body behaves as an inductor, as described by the zoomed box in Fig. 2.6, creating a resonator for each cavity. As a result, electromagnetic waves are excited by the cavities, which will interact with the circulating electron beam, causing bunching of the electrons. The generated electromagnetic waves are extracted by an antenna tapped into one of the cavities or by an open hole leading to a waveguide.

From the viewpoint of circulating electrons in the magnetron, they keep passing by a series of cavity openings, as if they travel along a very long multiple-cavity klystron [29]. In this view, electrons are "recycled" instead of reaching the collector after a single trip, which can explain the higher DC-to-RF efficiency typically exhibited by magnetrons compared to other vacuum devices. The oscillation frequency of magnetrons, which is determined by the cavity dimension and electron emission energy, is relatively lower than that of other devices described above. However, there have been efforts to push magnetrons toward THz operation, including a recent demonstration of a 210-GHz magnetron that utilizes a spatial harmonic [30]. Around 100 GHz, an output power larger than 10 kW can be achieved with magnetrons [31]. It is interesting to note that the magnetron is sometimes mentioned as a supporting example for the assertion that technical strength may significantly affect the outcome of military conflicts. During World War II, there were two types of radars competing, one based on magnetrons and the other on klystrons. There exists a notion that the superior performance of the radars based on the magnetrons critically affected the outcome in favor of the allies that had adopted magnetron-based radars. Ironically, afterwards, magnetrons have gradually faded out from radars mainly due to their persistent frequency instability issues, giving way to klystrons as well as TWTs as preferred signal sources for radars.

2.2 Diode Sources

Up to this point, THz sources based on vacuum electronic devices have been reviewed. Another main branch of electronic sources for THz operation is the solid-state devices. As mentioned earlier, the main advantages of solid-state THz

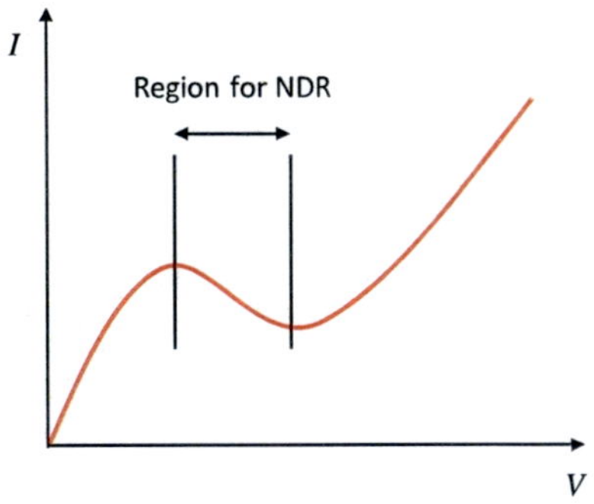

Fig. 2.7 Conceptual *I–V* relation of a diode with the negative differential resistance (NDR)

sources are their compact sizes as well as the low DC power dissipation, although they generate much lower output power than the vacuum sources. Solid-state THz sources can be based on either diodes or transistors. Diode-based sources will be discussed in this section, while transistor-based ones will be treated in the following section. Widely used diode-based sources are Gunn diodes, IMPATT (IMPAct ionization Transit Time) diodes, and resonant tunneling diodes (RTDs). They share a common main feature that the oscillation is attained based on the negative differential resistance (NDR). If NDR is established across a diode, *I–V* relation will show a region with a negative slope as is conceptually shown in Fig. 2.7. Then oscillation can be achieved with a proper termination of the device with a resonant circuit when biased correctly. The details of the physical principle to obtain NDR will differ for each diode, though, as will be described in this section.

2.2.1 Gunn Diodes

The Gunn diode is a two-terminal device with NDR originated from the Gunn effect. Its operation as an oscillator was first reported by Gunn in 1963 [32]. When the oscillation was first observed, the physical mechanism of the oscillation was not exactly understood, but Kroemer soon pointed out [33] that the behavior is consistent with a theory that had been described earlier by Ridley and Watkins [34] as well as Hilsum [35]. Formally known as Ridley-Watkins-Hilsum (RWH) theory, it describes how the NDR can be induced by the transfer of electrons between two adjacent energy valleys in the conduction band of III–V semiconductors, as briefly explained below.

Many III–V compound semiconductors widely employed for electronics, such as GaAs, InP, and GaN, share a following feature in the band structure. The main valley in the conduction band is located at the center of the *E–k* space (Γ point), and is neighbored by another valley in proximity. This neighboring valley is called the satellite valley. Its energy level is slightly higher than that of the main valley, as can be easily seen on the *E–k* diagram depicted in Fig. 2.8. The satellite valley usually shows a shallower curvature (larger radius of curvature) near the minimum point than the

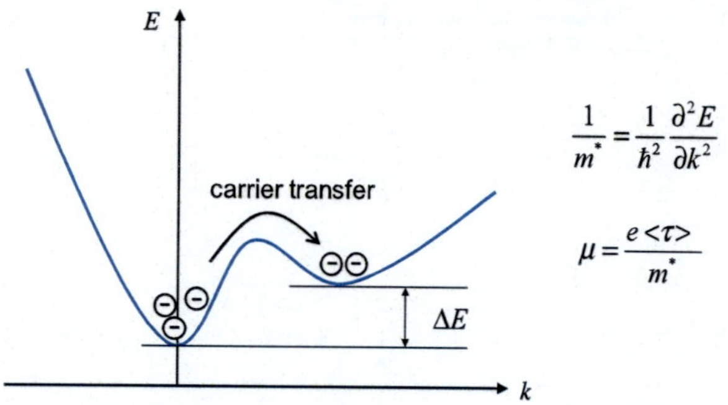

Fig. 2.8 Carrier transfer from the main valley to the satellite valley for some III–V semiconductors

main valley. This indicates a larger effective mass and thus lower mobility for electrons residing in that valley. When the semiconductor is in the equilibrium, most electrons remain in the main valley. When it is excited, typically by an external electric field from the bias voltage, the electrons will gain energy. If the acquired energy is large enough to exceed the energy separation between the two valleys (ΔE), most electrons will be transferred to the satellite valley. After the transfer is completed, the average electron mobility in the semiconductor is degraded, since the electrons in the satellite valley will have a larger effective mass and thus a lower mobility than in the main valley. For this whole process to take place, the energy band structure of a given semiconductor should satisfy following three conditions [35]: (1) ΔE should be significantly larger than the thermal energy at room temperature kT (otherwise electrons will enter the satellite valley even without an external electric field). (2) ΔE should be smaller than the band gap (otherwise the semiconductor will become highly conductive before the transfer occurs). (3) The satellite valley should show a lower mobility, larger effective mass, and higher density of states (otherwise the transfer will not cause mobility degradation). The semiconductors mentioned above, GaAs, InP, and GaN, satisfy all these conditions, while Si and Ge do not. Some III–V semiconductors, such as InAs, GaP, and InSb, do not qualify, either [36].

Now let us see what happens if a diode is made of such semiconductors that satisfy the conditions listed above. When the electric field established inside the device exceeds a certain critical value, the electron transfer will occur as electrons will obtain an energy larger than ΔE and the average electron mobility will be degraded. For an observer external to the device, this will appear as a sudden drop in the current flowing through the device when the bias across the device is increased beyond a certain voltage. This means that *I–V* curve will show a negative slope beyond this critical bias voltage, indicating that NDR is achieved across the device. By properly terminating the two electrodes of the device, typically with a resonant circuit, oscillation is attained. The intrinsic factors such as device dimension and doping profile, together with the resonant circuit, affect the oscillation frequency, which can reach a few hundred of GHz or higher.

Figure 2.9a shows the performance of selected Gunn diodes reported in the past years, in terms of the output power and the oscillation frequency for various materials and different harmonic outputs (The data points in this figure and other figures throughout this book, which show the recent performance trend, were

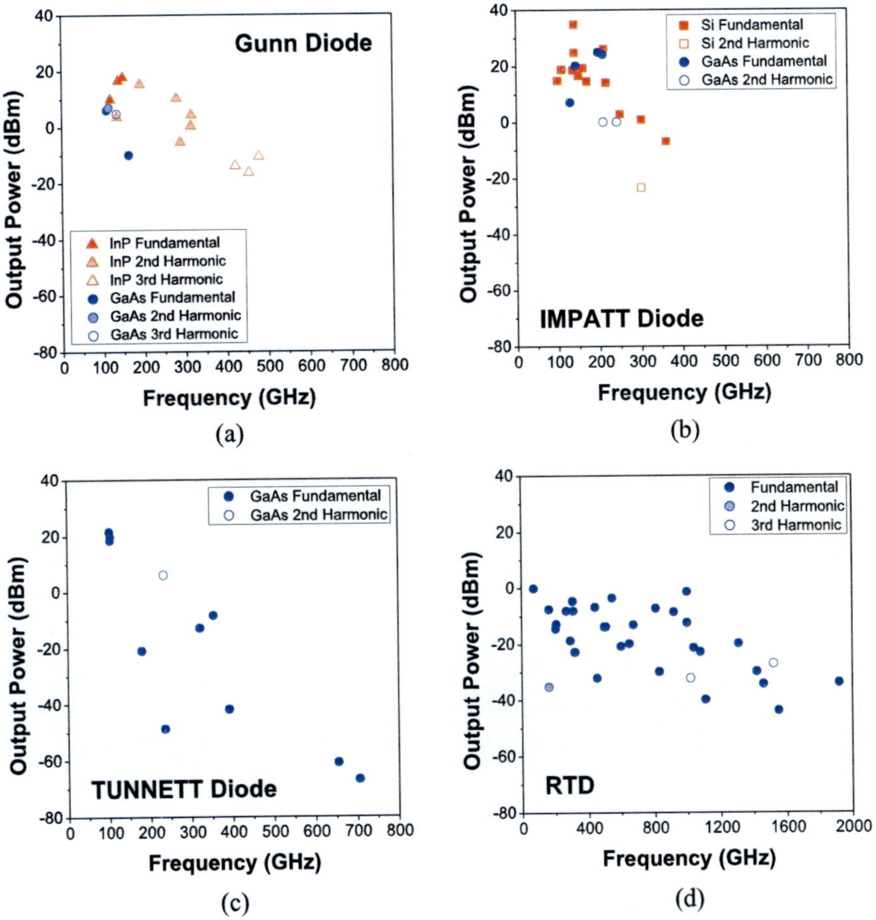

Fig. 2.9 Performance of recently reported diode-based signal sources: (**a**) Gunn diode, (**b**) IMPATT diode, (**c**) TUNNETT diode, (**d**) RTD

obtained directly from the original sources. Due to the large number, they are not individually referenced.). It shows that the output power of around 20 dBm can be achieved for a frequency range of 100–200 GHz range, and the oscillation frequency reaches nearly 500 GHz with output power of −10 to −15 dBm [37]. As of today, Gunn diodes are widely adopted for commercial purposes, and packaged products are available for various power and frequency ranges.

2.2.2 *IMPATT Diodes*

IMPATT (IMPAct ionization Transit Time) diodes attain NDR from time delays caused by the avalanche multiplication and carrier transit across the device. Unlike

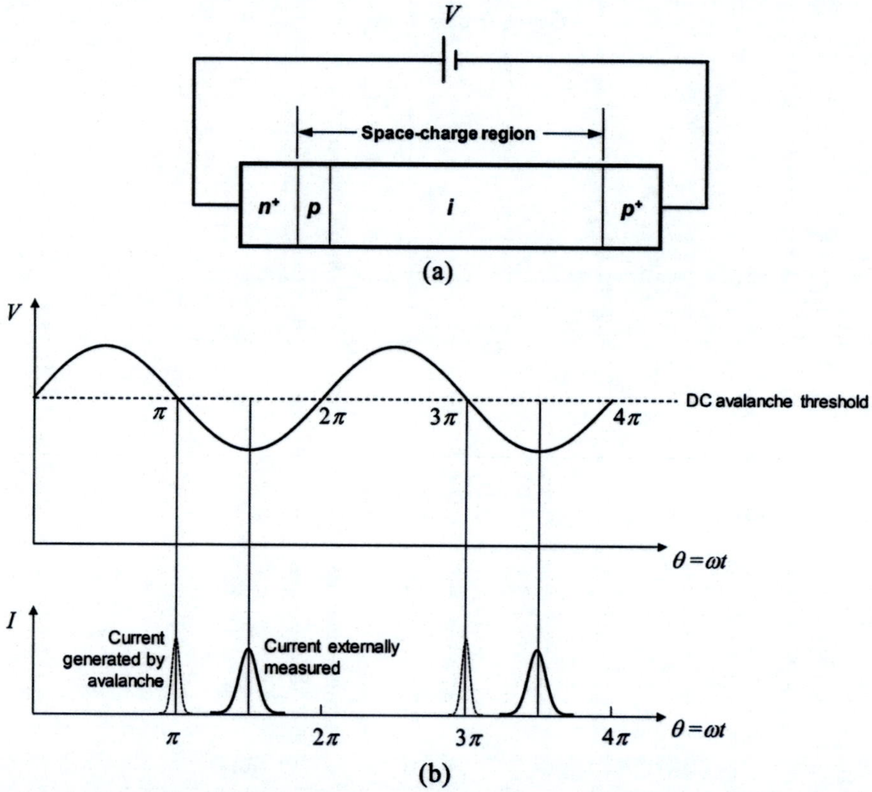

Fig. 2.10 (**a**) Structure of a typical IMPATT diode. (**b**) Applied voltage waveform and current profiles after avalanche breakdown and externally measured

Gunn diodes, which can be implemented on a bulk semiconductor, IMPATT diodes require a *p*-*n* junction in a semiconductor. The origin of IMPATT diodes can be traced back to the avalanche diode proposed by Read in 1958 [38]. The first operation was demonstrated by Johnston for a simple *p*-*n* diode and also by Lee for a more practical n^+-*p*-*i*-p^+ diode as proposed by Read (called Read diode), both in 1965 [39, 40]. The operation is described as follows.

The structure of a typical IMPATT diode is shown in Fig. 2.10a. It includes a highly doped injection region made of a n^+-*p* junction, neighbored by a lowly doped drift region (*i* region) and a contact layer (p^+ region). When a strong reverse bias is applied across the diode, carriers will be generated by the avalanche multiplication in the injection region around the n^+-*p* junction, leading to device breakdown. The carriers generated by the avalanche multiplication will be mostly holes, because the space charge region will be largely formed in the *p* region rather than the n^+ region due to its lower doping concentration in the n^+-*p* junction. The avalanche-generated holes will be drifted toward the p^+ terminal through the drift region and exit the device, establishing a diode current. As the avalanche multiplication is basically a chain reaction of impact ionization, there will be a finite time delay before a large number of carriers are generated. Besides, there will be an additional temporal delay for the generated carriers to transit across the drift region and reach the terminal. Hence, the total time delay will be composed of the two delay components.

Description above assumes a steady bias applied across the diode. Now let us consider a more practical case, in which a sinusoidal AC voltage signal is superposed on top of a DC bias that is very close to the breakdown voltage. During the positive cycle of the AC voltage, therefore, avalanche multiplication will be triggered as the total voltage will higher than the threshold for the avalanche process, leading to a sudden surge in the local hole current. The avalanche current, however, will quickly decay during the negative cycle that follows. Hence, the local hole current level near the junction will fluctuate with the same cycle as the applied AC signal, but with a finite phase lag. The generated holes will subsequently travel through the drift region before exiting the device at the p^+ contact region. The delay due to the drift will cause an additional phase delay. With a proper design of the drift region, the total phase delay between the diode current and the applied AC voltage can be tailored to match 180°, as shown in Fig. 2.10b [41]. In this case, NDR is induced across the diode terminals, which will open a possibility for oscillation with the diode. It is noted that the avalanche delay is a material property and thus strongly affected by the choice of the semiconductor employed for the diode. With a smaller avalanche delay, the hole generation can respond to a quicker transition of the AC signal, leading to a higher oscillation frequency. Similarly, a shorter drift region will help to increase the frequency. It is interesting to note that the avalanche delay is smaller for Si than GaAs, so that Si IMPATT diodes can operate faster than GaAs IMPATT diodes.

Instead of attempting to achieve a smaller avalanche delay for faster device operation, one may improve oscillation frequency by inducing a breakdown from a totally different mechanism that causes faster carrier generation. This is the idea behind the TUNNETT (TUNNEling Transit Time) diode, which is a variant of IMPATT diode [42]. It is well known that there are two mechanisms to cause a breakdown in a *p-n* junction: avalanche multiplication and tunneling. For a lower doping concentration near the junction, the former mechanism dominates, while the latter becomes dominant when the doping concentration is sufficiently high. For the generation of carriers to be injected into the drift region, TUNNETT diodes rely on the tunneling, as opposed to the avalanche multiplication for IMPATT diodes. Developing a tunneling current takes a shorter time than for the avalanche current, which results in a faster operation achieved with TUNNETT diodes. It is further helped by the suppressed dispersion effect that is available with the tunneling mechanism. Consequently, TUNNETT diodes tend to show a higher operation frequency and lower noise level than IMPATT diodes.

Figure 2.9b shows the reported performance of IMPATT diodes in terms of the output power and the oscillation frequency. It shows the output power exceeding 20 dBm for 100–200 GHz range, and the maximum frequency goes up close to 400 GHz [43]. On the other hand, a GaAs TUNNETT diode operating beyond 700 GHz with fundamental mode has been reported as Fig. 2.9c shows [44].

The output power of TUNNETT falls short of IMPATT diode output power, though. Also, it rapidly decreases with increasing frequency. Although experimental results are yet to appear, IMPATT diodes with an oscillation frequency exceeding 1 THz appears feasible from calculation [45].

2.2.3 Resonant Tunneling Diodes (RTDs)

RTDs achieve NDR by exploiting discrete quantized energy levels allowed for electrons confined in a quantum well. While Gunn diodes and IMPATT diodes are based on a single type of semiconductor, with either bulk or *p-n* junction, RTDs involve heterojunctions between semiconductors with different energy gaps to create quantum wells and barriers. The NDR due to the resonant tunneling through quantum wells in semiconductors was first observed by researchers in IBM in 1974 [46], soon after a prediction by calculation from the same group [47]. The operation principle is as follows.

When a diode is composed of a stack of semiconductor layers with different energy bandgaps, quantum wells will be formed in the conduction band. This is shown in Fig. 2.11a with a case of a single quantum well bounded by two energy barriers. Inside the quantum well, discrete energy levels for electrons will be created as predicted by the quantum theory. As the bias applied across the diode changes, the energy levels inside the quantum well will shift against the energy levels in the adjacent semiconductor regions across the barriers. When any of the discrete energy levels inside the quantum well aligns with the Fermi level of the outside region, a large current will flow across the barrier due to tunneling. On the other hand, when it is misaligned, the current will be small. Therefore, as the bias voltage is swept from low to high, there may exist a local reduction of the current level, which will happen when the increase in bias tends to further misalign the energy levels in that range. This will cause a negative differential resistance across the diode, which can be exploited to generate oscillation if proper bias voltage is applied. The oscillation frequency is affected by the electron dwell time in the quantum well and the transit time toward the terminal, while some external components, such as device parasitics, also affect the device speed. The reduction of these delay times by device optimization is the key path to RTDs with higher oscillation frequency.

Figure 2.9d shows the accumulated data points recently reported for RTDs. The oscillation frequency of RTD is known to be the highest among the diode-based oscillators, penetrating deep into the THz band. The highest oscillation frequency of RTD reported so far is 1.92 THz [48], while it is expected to further grow in view of the extensive recent research efforts, notably by the group in Tokyo Institute of Technology led by Asada. The output power of RTDs, on the other hand, is relatively smaller than those of Gunn or IMPATT diodes at low frequency range. However, there is a growing effort to increase the output power with device structure optimization as well as power combining with multiple RTDs.

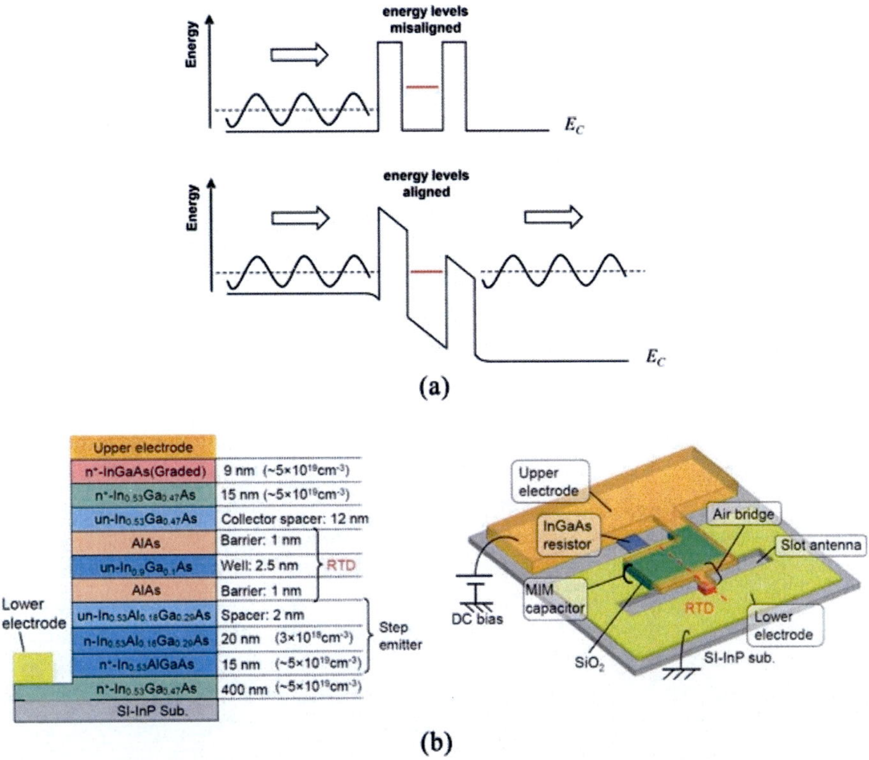

Fig. 2.11 (**a**) Conceptual energy band diagram of a typical RTD. In the upper case, the current does not pass through due to misalignment of the energy level in the quantum well. In the bottom case, current passes through owing to the energy level alignment. (**b**) Schematic of a typical modern RTD oscillator: semiconductor layer structure (left) and the entire device structure (right) [48]. (Copyright (2016) The Japan Society of Applied Physics)

The actual structure of modern THz RTD, which is integrated with a slot antenna for radiation, is depicted in Fig. 2.11b [48]. It is composed of multiple layers employing InGaAs and AlAs layers for narrow- and wide-band semiconductors, respectively, deposited on top of a semi-insulating InP wafer by a molecular beam epitaxy (MBE). The semiconductor layers are sandwiched by a pair of electrodes, which directly leads to the antenna. The dwell time and the transit time, which are the intrinsic delays, can be lowered by reducing the layer width of the RTD quantum well and the collector space layer shown in the figure, respectively. However, capacitance increase with the thinned layer may counteract the improvement with the consequent RC delay increase, which requires a careful optimization of the layer thicknesses. The extrinsic delay that arises from the parasitic components such as those from the link to the antenna needs also be reduced for increased oscillation frequency, which can be achieved through the optimization of device exterior as well as fabrication.

2.3 Transistor Circuit Sources

The signal sources based on diodes as described in the previous section benefit from an advantage over the signal sources based on vacuum device in that they are solid-state components and thus can be realized on a semiconductor substrate. Signal sources based on transistor circuits, as will be described in this section, provide further advantages. Most prominently, it allows possibility of integration with other circuits with different functions, digital circuits inclusive, enabling embedded signal sources in a monolithically integrated system. The generated signals can be conveniently delivered to other neighboring on-chip circuit components. Also, being an active device, transistors provide much more options for circuit implementation than passive devices would do, notably the availability of gain, which can help to further enhance output power after signal generation. The highly developed circuit design environment also helps efficient design processes. Realizing transistor-based signal source operating at the THz band, however, is certainly challenging and has long been limited due to the transistor performance not sufficient for such a high-frequency band operation. Nevertheless, the barrier is being lowered with the recent progress in semiconductor process technology for high-frequency operation, leading to a growing number of transistor-based THz signal sources being reported based on both III–V semiconductor technology and Si-based technology. This section will start with a brief introduction to the basics of oscillators, followed by the description on popular oscillator topologies widely adopted for THz application, such as LC cross-coupled oscillators, Colpitts oscillators, and ring oscillators. Also, additional topics of voltage-controlled oscillators (VCOs) and phase-locked loops (PLLs) will be covered with a focus on their THz operations.

2.3.1 Oscillator Basics

It will be useful to review the basics of oscillator principles from the circuit perspective before providing the details of the various oscillator topologies. There can be two views of oscillators: one-port and two-port views. For the one-port view, as is described in Fig. 2.12a, we assume a passive resonant circuit with a finite loss. Although oscillation may start inside the stand-alone resonator, it will not be sustainable as it gradually decays with the finite internal resistance. However, when the resonator output port is terminated with a network with a negative resistance, say R_1 (with a negative value), the loss of the resonant circuit will be compensated by the negative resistance if $R_1 = -R_2$, where R_2 is the resistance looking into the resonator. This will lead to a sustaining oscillation. One way to obtain the negative resistance is to employ a diode with a special structure as was discussed in the previous section. However, when transistors are available, the negative resistance can be obtained by *circuit techniques*, for which *conventional transistors* can be used with no special device-level modification. This is the primary

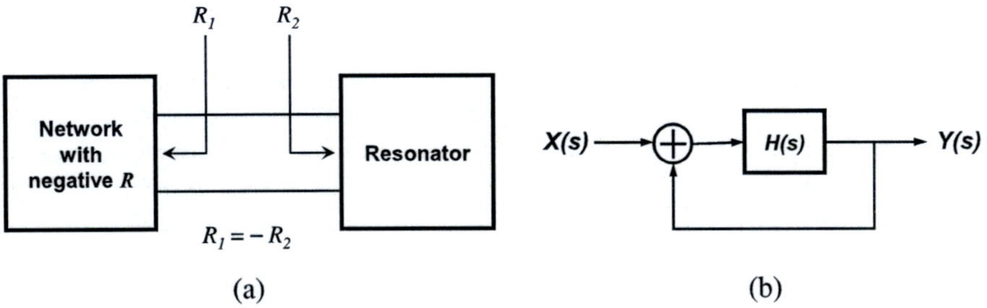

Fig. 2.12 (**a**) One-port view and (**b**) Two-port view of an oscillator

advantage of using active devices, or transistors, for realizing oscillators. Of course, for the oscillators operating at raised frequencies such as the THz band, those standard transistors employed in the circuit should operate at that high-frequency range. In fact, the lack of high-speed active devices used to be a major challenge for oscillator circuits in the past, but now transistor-based THz oscillators are a reality owing to the advance of high-frequency semiconductor technologies based on both Si and III–V compounds (see Fig. 1.5).

The two-port view is depicted in Fig. 2.12b. Assume a two-port network composed of a frequency-dependent core ($H(s)$) and a feedback loop, ported by an input node (X) and output node (Y). The transfer function of the two-port network is given by:

$$\frac{Y(s)}{X(s)} = \frac{H(s)}{1 - H(s)}. \tag{2.3}$$

Hence, when $H(s)$ becomes unity at a certain frequency, ω_o, the gain across the closed-loop network will be infinity. This means that any noise component inside the network that happens to have a frequency of ω_o can be picked up, amplified, and emerge at the output node with a very large amplitude, even without the input signal externally applied. Hence, oscillation is achieved. Of course, the output amplitude will not be infinite, as is suppressed by the nonlinearity of the network. As $H(s)$ is a complex number in general, the condition of $H(s) = 1$ can be divided into the magnitude and phase terms, leading to a pair of conditions:

$$\begin{aligned} &1. |H(s)| = 1 \\ &2. \angle H(s) = 0. \end{aligned} \tag{2.4}$$

These are called the Barkhausen's criteria for oscillation. The split of the original condition into the two conditions provides a useful view of the oscillation. According to these two conditions, the input signal (which may have originated from noise) applied to the two-port will appear at the output node without phase and magnitude change, and then be added to the incoming input signal after passing through the feedback. The modified input signal, now twice in magnitude, will go through the same path and be added to the incoming input signal again, leading to a tripled magnitude. This will go on and on, and the output signal will keep growing in its magnitude, until it reaches a stable magnitude as is limited by the circuit condition. Hence, oscillation is achieved. With this view in mind, consider a slight deviation from the Barkhausen's criteria, say, $|H(s)|$ being larger than unity, while the phase condition remains unchanged. With this modified condition, the two-port network will still develop an oscillation, presumably at a faster rate. Although this modified condition is not exactly the Barkhausen's criteria making Eq. (2.3) infinity, it will still lead to an oscillation. In fact, this is a preferred condition in most practical oscillator designs, as it provides a design margin over any uncertainty to guarantee oscillation start-up.

Now let us consider oscillator topologies widely adopted for high-frequency applications. Three most popular topologies for this purpose are LC cross-coupled oscillators, Colpitts oscillators, and ring oscillators.

2.3.2 *LC Cross-coupled Oscillators*

Figure 2.13a presents the schematic of a typical LC cross-coupled oscillator. Although the circuit shown is based on CMOS, it can also be implemented based on bipolar transistors. In those cases with bipolar transistors, though, decoupling capacitors are typically inserted at the feedback path between the base of one device and the collector of the opposite device, to allow different DC biases for the two nodes. The operation of LC cross-coupled oscillators can be explained by the either one-port view or two-port view described above. The one-port view is adopted here for a simple analysis as is given below.

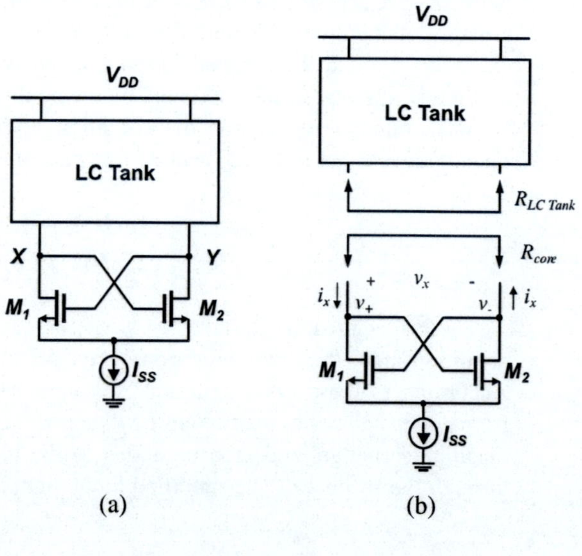

Fig. 2.13 (**a**) Schematic of a typical LC cross-coupled oscillator. (**b**) Input impedance looking into the oscillator core and LC tank

As depicted in Fig. 2.13b, let us divide the oscillator into two parts: the core and the LC resonator. According to the one-port view descried earlier, a negative resistance should be seen when looking into the core from the LC resonator side. This can be proved with a simple circuit analysis, which is presented here to emphasize how a simple active network can generate a negative resistance. To obtain the resistance looking into the oscillator core, R_{core}, let us apply a test voltage v_x across the drain nodes of the differential pair. Then, v_x can be expressed as

$$v_x = v_+ - v_- = v_{\mathrm{g2}} - v_{\mathrm{g1}} = v_{\mathrm{gs2}} - v_{\mathrm{gs1}}. \tag{2.5}$$

The current induced by the test voltage will be equal to the drain current of M_1 and the negative of the drain current of M_2. That is:

$$i_x = i_{\mathrm{D1}} = g_{\mathrm{m1}} v_{\mathrm{gs1}} = -i_{\mathrm{D2}} = -g_{\mathrm{m2}} v_{\mathrm{gs2}}. \tag{2.6}$$

Assuming $g_{\mathrm{m1}} = g_{\mathrm{m2}}$ $(=g_{\mathrm{m}})$, which is valid for most integrated differential pairs, the test voltage v_x can be related to the induced current i_x as:

$$v_x = -\frac{i_x}{g_{\mathrm{m2}}} - \frac{i_x}{g_{\mathrm{m1}}} = -\frac{2 i_x}{g_{\mathrm{m}}}. \tag{2.7}$$

Hence, R_{core} naturally bears a negative value as:

$$R_{\mathrm{core}} = \frac{v_x}{i_x} = -\frac{2}{g_{\mathrm{m}}}. \tag{2.8}$$

This is a good example of how a negative resistance can be readily achieved with standard transistors with a simple circuit topology, not relying on novel device-level structures or phenomena. Note that the analysis given here is a small-signal analysis, and thus the obtained negative resistance is the small-signal negative resistance, which is basically equal to the negative differential resistance (NDR) mentioned earlier by nature. The operation of LC cross-coupled oscillators can be analyzed with the two-port view as well, treating the circuits as a cascaded two-stage amplifiers with reactive loads and a feedback loop, which is not elaborated here.

LC cross-coupled oscillators are widely employed for various RF applications owing to its simple topology and frequency stability especially when compared with ring oscillators. Based on a differential topology, they are inherently robust against common-mode noise and well suited for differential systems. If connection with single-ended systems is needed, however, they may require a balun that converts the differential output signal to a single-ended signal. Alternatively, the signal from one output node of the two oscillator differential output nodes can be taken while the other output node is terminated with a proper load, but with a risk of breaking the symmetry in the circuit. One downside of LC cross-coupled oscillators (or any LC oscillators that comprise inductors and capacitors) is the large area occupied by the on-chip inductors, especially for designs intended for relatively low-frequency

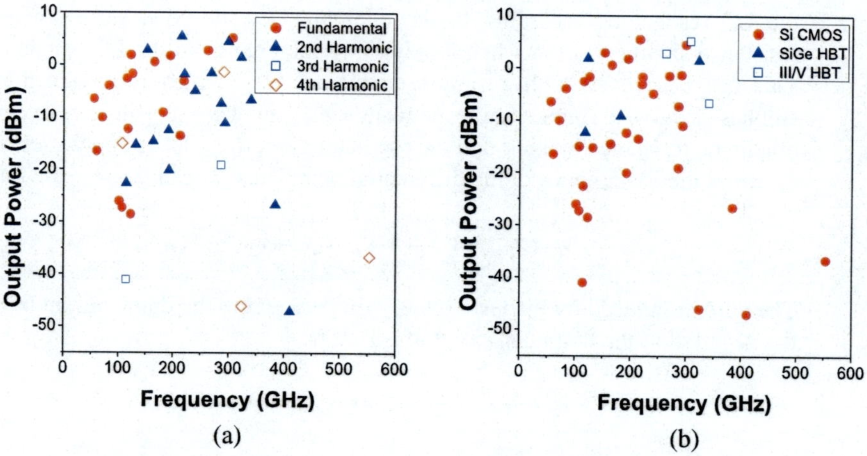

Fig. 2.14 Performance of recently reported LC cross-coupled oscillators operating beyond 100 GHz: (**a**) Plotted for various signal harmonics, (**b**) Plotted for different technologies

bands. A relief for high-frequency applications, including THz ones, is that the area issue becomes less stringent with increasing frequency, since the required inductance and thus the inductor size become small. High-frequency operation also allows chances for transmission lines to replace inductors, which may also help to reduce the circuit area with the reduced wavelength.

There are still challenges for developing LC cross-coupled oscillators operating at the THz band. One is the precise control of the small capacitance and inductance values employed for the LC tank to increase its resonance frequency. The required smaller component sizes are favored for chip size reduction, but, on the other hand, even a tiny change in the device dimension may result in a great impact on the component values. Also, the effect of the parasitic components becomes increasingly significant as the frequency rises. The limited availability of sufficient g_{m} of transistors at high frequencies, which is needed to obtain enough negative resistance, also deters the effort to push the oscillation frequency up to the THz band, although the recent development of high-frequency semiconductor technologies is gradually relaxing this constraint. A popular approach of taking the harmonic signals as the output (will be discussed in Sect. 2.4.1) for elevated frequency will inevitably lower the output power, which is also one of the challenges for THz operation of the LC cross-coupled oscillators.

Despite the challenges, there have been a great number of reported LC cross-coupled oscillators operating in the THz band. Figure 2.14a shows the performance of recently reported LC cross-coupled oscillators operating near or beyond 100 GHz, in terms of the output power and the oscillation frequency. A great portion of the reported oscillators employ harmonic techniques to boost the oscillation frequency (will be discussed in Sect. 2.4). Also, some data points are for the circuits with power combining techniques (will be discussed in Sect. 2.5.2). Figure 2.14b presents the same set of data again, now categorized by the technology they are based on for technology comparison. It is clear that those circuits based on III–V compound semiconductors such as GaAs and InP show higher frequency as well as larger output power. However, a growing number of THz oscillators based on Si technologies, both Si CMOS and SiGe HBT (heterojunction bipolar transistor) technologies, are being reported, with SiGe HBT oscillators tending to exhibit slightly better performance than Si CMOS ones.

2.3.3 Colpitts Oscillators

The second oscillator type to be considered is Colpitts oscillators. The origin of Colpitts oscillators can be dated back to the early twentieth century, around the time when the oscillators based on active devices (vacuum tubes at that time) with positive feedback first appeared [49]. The basic Colpitts oscillator is an oscillator based on a single active device, with a feedback made of a capacitive voltage divider. Hence, the original form of Colpitts oscillators is a single-ended circuit by nature, as opposed to the LC cross-coupled oscillators that are based on differential topology. However, with the advent of integrated circuits, for which the increase in the active device (transistor) count is not much a design concern anymore, the differential version of Colpitts oscillators became also popular. The schematic of a typical single-transistor single-ended Colpitts oscillator is shown in Fig. 2.15a. Note that Colpitts oscillators are also a variant of LC oscillators since they also rely on inductance and capacitance to form a resonator, while often the term "LC oscillators" specifically refers to LC *cross-coupled* oscillators.

To examine how oscillation can be attained with Colpitts oscillators, let us first take a qualitative view on the structure given in Fig. 2.15a. If the voltage at the node X is increased, then the voltage at node Y and thus the source voltage will be accordingly increased, which will lead to a reduction in the drain current. This will

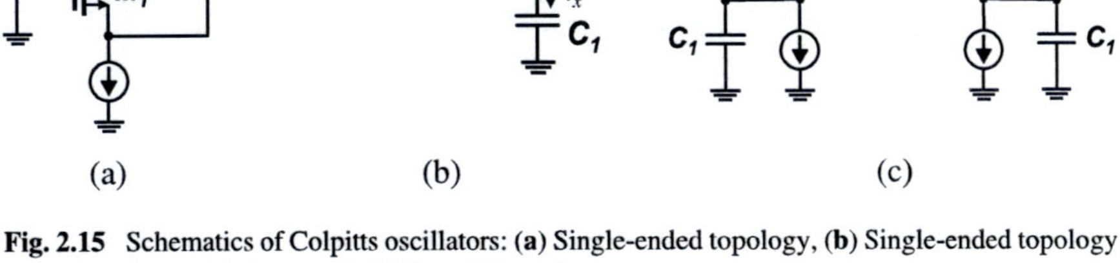

(a) (b) (c)

Fig. 2.15 Schematics of Colpitts oscillators: (**a**) Single-ended topology, (**b**) Single-ended topology seen from the load inductor, (**c**) Differential topology effectively further increase the voltage at X as the voltage drop across the load diminishes, resulting in a positive feedback. This will cause oscillation. One may argue that a positive feedback can be achieved even without the capacitive division, questioning the need for the capacitive division. The truth is that, without the capacitive division (meaning without C_2), the node X will see a resistance equivalent to $1/g_m$, which is not a very large value in typical cases. This will degrade the Q-factor of the LC load. Such a situation is obviously not favored and can be avoided by the adoption of the capacitive division that will transform $1/g_m$ to a larger value, or $\sim(1 + C_1/C_2)^2/g_m$ for the configuration in Fig. 2.15a [50]. This is the main idea behind Colpitts oscillators.

As an alternative way to understand the operation of Colpitts oscillators, let us take a quantitative view on the oscillator, in which we will obtain the impedance of the oscillator core seen from the load inductor. For this, the oscillator is redrawn as Fig. 2.15b, and the impedance is calculated as the ratio of v_x and i_x. The current i is the sum of i_1 and i_2, which are the drain current of M_1 and the current flowing across C_2, respectively. Noting that i_1 is driven by v_{gs} of M_1, which is given by the voltage across C_1 with the sign switched, following relations can be obtained for the current components:

$$i_1 = g_m v_{gs} = g_m \left(-\frac{i_x}{C_1 s} \right), \tag{2.9}$$

$$i_2 = (v_x + v_{gs}) C_2 s = \left(v_x - \frac{i_x}{C_1 s} \right) C_2 s, \tag{2.10}$$

$$i_x = i_1 + i_2 = -g_m \left(\frac{i_x}{C_1 s} \right) + \left(v_x - \frac{i_x}{C_1 s} \right) C_2 s. \tag{2.11}$$

By rearranging the terms and assuming a sinusoidal signal so that $s = j\omega$, the impedance Z_{in} can be obtained:

$$i_x \left(1 + \frac{g_m}{j\omega C_1} + \frac{C_2}{C_1} \right) = j\omega v_x C_2, \tag{2.12}$$

$$Z_{in} = \frac{v_x}{i_x} = -\frac{g_m}{\omega^2 C_1 C_2} - j\frac{1}{\omega} \left(\frac{1}{C_1} + \frac{1}{C_2} \right). \tag{2.13}$$

The result clearly shows that a negative resistance is obtained, which is connected in series with C_1 and C_2. Hence, by terminating the input with an inductor, which is actually the case for Colpitts oscillators, the reactive and the (negative) resistive components will be compensated by the inductance and the parasitic resistance of the inductor, respectively. This will trigger oscillation.

The circuit considered above is based on the common-gate (or common-base for bipolar version) topology. However, the common-drain (common-collector) topology can also be adopted, for which the capacitive division is made between the gate (base) and the ground. Also, for each of these topologies, differential versions can be implemented. Figure 2.15c shows a differential Colpitts topology of the common-gate case. For the differential cases, a phase difference of 180° needs to be maintained between the two transistors that form the differential pair, which is typically achieved through a capacitive or inductive coupling across the two branches. It is noted that for some Colpitts oscillators, the capacitive divider is realized by parasitic capacitance of transistors, without employing separate devoted capacitors. This is especially true for oscillators operating at high-frequency ranges, where small device capacitance value is sufficient to generate reactance required for the capacitive division. It is to be mentioned that a variant of the Colpitts oscillator, in which another capacitance is inserted in series with (sometimes in parallel with) the load inductor in the basic Colpitts topology, bears a different name, Clapp oscillator. As proposed by Clapp in 1948, this version of Colpitts oscillator is known to provide a much better frequency stability [51] with an additional degree of freedom for design obtained from the newly added capacitance component, and often adopted for high-frequency applications.

For the designers of high-frequency LC oscillators, one of the first choices they need to make is the topology of the oscillator, which typically narrows down to either LC cross-coupled oscillators or Colpitts oscillators. Although there are various performance aspects to be considered for the decision, the phase noise is usually at the top of the list. To make the choice difficult, there is no obvious winner in terms of the phase noise performance. Based on the analysis in [52], LC cross-coupled oscillators show a superior phase noise performance than differential Colpitts oscillators. This observation has been confirmed by some works that followed, but the comparison was limited for oscillators based on CMOS technologies only. Because the way the device-level noise affects the phase noise of the overall oscillator is different for LC cross-coupled oscillators and Colpitts oscillator (Colpitts with more noise from device but less conversion to phase noise; LC cross-coupled with less noise from device but more conversion to phase noise [52]), the preference may turn out different for different device types employed.

For example, bipolar transistors exhibit much better $1/f$ noise characteristics than MOSFETs, mainly because the main current path in (vertical) bipolar transistors stays away from the interface usually full of traps, while (lateral) MOSFET current runs just beneath the oxide–semiconductor interface, making it vulnerable to carrier trap/detrap. Also, bipolar transistors show a much larger transconductance g_{m} than MOSFETs, a direct result of the exponential relation between the current and base bias. These device properties certainly make difference in phase noise generation in oscillators. In fact, in another comparison study for the two types of oscillators, which was made in simulation based on bipolar transistor technology, it was reported that differential Colpitts oscillators show better phase noise performance than LC cross-coupled oscillators [53]. This result is consistent with the fact that Colpitts oscillators are particularly popular with bipolar transistor technologies. The relative performance may depend on the oscillation frequency range as well. The comparison studies mentioned above are for frequencies below 10 GHz. For operations in high-frequency bands, including the mm-wave and THz bands, there is a

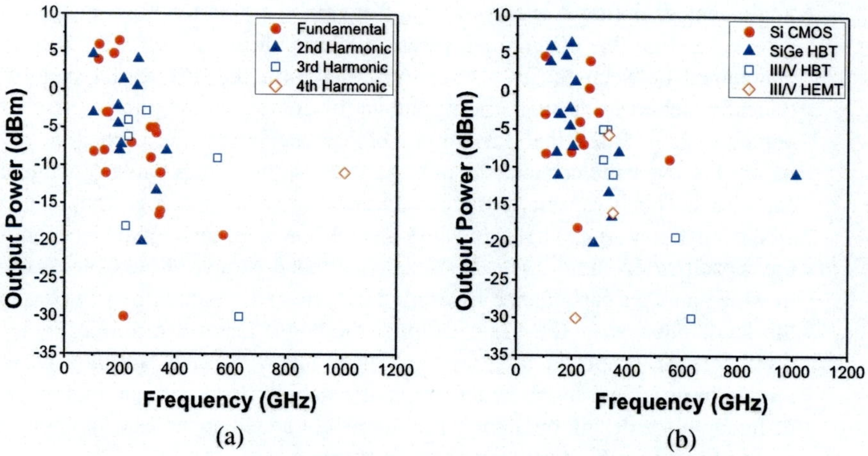

Fig. 2.16 Performance of recently reported Colpitts oscillators operating beyond 100 GHz: (**a**) Plotted for various signal harmonics, (**b**) Plotted for different technologies

widely spread notion that Colpitts oscillators exhibit better performance in terms of the phase noise. Despite this apparent complication in the comparison, it is commonly accepted that the start-up condition is more stringent for Colpitts oscillators, partly due to the higher g_m needed for Colpitts than LC cross-coupled oscillators. This has prompted some techniques developed to mitigate this issue [54].

The challenges for realizing THz Colpitts oscillators are similar to those for LC cross-coupled oscillators as discussed above, since they are both based on LC resonators. Again, these challenges are gradually being addressed, leading to a growing number of Colpitts oscillators operating beyond 100 GHz. Figure 2.16 shows the performance of recently reported high-frequency Colpitts oscillators, again on the output power—oscillation frequency space. The highest oscillation frequency reported till today is 1 THz, which is from a 42-component Colpitts oscillator array based on SiGe HBT with a combined output power of −10.9 dBm [55].

2.3.4 Ring Oscillators

The third type of oscillator to be discussed is the ring oscillator. Ring oscillators do not necessarily contain an LC resonator, a clear contrast to the two oscillator types discussed so far. Instead, the delay inherent to inverter stages is exploited to generate the phase shift that is required for the phase inversion in the two-port view of the oscillator as was discussed above. A typical configuration of a ring oscillator is shown in Fig. 2.17a. It is composed of a series of inverters cascaded, where the output of the last stage is fed back to the input of the first, thus forming a loop or a ring. The intrinsic delay of a single inverter is π (or 180°). If there is an extra time

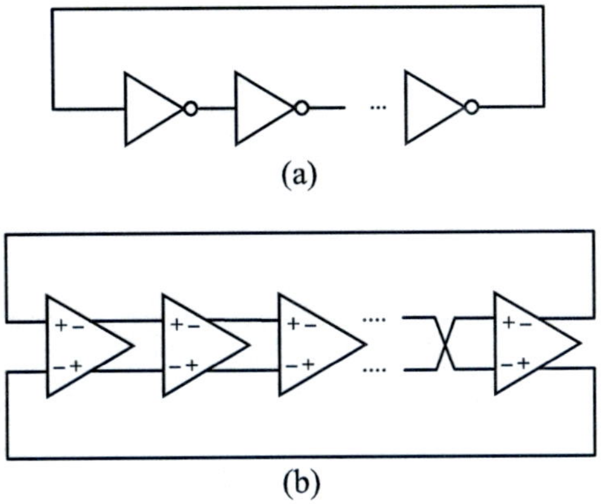

Fig. 2.17 A typical configuration of a ring oscillator: (**a**) Single-ended topology, (**b**) Differential topology

delay of τ per each inverter stage, which is basically a gate delay due to the finite RC time constant, the total delay around the loop will be equal to $n(\pi + \phi)$, where n is the total number of the inverter stages around the loop and ϕ is the phase shift corresponding to the time delay τ. If n is an odd integer and $n\phi = \pi$, the total phase shift around the loop will be an integer multiple of 2π. Hence, for every round trip of a signal around the loop, the signal will be added in-phase and the signal strength will grow, leading to an oscillation. Based on the relation between ϕ and τ as follows,

$$\phi = 2\pi \cdot \frac{\tau}{T}, \tag{2.14}$$

where T is the period of the signal, the oscillation frequency is simply given as:

$$f_0 = \frac{1}{T} = \frac{\phi}{2\pi} \cdot \frac{1}{\tau} = \frac{1}{2n\tau} \quad \text{since} \quad \phi = \frac{\pi}{n}. \tag{2.15}$$

In most high-frequency applications, n is kept as the smallest number available, or three, to maximize the oscillation frequency for a given set of design parameters (certainly, $n = 1$ is not acceptable because a single stage does not form an oscillator loop). Hence, it is clear that the reduction of τ is a key to the successful high-frequency and THz operation of ring oscillators. For most advanced semiconductor technologies of today, the gate delay τ of around 2 ps is obtained [56]. With $n = 3$, this would result in f_0 of 167 GHz, indicating the feasibility of THz operation of ring oscillators. The frequency can be pushed even higher with harmonic techniques applied, which will be discussed later in Sect. 2.4.1. The constraint of n being an odd integer can be relaxed if a differential version of ring oscillators is considered. As shown in Fig. 2.17b, if the output wiring is crossed for one stage in a differential ring

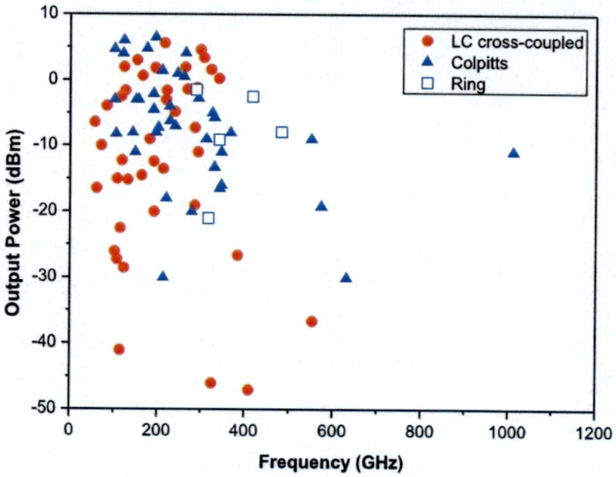

Fig. 2.18 Performance comparison of three types of transistor-based oscillators: LC cross-coupled, Colpitts, and ring oscillators. The data points include all the harmonic numbers and process technologies (Si CMOS, SiGe HBT, III–V, HEMT, III–V HBT)

oscillator, an even number of stages will satisfy the oscillation condition described above.

Ring oscillators benefit from the small area owing to the absence of the bulky on-chip inductors that are typically needed for LC oscillators. However, for ring oscillators, the oscillation frequency is not very precisely predictable (due to the uncertainty in the gate delay, etc.), and the phase noise is much worse than the oscillators with LC resonators. When it is injection-locked with a much more stable signal source, however, those issues are largely resolved, since the oscillation frequency and the phase noise are now dictated by the signal source that locks the ring oscillator rather than the ring oscillator itself. For this reason, ring oscillators are often employed inside a locked circuit, such as phase-locked loops (PLLs) that will be discussed in Sect. 2.3.6.

Figure 2.18 shows recently obtained performance of ring oscillators along with those of LC cross-coupled and Colpitts oscillators in comparison. As the plot shows, the performance of ring oscillators appears to fall slightly behind those of LC resonator-based oscillators based on the given data points. Remember, though, they have their own attractive features, such as the small area as just mentioned above. Figure 2.18 also provides a direct comparison between LC cross-coupled oscillators and Colpitts oscillators (the data points are based on Figs. 2.14 and 2.16.) The two types of oscillators are almost on par in their output power performances, while higher operation frequencies have been achieved with Colpitts oscillators.

2.3.5 Voltage-Controlled Oscillators (VCOs)

Many applications of oscillators benefit from the tuning of the oscillation frequency. For communication applications, for example, the channel or band selection would require the tuning capability. For imaging applications, frequency tuning would allow imaging over different frequencies, which is known as the spectral imaging. There is a need from a highly practical point of view as well: when the fabricated oscillator shows an oscillation frequency off the target, it can be still recovered if frequency tuning is available. Finally, frequency tunability is indispensable for an oscillator to be employed for PLLs, which will be discussed in more detail in the next subsection.

Frequency tuning can be realized with various approaches. For oscillators based on LC resonators, such as LC cross-coupled oscillators or Colpitts oscillators as introduced earlier, the oscillation frequency is determined by the resonant frequency of the resonator ($f_{osc} \propto 1/\sqrt{LC}$). Hence, the tuning of inductance or capacitance will lead to frequency tuning. For oscillators driven by the gate delay, such as in ring oscillators, it is the RC time constant that dictates the oscillation frequency ($f_{osc} \propto 1/RC$). Therefore, the control over resistance or capacitance values will enable the frequency tuning. Despite these various options, the tuning of the oscillation frequency of most oscillators is predominantly achieved by capacitance variation. Hence, a capacitor with variable capacitance, or varactor, is a core component for oscillators with frequency turnability. Because the varactor tuning is accomplished by *voltage* tuning in most cases, such frequency-tunable oscillators are called voltage-controlled oscillators (VCOs). VCOs can be implemented with most types of oscillators, including the three topologies discussed previously.

Varactors can be realized by various methods with semiconductor technologies. The most straightforward approach is to make use of a *p-n* junction, which is readily available in conventional semiconductor technologies. It is well known that the capacitance (per unit area) of a *p-n* junction is given by $\varepsilon_0\varepsilon_s/x_d$, where ε_0 is the permittivity of vacuum and ε_s is the relative permittivity or the dielectric constant of the semiconductor, and x_d is the depletion width of the *p-n* junction. For one-sided step junction (where doping concentration difference between the *n*- and *p*-region of the *p-n* junction is so large that the depletion region is formed mostly in one side of the junction only), which is the case for many practical *p-n* junctions inside semiconductor devices, the depletion width is given as:

$$x_d = \sqrt{\frac{2\varepsilon_0\varepsilon_s(V_{bi} + V_r)}{qN}}, \tag{2.16}$$

where q is the elementary charge, N is the doping concentration (of the lowly doped side of the *p-n* junction), V_{bi} is the built-in potential of the *p-n* junction, and V_r is the reverse bias across the junction. Hence, the capacitance is given as:

$$C = \sqrt{\frac{\varepsilon_0\varepsilon_s qN}{2(V_{bi} + V_r)}}. \tag{2.17}$$

Thus, capacitance varies over V_r and the junction can be used as a varactor. This is a popular approach to realize a varactor, especially for bipolar transistor technologies, in which the base–collector junction is typically utilized for this purpose.

However, a downside of this type of varactor is the small tuning range. V_{bi} is already close to 1 V for typical *p-n* junctions in Si technology and may exceed 1 V for GaAs or InP technologies, whereas V_r is limited by a maximum voltage allowed for a given technology. Assuming $V_{bi} = 800$ mV and the maximum allowed V_r of 2 V, a simple calculation reveals that the capacitance tuning range is around 87%. This is the ideal case and the actual tuning range will be smaller, typically down to around 50%, due to fixed capacitance components of the device interior and nonuniform doping profile. Besides, when it is embedded in a circuit, fixed capacitance from other neighboring circuit elements will additionally limit the circuit tuning range. Also, the oscillation frequency has a square-root relation with the capacitance for LC oscillators ($1/\sqrt{LC}$), further suppressing the available frequency tuning of the oscillators.

There is a variant of the *p-n* junction varactor, which is devised to overcome the issue and boost the tuning range. Called the hyperabrupt junction varactor [57], it employs a doping profile tailored for a larger capacitance variation for a given voltage tuning. To be more specific, it has a retrograded doping concentration profile (lower doping when farther away from the junction), which causes a faster change in depletion width (and thus capacitance) with voltage change than the uniform doping case. As a result, capacitance tuning can be significantly improved, by a factor of 2 or more when compared to the conventional junction varactors. However, this approach would require an additional ion implantation step dedicated to the varactor formation, adding to the total fabrication cost. This is what you need to pay in exchange of performance improvement. Note that the conventional junction varactors are basically "free," as they make use of existing junctions formed for other purposes.

Widely adopted in Si CMOS technologies is the accumulation-mode MOS varactor [58, 59]. The capacitance–voltage (*C–V*) curve of MOSFETs shows a wide capacitance variation over the voltage across the gate and the source/drain (Fig. 2.19a). This is because the width of the depletion region, formed just beneath the oxide layer, changes with the gate voltage V_g, leading to a variation in the capacitance. This depletion capacitance (C_d) is in series connection with the oxide capacitance (C_{ox}), resulting in the total gate capacitance ($C_{tot} = C_d \parallel C_{ox}$). As C_d decreases with increasing V_g, so does C_{tot} as shown in Fig. 2.19a with a lower curve (denoted as high frequency). With this *C–V* curve given, any MOSFET can serve as a varactor in principle. However, the MOSFET *C–V* relation shows a frequency dependence, exhibiting a "snap-back" behavior if the voltage changes slowly or at low-frequency operation. In this case, the capacitance returns back to the "high" state ($= C_{ox}$), although the MOSFET is still in the inversion mode with a sufficiently large V_g. This is due to the junction structure of the conventional MOSFETs: *n*-type source/drain regions formed on *p*-type substrate for NMOSFETs (or *p*-type source/drain regions formed on *n*-type substrate for PMOSFETs). When the voltage change is slow with this type of device in the inversion mode, the depletion width (and thus the total space charge in the depletion region) does not respond to the change, because a sufficient time is allowed for the inversion charges to be generated in the source/drain region and supplied to the channel, which absorbs the effect of the

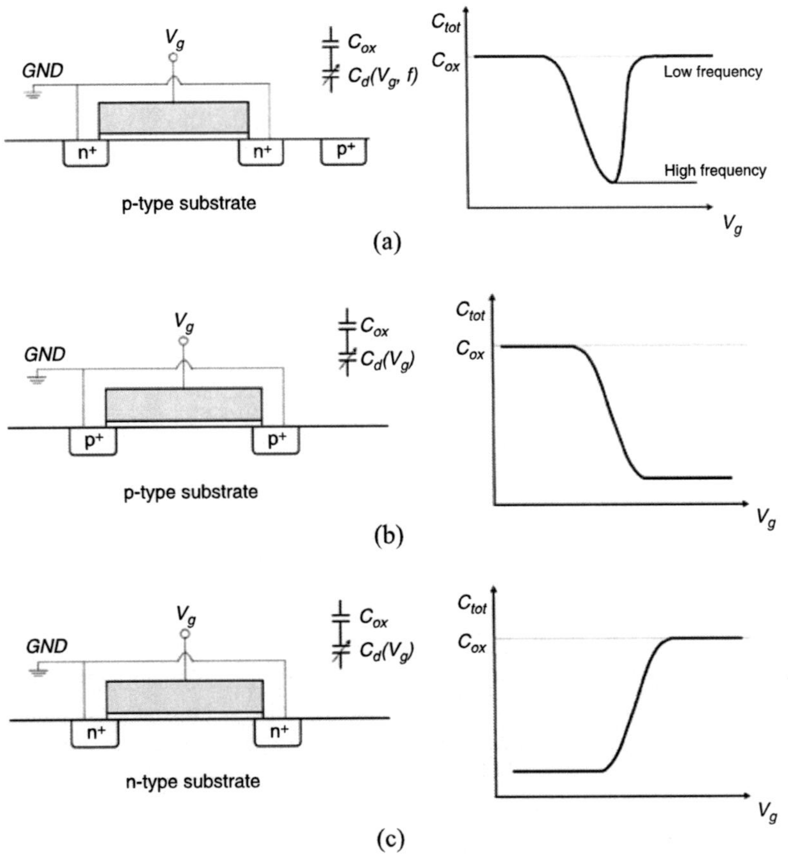

Fig. 2.19 (**a**) Structure of a conventional *n*-type MOSFET and its *C–V* curve. Note in the *C–V* curve that the frequency-dependent split between low- and high-frequency states is bimodal, without intermediate states. (**b**) Structure of a *p*-type accumulation-mode MOS varactor and its *C–V* curve. (**c**) Structure of an *n*-type accumulation-mode MOS varactor and its *C–V* curve

voltage change. In view of the definition of capacitance, the ratio of charge variation over voltage variation, there is no capacitance associated with the depletion region in this case as the depletion charge is fixed at this low-frequency condition. Hence, C_d does not contribute to C_{tot} at low-frequency operation, which leaves C_{ox} the only capacitance component of C_{tot}, leading to the high state of the capacitance. In contrast, in high-frequency operation, the fast change of voltage would not allow this process, and thus both C_d and C_{ox} contribute to the total capacitance C_{tot}, maintaining the low state of capacitance throughout the inversion mode. It is noted in the *C–V* curve that the frequency-dependent split between low- and high-frequency states is bimodal, without intermediate states in between.

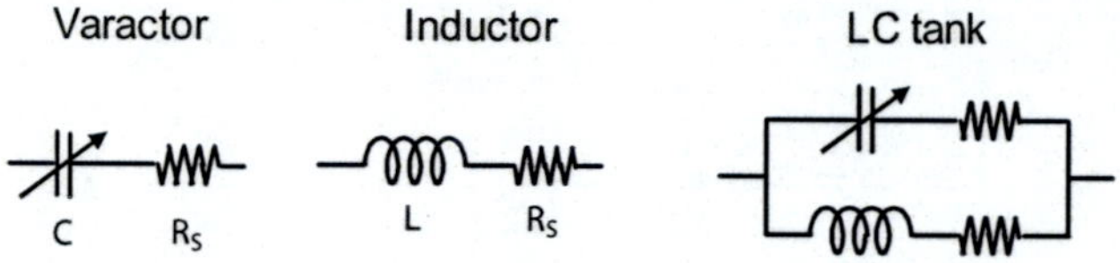

$$Q_{\mathrm{var}} = \frac{1}{\omega C R_S} \qquad Q_{ind} = \frac{\omega L}{R_S} \qquad \frac{1}{Q_{tot}} = \frac{1}{Q_{\mathrm{var}}} + \frac{1}{Q_{ind}}$$

Fig. 2.20 Simple circuit model and Q-factor of a varactor, inductor, and LC tank

This troublesome snap-back behavior at low frequencies, which limits the allowed tuning voltage range, can be eliminated with the accumulation-mode MOS varactors. In the accumulation-mode MOS varactors as shown in Fig. 2.19b, c, the source/drain region and the substrate are now formed with the same polarity of semiconductor. For a p-type (n-type) accumulation-mode MOS varactor, p^+ (n^+) source/drain is formed on a p-type (n-type) substrate. In the inversion mode, the inversion charges are not generated fast enough to instantly respond to the AC voltage change even at low-frequency operation, due to the absence of the source/drain region that would supply the inversion charge (source/drain supplies the accumulation charge, not inversion charge, in the given structure), an outstanding difference from the conventional MOSFET case. As a result, the depletion capacitance C_{d} continues to contribute to the total capacitance C_{tot} at the inversion mode, avoiding the snap-back behavior. This makes the accumulation-mode MOS structure an attractive option for varactor applications. A typical capacitance variation for the accumulation-mode MOS varactor reaches as high as 3, or a tuning range of around 200%, much higher than what p-n junction varactors would provide. The large capacitance ratio, as well as the favorable C–V profile, has made the accumulated-mode MOS varactor a highly popular varactor solution in Si CMOS technologies.

Up to this point, various types of varactors have been reviewed with their respective pros and cons. One common behavior that is found from all types of varactors (in fact, all types of capacitors) is the decreasing Q-factor with increasing frequency. This can be easily understood from the simple circuit model of varactors as shown in Fig. 2.20, which is composed of C (main capacitance component) and R_{S} (parasitic series resistance). For this configuration, the Q-factor is given as $1/(\omega C R_{\mathrm{s}})$. This relation clearly explains the decreasing trend of the Q-factor. For actual devices, the trend will slightly deviate from this pure inverse relation due to other components neglected in the circuit model, but the decreasing trend will be maintained at least up to the self-resonance-frequency (SRF) of the device. This can be compared with the inductor case, which is also shown for comparison. For inductors, the Q-factor increases with frequency with a linear relation, as is given by $\omega L/R_{\mathrm{s}}$. Again, the actual behavior of on-chip inductors will be altered by other components omitted in the model, but the overall trend will be still valid. The low Q-factor of varactors is known as a major problem in VCOs, especially for those oscillators operating at high frequency, including the THz band. The total Q-factor

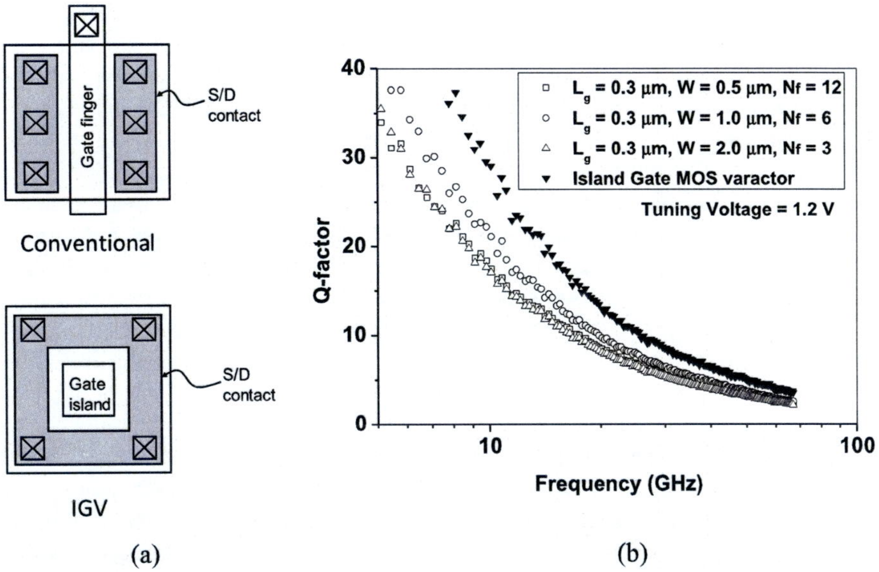

Fig. 2.21 Comparison of the island-gate varactor (IGV) and conventional MOS varactor: (**a**) Layout (only the unit device is shown), (**b**) Measured performance (IGV is for $20 \times 0.3\ \mu m \times 0.3\ \mu m$ device). (© 2009 IEEE [60])

of an LC tank composed of a capacitor and an inductor will be determined by $1/Q_{LC} = 1/Q_{var} + 1/Q_L$. Hence, the Q-factor of the LC tank is limited by the Q-factor with the lower value, which is mostly the Q-factor of the varactor at high-frequency operation. In fact, at around 100 GHz, the Q-factor of typical varactors drops down well below 10. This makes the design of VCOs more challenging than fixed-frequency oscillators for high-frequency operation. It also leads to degraded phase noise performance of oscillators when varactors are included.

There have been attempts to improve the Q-factor of varactors. One approach is introduced here. The proposed varactor, called the island-gate varactor (IGV), adopts a different shape of gate electrode than conventional MOS varactors, which helps to reduce the series resistance [60, 61]. The conventional MOS varactors inherit the finger-shaped gate of MOSFETs, which would allow the minimum gate length that is required for the improved speed of transistors. In IGVs, where the minimum gate length is not necessarily desired, the gate adopts a square shape, as is illustrated in Fig. 2.21. With this structure, source/drain current will access the varactor core from four sides, rather than two sides as is for the finger-shaped gate, leading to a reduction in the source/drain access resistance. Further, the gate contact can be made directly on top of the gate capacitance for IGVs, which will lead to a reduced gate access resistance. Note that for conventional MOS varactors, the gate contact is made off the finger area. The remote gate contact necessitates gate current flowing along the narrow gate finger, resulting in a larger gate access resistance. A comparison study based on actually fabricated devices showed that IGVs indeed show reduced series resistance and significantly improved Q-factor [60, 61]. The study was extended to build and compare VCOs based on IGVs and the conventional MOS varactors [62], which showed improved phase noise for the VCO employing IGV, as expected.

There have been efforts to build VCOs *without* varactor as well. Various techniques have been reported for such varactor-less VCOs, which are in many cases based on variable inductors in principle although the details may vary. In an approach named magnetic tuning [63], a transformer-capacitor tank is employed, in which one of the two windings of the transformer forms a loop in series with a capacitor. For this resonator structure, the resonant frequency is determined by the self and mutual inductances of the transformer and the capacitance of the capacitor, as well as the ratio of the *currents* in the two windings of the transformer. Hence, by controlling the current ratio, a frequency tuning is achieved when it is employed in an oscillator. The tunable oscillator built in this way is in fact a current controlled oscillator, or ICO, as opposed to VCO. Another example of the current controlled oscillator was reported in [64]. In this work, a tunable inductor is realized by terminating the base of an emitter follower with an RL degeneration network. With this structure, an effective inductance is obtained when looking into the emitter node of the transistor, the value of which is controlled by the current level of the emitter follower. Hence, the block works as a current-tuned inductor. By combining the tunable inductor with a transformer in the differential topology, an oscillator operating at 200 GHz was realized with a frequency tuning range of 3.5%. Yet another study reports a frequency tuning realized by tuning the transconductance. By properly arranging RLC network with a dependent current source driven by a transconductance, an inductive impedance can be established across the current source. The effective inductance depends on the transconductance, and thus an inductance tuning is achieved by controlling the transconductance, which can be readily obtained by controlling a transistor bias current. A tunable oscillator was realized with the tuning component showing a 23.6% tuning near 26 GHz [65].

2.3.6 Phase-Locked Loops (PLLs)

Stand-alone oscillators implemented by on-chip semiconductor devices typically show much worse frequency stability and phase noise than off-chip high-Q oscillators such as crystal oscillators. One highly popular technique to enhance the performance of the low-Q oscillators is to lock them with another high-Q oscillator by forming a loop, for which phase-locking is widely adopted. Such phase-locked loops (PLLs) are employed for a vast wide range of applications that need highly stable signal source. As is depicted in Fig. 2.22, a typical PLL is composed of a VCO, frequency divider chain, phase/frequency detector (PFD), charge pump, and loop filter. (Sometimes this particular PLL topology is called the Charge Pump PLL (CPPLL) because of the presence of the charge pump. It is also called Type II PLL, as opposed to Type I PLL that does not include a charge pump and employs a

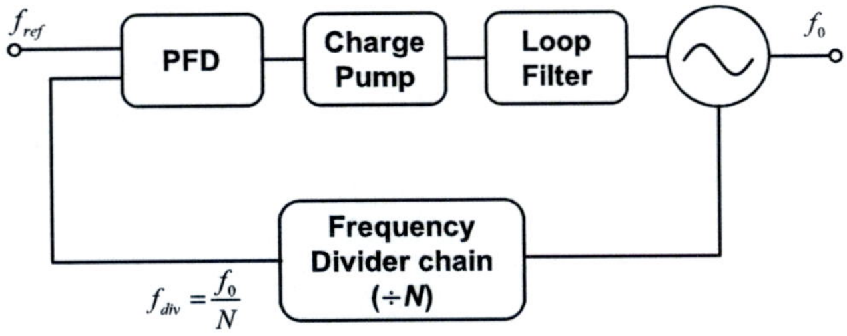

Fig. 2.22 Block diagram of a typical PLL

phase detector (PD) instead of PFD.) A detailed description of PLLs can be found elsewhere, and we will here focus on the issues of realizing high-frequency PLLs, which relates the high-frequency components in PLLs, say VCO and the front stages of the frequency divider chain.

The key principle of PLL operation is simple: if there occurs a drift in frequency (and thus the phase) with the embedded oscillator, it is immediately detected by the PFD and a restoration process is triggered along the loop, which would bring the frequency back to the original frequency. For this to happen, the oscillator needs to have a tuning mechanism installed, preferably by voltage adjustment, which requires the oscillator to be a VCO. Since the output of the VCO is the output of the PLL, the issues of developing PLLs operating at higher frequency bands, including the THz band, inherit those for high-frequency VCOs. These issues related to VCOs have already been discussed above and will not be repeated here. Instead, a review of *frequency dividers* operating at high frequencies will be presented below.

Frequency dividers (also called prescalers) operating at low frequencies can be readily realized with digital logic circuits, for which CMOS latches with current-mode logic (CML) are widely adopted. The *digital CML frequency dividers* generally show a wide locking range, and its operation frequency can be stretched up to quite high frequencies, nearly 100 GHz as indicated by a recent result [66]. However, such a frequency extension with digital frequency dividers typically accompanies a rapid increase in the DC power dissipation. For example, the 94-GHz frequency divider reported in [66], which is basically a flip-flop made of two CML latches, consumes 65 mW. This is quite large and comparable to the power consumed by a VCO in PLLs. The combined overall power would lead to an excessive PLL power dissipation. Thus, from a practical viewpoint, the frequency dividers for high-frequency PLLs, especially those located at the front-end of a divider chain, need to take a different topology suited for higher frequency operation.

A *dynamic divider* is one of the favored high-frequency solutions. Also called Miller divider after the name of the inventor [67] (sometimes it is called the regenerative divider as well, reflecting the regenerative nature of the operation), it is basically a mixer followed by a low pass filter, the output of which is fed back to the input (see Fig. 2.23). With this configuration, a signal with a half frequency of the

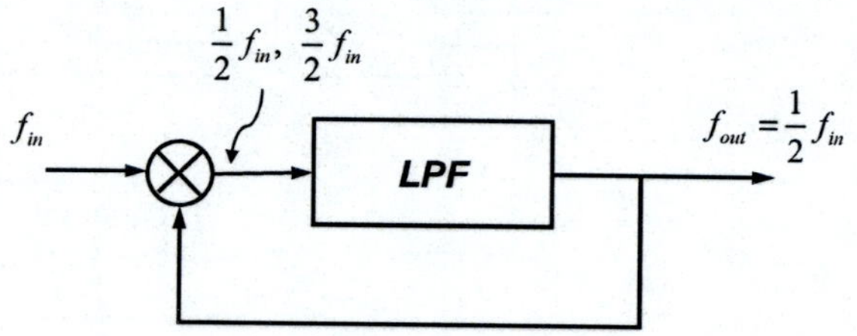

Fig. 2.23 Block diagram of a dynamic frequency divider (also known as Miller divider or regenerative divider)

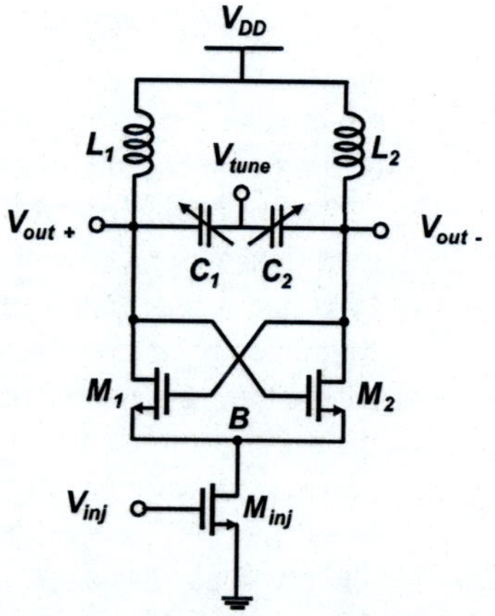

Fig. 2.24 Schematic of an example injection-locked frequency dividers (ILFDs). This circuit is based on an LC resonator-type oscillator while a ring-type oscillator can also be employed as the core of ILFDs

input ($f = f_{in}/2$) can be sustained at the output, while another output frequency of the mixer ($f = 3f_{in}/2$) is filtered out by the LPF. Although a higher division ratio can be attained with an extra component inserted in the loop, a division ratio of 2 is generally obtained from a dynamic divider as is achieved with the standard structure shown in Fig. 2.23. The operation frequency of dynamic dividers easily enters the THz band. With a SiGe HBT technology, a dynamic divider with an operation frequency beyond 200 GHz has been reported [68], which showed a wide locking range as well that exceeds 150 GHz. With an InP HBT technology, a dynamic divider has reached beyond 500 GHz [69]. Although dynamic dividers are preferred over the digital dividers (sometimes called static dividers) for high-frequency operation, it still suffers from the power dissipation issue. The 217 GHz operation in [68] required a DC power dissipation of around 200 mW including the buffer circuits.

From a power dissipation perspective, *injection-locked frequency dividers* (ILFDs) are the most attractive divider option for high-frequency applications [70]. As can be seen from an example circuit in Fig. 2.24 [71], it is an oscillator with an injection port, through which the input locking signal can be injected. Hence, it is basically an injection-locked oscillator. When the injection signal frequency is a multiple harmonic of the free-running frequency of the oscillator, the circuit functions as a frequency divider. (Likewise, the circuit may operate as a frequency multiplier if the free-running frequency is a multiple harmonic of the injection signal frequency.) One favorable aspect of ILFDs is that a frequency division ratio higher than 2 can be easily attained. In fact, one can obtain different division ratios for a given ILFD without changing the circuit, by simply applying injection signals with different harmonic numbers. Higher division ratios will save power and area of the divider chain, although the operation becomes more challenging as the division ratio grows. While a division ratio of 2 or 3 are still most popular choices for ILFDs, higher ratios have also been adopted and reported [72]. The operation frequency of ILFDs has reached 300 GHz with a Si CMOS technology [73]. The DC power required to drive the 300-GHz ILFD was only 9.2 mW, a tiny fraction of those needed for digital dividers or dynamic dividers with even lower operation frequencies. High-frequency PLLs will greatly benefit from this type of divider for reduced operation power.

One major drawback of the ILFDs, however, is the narrow locking range. The locking range of the above 300-GHz ILFD is less than 10 GHz, or around 3% of the operation frequency. A narrow locking range of dividers increases the chance of lock failure in the entire loop of PLLs. Also, a narrow locking range would require a significantly increased input power even for a slight mismatch between the (harmonics of) input signal frequency and free running frequency. This imposes a huge challenge on circuit design for its preciseness, which will also be affected by the device model accuracy and fabrication variation. The adoption of a VCO rather than a fixed-frequency oscillator for the divider core to enable locking frequency *tuning* may help in some cases (the resultant improved locking range with voltage tuning is sometimes called the *operation range*, as opposed to the *locking range* obtained without tuning). However, such a voltage tuning is not a practical option for ILFDs embedded in a PLL. There have been many reports to improve the locking range of ILFDs, most of which can be reduced to the efforts to lower the Q-factor of the oscillator core and its current level, following the implication by the relation for locking [74]:

$$\omega_{\mathrm{L}} \propto \frac{\omega_0}{2Q} \cdot \frac{2}{\pi} \cdot \frac{I_{\mathrm{inj}}}{I_{\mathrm{osc}}}, \tag{2.18}$$

where ω_{L} is the locking range, ω_0 is the oscillation frequency of the ILFD core, I_{inj} is the injection current, and I_{osc} is the oscillation current. Another approach to overcome the narrow locking range of ILFDs and ensure the locking with the adjacent VCO in a PLL is to employ the identical topology for the ILFD and the VCO [75]. With this strategy, the two circuits will be affected in a similar way with external variations, which may arise from device model, circuit simulation, and fabrication issues, and so forth, possibly retaining the frequency alignment between the ILFD and the VCO.

Up to this point, we have observed a variety of issues present in the high-frequency components to be included in PLLs, i.e. VCOs and frequency dividers. Despite all these issues and related challenges for these components, researchers

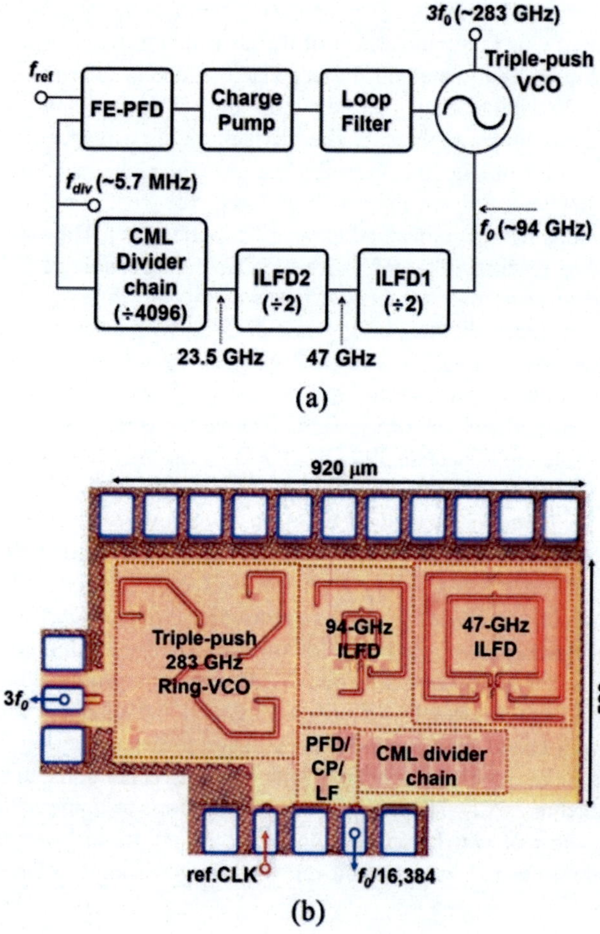

Fig. 2.25 (**a**) Block diagram of a CMOS 283-GHz PLL. (**b**) Chip photo of the fabricated PLL. (© 2018 IEEE [77])

have found ways to operate PLLs at elevated frequencies including the THz band. A growing number of PLLs working at the THz band are being reported, including the one that operates at 560 GHz based on a Si CMOS technology [76]. Another THz PLL to be introduced here as an example is shown in Fig. 2.25 [77], which is a Si CMOS PLL operating around 280 GHz. For the VCO, a triple-push topology based on ring oscillators is employed. The frequency divider chain converts 283 GHz down to 5.8 MHz with a division ratio of 16,384, in which the first two dividers are realized with ILFDs while CML static dividers complete the remaining division. Owing to the availability of the fundamental signal from the embedded triple-push VCO (f_0), which is one third of the PLL output frequency ($3f_0$), the operation frequency and the total division ratio for the divider chain is relaxed by taking f_0

as the input signal frequency to the chain. With a total power dissipation of 114 mW, about a half coming from VCO and one-fifth from the two ILFDs, the PLL exhibits an output power of −22.5 dBm.

2.4 Frequency Enhancement Techniques

As the required operation frequency of high-frequency systems keeps increasing, it may approach the cutoff frequency of the transistors being used. Hence, achieving fundamental-mode oscillation at the target frequency is increasingly difficult, which mainly arises from the limited transistor gain available around the frequency. As briefly mentioned above, a widely accepted technique to overcome this limit is to extract an internal *harmonic* signal of the oscillator as the output signal. Another technique is to *multiply* the output frequency of the oscillator by placing a frequency multiplier next to it. Both the techniques have their own pros and cons. In this section, these two techniques are introduced and discussed.

2.4.1 n-push Techniques

The technique to extract the *n*th harmonic signal internal to the oscillator is widely referred to as *n*-push technique [78, 79]. The oscillators employing the technique are called the *n*-push oscillator. The most widely adopted are the push-push (2-push) [80, 81] and the triple-push (3-push) techniques [82, 83]. There have been reports on oscillators with higher order harmonics as well [84]. A brief operation principle of these techniques is introduced below based on Fig. 2.26. A push-push oscillator is composed of two oscillator cores as shown in Fig. 2.26a, which are locked with a 180° phase difference. At the symmetric nodes, therefore, the fundamental-mode signal will be canceled out due to the anti-phase relation, leading to a negligible fundamental output signal when the output is taken at one of the symmetric nodes. In contrast, for the second harmonic signals, the phase difference between the two cores is doubled and becomes 360°, as indicated by the phase diagram in the figure. This leads to the addition of the two signals at the output of the push-push oscillator. A similar principle applies to triple-push oscillators as well. They are composed of three oscillator cores as shown in Fig. 2.26b, each locked with a phase difference of 120° against the adjacent core. At the symmetric nodes, the fundamental mode and the second harmonic signals vanish by cancelation, while the third harmonic signal adds up, which is taken as the output signal of the oscillator. For the general case of *n*-push oscillators, all the signals with harmonic number less than *n* (including the fundamental signal) are eliminated at the symmetric node, while the superposed *n*th harmonic signal is taken as the output, as indicated by Fig. 2.26c. In theory, *n* can be as large as desired. In practice, however, *n* cannot be too large, since the signal power

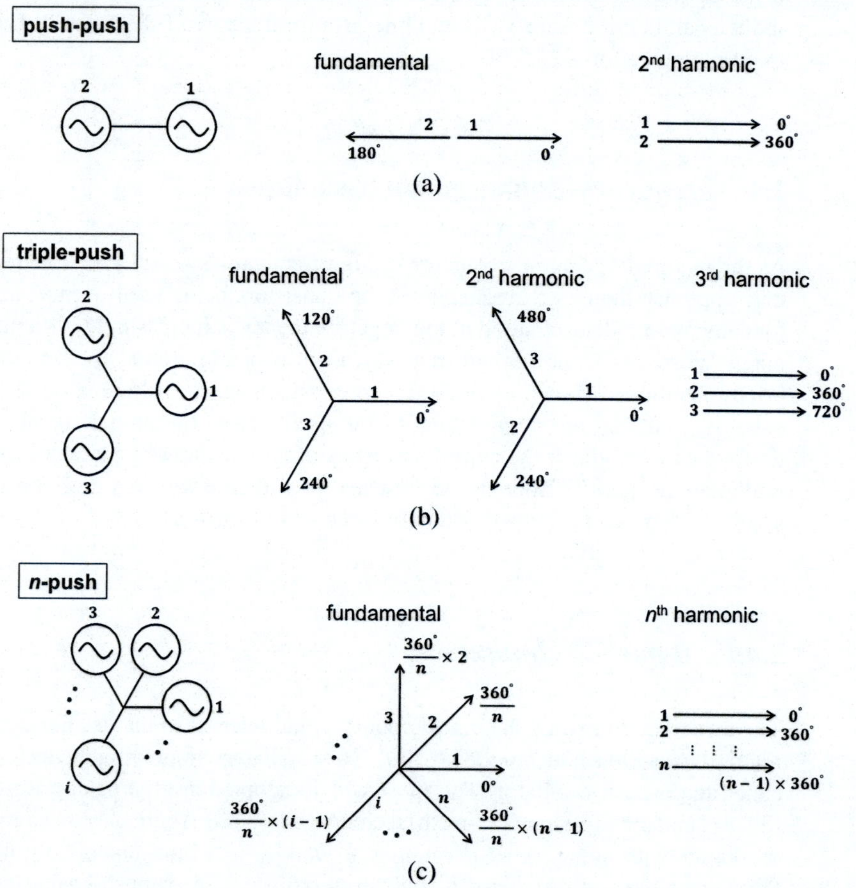

Fig. 2.26 The operation of n-push oscillators. The conceptual diagram is shown together with the phase of the fundamental mode and the higher-order modes: **(a)** Push-push oscillator ($n = 2$), **(b)** Triple-push oscillator ($n = 3$), **(c)** n-push oscillator

drastically reduces as the harmonic number grows. This is why most n-push oscillators are realized with $n = 2$ or 3 as mentioned earlier.

For the design of n-push oscillators, it is essential to maintain the symmetry in both schematic and layout, since any broken symmetry may result in incomplete suppression of the undesired harmonics. Also critical is to maintain the required phase difference between the adjacent oscillator cores. For push-push oscillators, the required phase difference can be achieved with a relative ease, because a differential configuration naturally shows a phase difference of 180°, as is employed in the example circuit presented in Fig. 2.27a. In this SiGe HBT push-push oscillator oscillating at 320 GHz, Q_1 and Q_2 of the differential pair function as the two oscillators cores [85]. At any symmetric point of the differential circuit, the fundamental signal vanishes and the second harmonic signal adds up. The collector

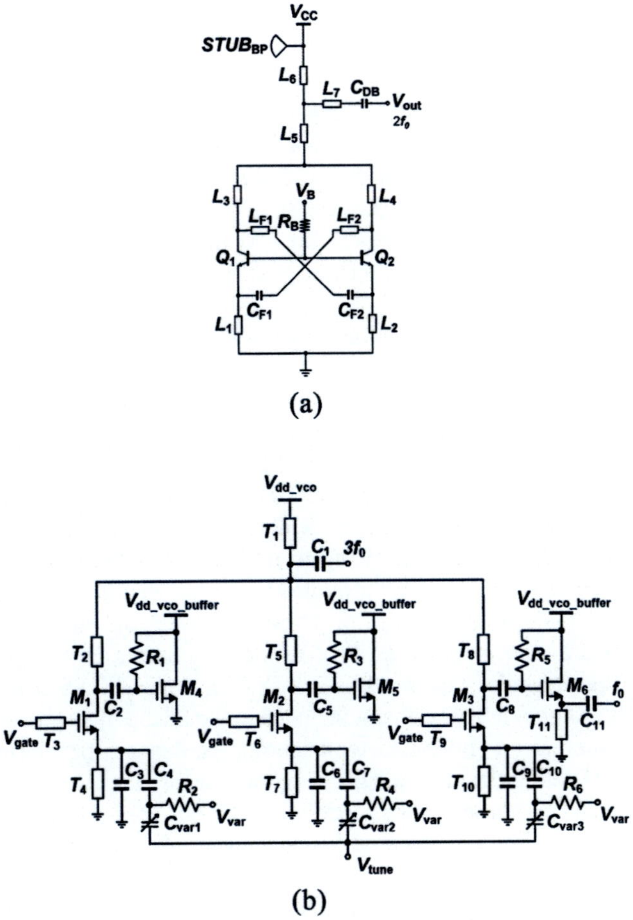

Fig. 2.27 Examples of n-push oscillators: (**a**) 320-GHz SiGe HBT push-push oscillator (© 2015 IEEE [85]), (**b**) 280-GHz CMOS triple-push oscillator (© 2015 IEEE [87])

(or drain) node is a popular choice for output signal extraction as is the case for this example circuit. For triple-push oscillators, a three-stage ring oscillator is widely employed to generate the third harmonic signal, because a phase difference of 120° between adjacent stages is enforced in this topology [86]. In Fig. 2.27b, an example of triple-push oscillators based on Colpitts topology is introduced [87]. In this oscillator operating at 280 GHz, the third harmonic output is taken at the common drain node of the stages, which is supposed to be symmetric. It can be seen in the schematic that the fundamental signal, with one-third of the output signal frequency, is also extracted from the oscillator in addition to the third harmonic signal. It flows into the frequency divider chain that follows, which would save a divide-by-three frequency divider. While the fundamental signal exits, the circuit through the buffer of the third stage, dummy buffers are attached to the first and second stages as well, to retain the symmetry that is critical for the n-push operation.

2.4.2 Frequency Multipliers

For most of electronic systems, linearity is the highly preferred characteristic as most of the nonlinear effects are malignant from a general system point of view. There are some components of exception, however, which benefit from the nonlinearity and exploit it. Mixers and frequency multipliers are well-known examples of these types of circuits. The former operates on two inputs and the latter on a single input. Frequency multipliers are discussed in this section, while mixers will be one of the topics in Chap. 3.

Probably the simplest nonlinear device is the diode, as its output response is given as a nonlinear function of the input, typically an exponential function. Hence, if followed by an appropriate bandpass filter that suppresses the fundamental and unwanted harmonics, a diode can operate as a frequency multiplier as its nonlinearity inherently leads to the generation of harmonic signals. In fact, diodes indeed have long been used as useful solid-state frequency multipliers. In particular, Schottky barrier diodes (SBDs) have been widely adopted for high-frequency applications, preferred over *p-n* diodes, owing to its faster switching time [88–90]. The faster operation of an SBD can be attributed to the fact that it is a majority-carrier device. Being a majority-carrier device (typically based on *n*-type semiconductor with electrons as the majority carrier), it does not suffer from the long (dis)charging time that degrades the speed as is found in minority-carrier devices such as *p-n* diodes. Although SBDs are not included in the lineup of standard devices offered by most commercial semiconductor technologies, they can be realized with a mix-and-match of various layers available from those technologies. This requires customized layout and in-house device modeling [91], but there have been reports on THz frequency multipliers based on SBDs built in this way based on both CMOS [92] and SiGe BiCMOS technologies [93]. They usually employ a single SBD or a pair of SBDs with an anti-parallel connection as the nonlinear component, followed by a filter that suppresses the fundamental frequency f_0, often based on a $\lambda_0/4$ open stub. However, the conversion gain of diode-based multipliers is typically lower than what is obtained with transistors due to the lack of the device gain. Also, circuit topology options are rather limited with diodes. This brings our attention back to transistor circuits for frequency multipliers.

Transistors are also nonlinear devices by nature and thus can be employed for frequency multipliers. The detailed characteristics may vary depending on whether they are bipolar or FET type, because the response of the former is characterized by the exponential relation while that of the latter is generally given by the quadratic relation. Nonetheless, the nonlinearity will be observed in both cases. It should be noted, however, that the nonlinearity alone will not lead to frequency multipliers with acceptable performance. The desired properties include high output power, small conversion loss, and the suppression of undesired harmonics, which can be achieved with properly tailored and optimized circuits. The availability of the device gain and various circuit techniques with transistors will certainly help, leading to richer options for the realization of frequency multipliers.

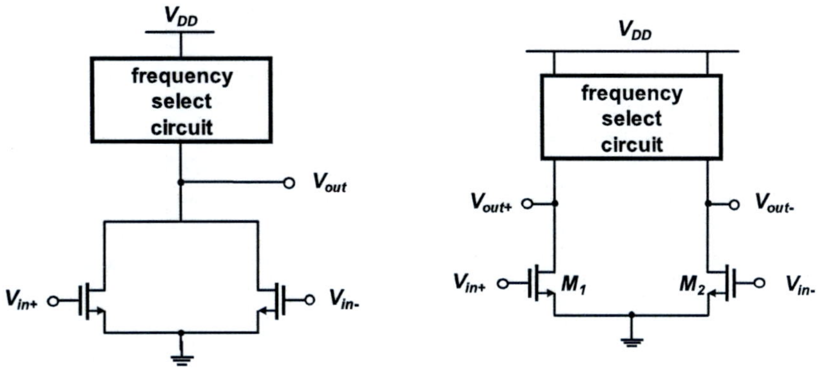

(a) (b)

Fig. 2.28 Conceptual schematics of nonlinearity-based frequency multipliers with differential topology: (**a**) Even-harmonic multiplier, (**b**) Odd-harmonic multiplier

Widely adopted transistor-based frequency multiplier topology is the differential pair, with the load network employing a stub (open or short) or LC tank resonator to suppress the unwanted harmonics. A popular circuit realization example is shown in Fig. 2.28. For even harmonics (Fig. 2.28a), a single-ended output signal can be taken from the common node of collector (or drain). Some readers may wonder if there is any signal detected at the common node, because it is known to be a virtual ground point. They are advised to remember that the virtual ground is a concept that belongs to the small-signal analysis. A frequency multiplier is a large-signal circuit driven by input signals large enough to trigger the nonlinear behavior. In such a situation, the common node is not a perfect ground anymore and even-harmonic signals will develop at the point, which can be taken as a desired output. This is similar to the case for the push-push oscillators in which the output is taken at the common collector (drain) node as discussed earlier. For odd harmonics (Fig. 2.28b), a differential output can be taken from the individual collector (drain) nodes of the differential pair. This is basically a differential amplifier, simply driven by a large input signal so that harmonic signals can be generated at the output due to nonlinearity, which will be helped by the load network that filters out undesired harmonic components.

For the examples shown in Fig. 2.28, the input and output configurations are pre-fixed (as single-ended or differential) by the needs from the circuit operation, while desired configuration may differ depending on the configuration of the preceding or following circuit block. Baluns can be employed if a conversion between configurations is needed. While the given configurations can be used for any even or odd harmonic numbers in theory, $\times 2$ (doubler) and $\times 3$ (tripler) multiplications are often practiced. Higher-order multiplication can be simply realized by cascading low-order multipliers, in which amplifiers are often inserted to boost up the signal power, leading to a circuit often referred to as the amplifier-multiplier chain (AMC).

In AMCs, the last stage is typically made of a frequency multiplier, often preceded by a driving amplifier, enabling the final frequency boost beyond the maximum operation frequency of the amplifiers available. A single-stage higher-order multiplier is also possible [94], but typically with a compromise in the performance. While the examples in Fig. 2.28 takes on the differential configuration, single-ended multipliers are also a popular option [95]. It is interesting to note that such single-ended topologies are often found with III–V technologies, while the differential approaches are more popular with Si-based technologies. This tendency is not specific to frequency multipliers only but applies to most high-frequency circuit components in general, partly due to the traditional individual component-based design practices among III–V community. It can also be attributed to the larger (lossy) substrate noise of Si circuits that calls for greater needs for common-mode noise rejection.

There are various other ways to implement frequency multipliers. When a wideband operation is required, traveling-wave frequency multiplier can be employed [96, 97]. The structure is based on the artificial transmission line constructed with transistor parasitic capacitance and line inductance (sometimes with additional reactive components), which is a hallmark feature of the traveling-wave amplifier (also known as the distributed amplifier). This enables a wideband characteristic for frequency multipliers as is the case for traveling-wave amplifiers. There are various ways to apply the artificial transmission line for the frequency multipliers. Input signal can be injected at both ends of the line while the output is extracted from each transistor stage and then combined [96], or input signal is applied at one end while the output is extracted from the other end [97]. Another way to implement a frequency multiplier is to simply employ a mixer and apply the input signal to both RF and LO nodes. With this configuration, the second harmonic signal will appear at the mixer output, rendering the circuit to work as a frequency doubler. While the operation is straightforward, the saturation by the DC offset may affect the performance with this approach. When frequency multiplication at high power level is required, a grid frequency multiplier based on the quasi-optical method (to be discussed in the next section) can be used. A frequency doubler with a grid structure operating at 1 THz was reported with an input power of 47 W at 500 GHz, generating an output power of 24 mW in a pulse mode operation [98].

There are also frequency multiplier types that do not rely on the nonlinearity or mixing, one popular example being the injection-locked frequency multiplier (ILFM) [99]. Its operation is similar to the injection-locked frequency dividers (ILFDs) discussed above. The only difference is that, for ILFMs, the output frequency is the integer-*multiple* of the injected input frequency (sub-harmonic injection), whereas, for ILFDs, the output frequency is the integer-*division* of the input frequency (super-harmonic injection). Therefore, in ILFMs, subharmonic injection signal triggers oscillation near (but not necessarily exact to) the free-running frequency of the core oscillator, which is taken as the output signal. For this purpose, the n-push topology is nicely suited as the core oscillator, with the inject signal corresponding to the fundamental signal and the output signal to its nth harmonic. For ILFMs, the output power is determined by the bias condition of the circuit rather

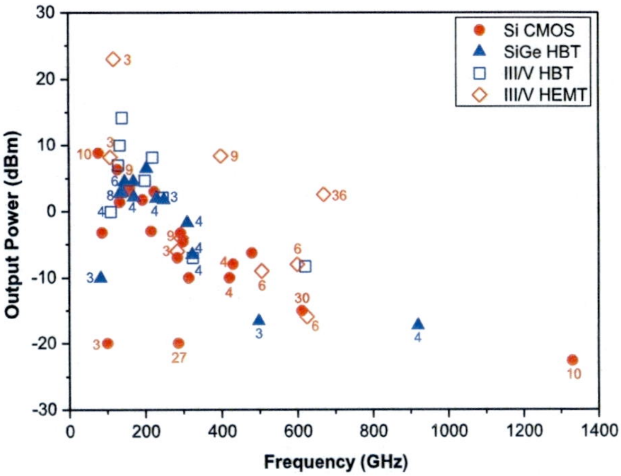

Fig. 2.29 Performance of recently reported frequency multipliers with various technologies. The numbers beside each symbol indicate the multiplication factor. The data points without a number are for frequency doublers (×2)

that the input power level as is often the case for conventional frequency multipliers, which raises chances for high output signal power. Also, a better harmonic suppression is expected for ILFMs, since a small injection signal can trigger the oscillation in the circuit.

A major challenge for THz frequency multiplier is to ensure sufficient output power, which in turn requires both low conversion loss and the availability of high input power level for most multipliers (note that input frequency is also high for THz multipliers). Filtering undesired harmonics at the high frequency will be also more difficult than with lower frequency cases. Figure 2.29 shows the reported performance of frequency multipliers operating beyond 100 GHz in terms of the output power and frequency with different technologies. The numbers beside each symbol indicate the frequency multiplication factor (data points without numbers are for frequency doublers (×2)). As can be seen from the plot, operation beyond 1.3 THz has been reached, which was obtained by a cascade of a ×5 and ×2 Si CMOS multipliers [100]. Output power over 1 mW (0 dBm) has also been obtained beyond 600 GHz with a ×36 InP HEMT multiplier chain [101]. As for the noise, it is generally known that frequency multipliers will degrade (increase) the phase noise by 20log*N* dB for the ideal cases, where *N* is the multiplication factor. The reported phase noise degradations with frequency multipliers operating near the THz band are quite close the ideal value [102, 103], indicating the dominance of the input phase noise in usual cases.

2.5 Power Enhancement Techniques

Although THz signal generation with electrical means became feasible owing to various techniques developed as discussed earlier, the output power level from a stand-alone signal source is not large enough. As many of potential THz applications prefer high output power, the lack of high-power sources is one of the major bottlenecks for the THz applications. From this viewpoint, the development of effective power enhancement techniques for THz signals is as critical as that of individual signal sources. Two power enhancement techniques can be considered: amplification and power combining. Each of these will be discussed in this section.

2.5.1 Amplifying Techniques

Amplification is the most effective way to boost the signal power and amplifiers are an indispensable component in most electrical systems. The amplifier type that is used to amplify and maximize the output power of a system is loosely called the power amplifier (PA). At low-frequency bands, various types of amplifiers have been developed to meet the key performance specs of power amplifiers such as output power, efficiency, as well as gain. In particular, achieving a high efficiency is critical for the increasingly popular mobile applications, for which the envelop tracking technique or Doherty amplifiers are currently widely employed. In the THz regime, however, amplifiers barely obtain a meaningful gain, since the operation frequency is close to the maximum operation frequency of the devices employed, typically dictated by the maximum oscillation frequency f_{max}, or the frequency on which device power gain reduces to unity. As such, the general strategy for THz amplifier development is to ensure the required gain first, and then meet the required output power, optionally with power combining if needed. Efficiency is of less priority for amplifiers operating at the THz band yet, and a high-efficiency THz amplifier would be regarded "luxury" at this stage of development. For the same reason, THz amplifiers reported so far are exclusively linear amplifiers.

So how to ensure a gain at THz? Obviously, the best strategy is to have access to high performance process technology. Probably the biggest factor that delayed the birth of THz amplifiers was the lack of a proper semiconductor technology that would support the gain required at the THz band. This rather fundamental issue was first addressed by III–V compound semiconductor technologies such as HEMT (high-electron mobility transistor) or HBT (heterojunction bipolar transistor) technologies, now both boasting f_{max} in excess of 1 THz [104, 105], which allows a sufficient gain deep into the THz regime. There have been impressive improvements in Si technologies as well, with CMOS and SiGe HBT technologies exhibiting f_{max} of close to 500 GHz [106] and over 700 GHz [107], respectively. Owing to these developments in the technology side, the designers now have a privilege to choose their favorite from the lineup of technologies available for the THz amplifier design depending on the application needs.

When the technology selection is done (hopefully affordable to the designers), the next is to choose the amplifier topology. Three basic options are widely used for THz amplifiers: common-emitter (CE) (or common-source (CS)), common-base (CB) (or common-gate (CG)), and cascode. While each topology has pros and cons from various aspects and needs, the focus will be maintained on the gain below. For

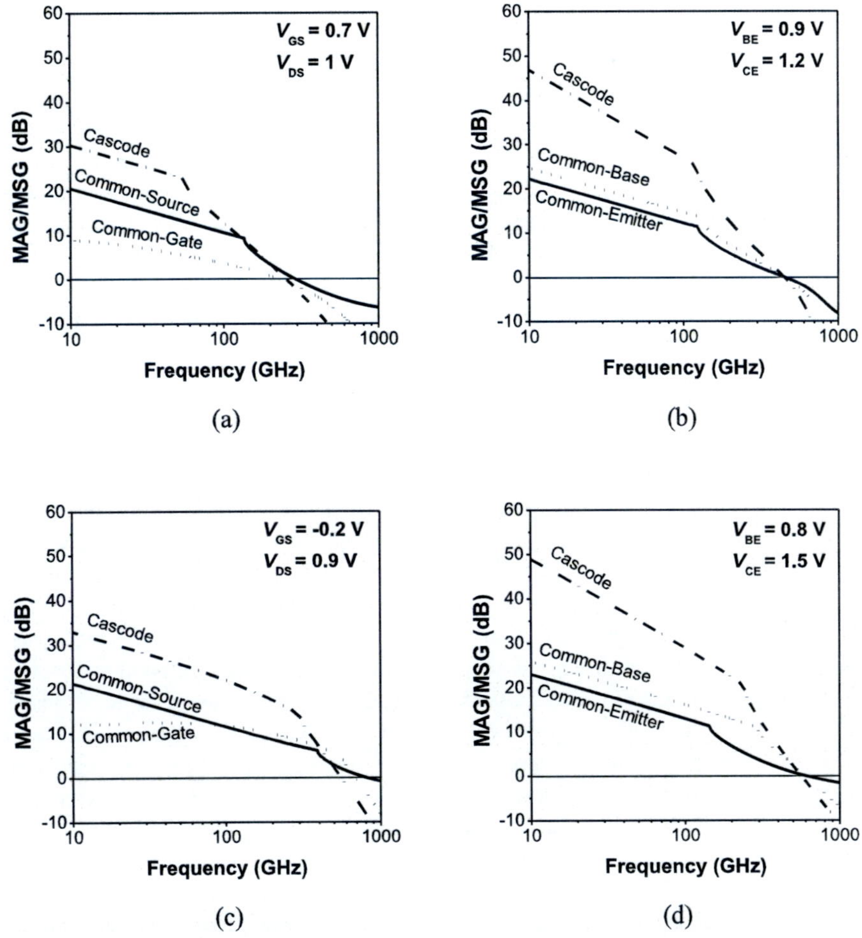

Fig. 2.30 MAG/MSG of various configurations for four different technologies based on device model: (**a**) 28-nm CMOS (A_G = 28 nm × 1 μm × 15), (**b**) 130-nm SiGe HBT (A_E = 130 nm × 1 μm × 3), (**c**) 80-nm InP HEMT (A_G = 80 nm × 20 μm × 2), (**d**) 250-nm InP HBT (A_E = 250 nm × 1 μm × 5)

conventional design approaches, the maximum available gain/maximum stable gain (MAG/MSG) dictates the upper limit of the gain available for the amplifier. Figure 2.30 compares MAG/MSG of those topologies for four example technologies: 28-nm Si CMOS, 130-nm SiGe HBT, 80-nm InP HEMT, 250-nm InP HBT. Of course, the characteristics highly depend on the device size and bias point, but we assume a typical size with an optimal bias point based on the device model provided by the process design kit (PDK). Also, the technologies compared in the figure may not be representatives of the given technology node and type, but they will still serve the purpose of comparison. As is indicated by the plot, cascode topology shows the largest gain, which is well expected considering its two-stage configuration. For bipolar transistors, CB tends to show a larger gain than CE, while the opposite is exhibited by FETs (CS shows a larger gain than CG). Still, most of the devices show a gain well beyond 100 GHz, enabling their application to THz amplifiers. It would be worthwhile to note that there is a stability issue with bipolar transistors in CB configuration, partly related to its high gain. The issue is aggravated by the uncertainty in the base inductance value in the device model, which has emerged as a major concern and prompted some efforts to mitigate the effect [108, 109].

Can we get a gain higher than MAG/MSG? Yes, we can, but with an extra effort in the circuit design. Consider the unilateral power gain U that is higher than MAG/MSG below f_{max}, which offers an opportunity to attain a gain higher than MAG/MSG. The readers are referred to the definition of U, which is the maximum gain available for a two-port when the two-port is *unilateral* or $S_{12} = 0$. Hence, if an amplifier is modified to be unilateral, or "unilateralized" (sometimes called "neutralized"), the gain may reach U for an ideal case (no loss, perfect input/output match), which exceeds MAG/MSG. For transistor amplifiers, this can be effectively achieved by canceling out the gate-drain capacitance C_{gd} for FETs or base-collector capacitance C_{bc} for bipolar transistors by adding extra components, as those capacitances are the main bilateral path between input and output. However, in most cases, the difference between U and MAG/MSG is not significant near f_{max}, as U and MAG/MSG eventually merges at f_{max} (this merge at zero gain point allows f_{max} to be defined based on either MAG/MSG or U). Therefore, the extra loss caused by the extra network may negate the extra gain attained with the unilateralization near f_{max}. It is interesting to note, however, that U is the maximum gain that can be obtained from a given two-port network assuming internal extra components can be added to the network besides the input/output matching (note that MAG/MSG is the maximum gain achieved with optimum input/output matching only that are *external* to the network). An extension of this idea was proposed in [110]. By embedding an amplifier inside a carefully tailored linear lossless passive network, the amplifier can achieve a gain higher than U. This new upper limit, dubbed the maximum achievable gain or G_{max}, can be as high as four times the value of U in the optimal case, leading to a potential gain boost up to 6 dB. This is a highly attractive option for THz amplifiers that are always hungry for gain and has been successfully applied to amplifiers operating beyond 200 GHz [111, 112].

There are other factors to be discussed for THz amplifier designs. For each of the three basic topologies mentioned earlier, both single-ended and differential configurations can be implemented. Besides the well-known advantage of common-mode noise rejection, differential topology also provides a highly valuable virtual ground. This is especially critical for THz applications, since the path to the external ground always causes extra inductance, which becomes increasingly problematic with increasing frequency as the reactance linearly grows with frequency ($X = j\omega L$). The availability of the virtual ground also obviates the need for bypass capacitors, which are necessary components for single-ended circuits but with known issues at high frequencies. The large capacitance required for bypassing typically tends to involve a low self-resonance frequency (SRF). For example, bypass capacitors of around 1 pF, a typical value for bypassing in high-frequency circuits, would result in SRFs lower than 100 GHz with conventional Si technologies [113]. This implies that circuits operating beyond this frequency will not have suitable RF bypassing because the capacitor converts to an inductor at the operation frequency. This calls for smaller bypass capacitors with higher SRF, which in turn leads to insufficient bypassing. For these reasons, differential amplifiers are highly favored in many applications, especially with Si technologies as briefly mentioned earlier. On the other hand, single-ended structures have their own strengths, too. For example, they benefit from its simpler layout and small size as well as lower power dissipation. Also, the compatibility with single-ended on-chip antennas and test interfaces helps in many practical situations, for which differential amplifiers would need extra baluns.

Up to this point, the issues for THz amplifiers were discussed mostly in relation to the circuit topologies. There are other aspects that need to be addressed for THz amplifiers. One major challenge for successful development of amplifiers operating at this high frequency is the availability of the accurate device model, both active and passive. The accuracy of the device models provided by the foundry is not guaranteed for the high frequencies that belong to the THz band. For this reason, in-house modeling is often carried out to reduce the risk caused by possible model mismatch. A typical RF model for active devices is composed of two parts: core model for the intrinsic and parasitic components for the extrinsic part of the device. While the core model is seldom touched by the circuit designers, improvements can be still made by the users based on simple DC measurements. Modification is often made for the parasitic components that are typically composed of series resistors for electrodes (with occasional inclusion of series inductors), inter-electrode capacitors, and a substrate network to account for the intra-device substrate coupling. The component values can be modified, with an optional change in the parasitic network topology itself, in a way to minimize the model-measurement discrepancy over a wide range of bias, which is typically based on repeated RF measurements. If the scalability of the original model is retained in such modification, it will be even more useful in the actual design.

The accuracy of the passive device model is as critical as the active device model. One relief for passive device modeling (and modification) comes from the availability of EM (electromagnetic) simulators. The results from EM simulation can be used to modify or develop passive device models. Alternatively, it can be directly used in circuit simulation bypassing the modeling step. To assure accurate EM simulation, however, it is critical to have precise dimension and material parameters of the technology. They are routinely provided by the foundry, but it is still desirable to check its accuracy with actual measurements with calibration. It is also a good practice to cross-check the results from multiple EM simulators for further verification. These issues and solutions related to the device models are in fact generic and they can be applied not only to amplifier design but any type of THz circuit designs.

Figure 2.31 shows the recently reported performance of THz amplifiers in terms of the gain versus frequency. As is indicated by the plot, an amplifier operating beyond 1 THz has been recently achieved based on a 25-nm InP HEMT technology,

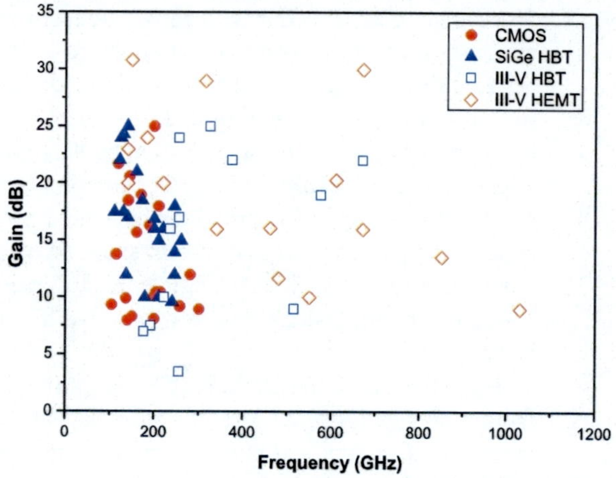

Fig. 2.31 Performance of recently reported THz amplifiers in terms of the gain

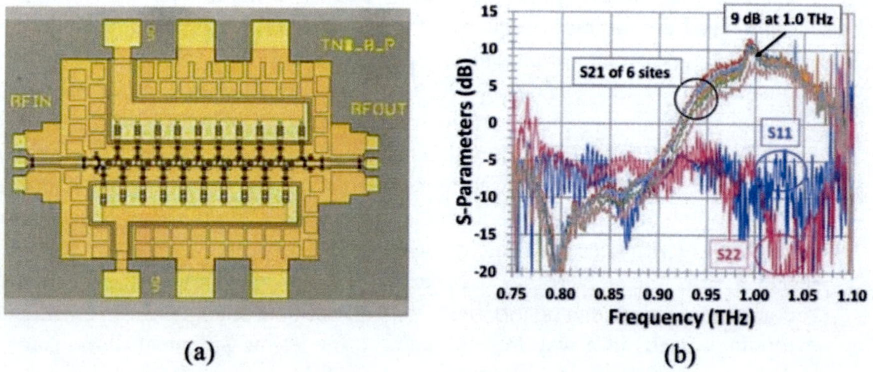

Fig. 2.32 One-THz InP HEMT amplifier: (**a**) Chip photo, (**b**) S-parameters. (© 2015 IEEE [105])

which is a great milestone for the semiconductor-driven solid-state electronics [105]. This 10-stage CS amplifier shows a peak gain of 9 dB at 1 THz and maintains gain up to 1.1 THz. The gain per stage is only 0.9 dB, a clear example of the challenges in developing amplifiers at the THz band. Figure 2.32 shows the chip photo and the *S*-parameters obtained with this THz amplifier. Amplifiers with InP HBT technologies also reached an operation frequency exceeding 600 GHz [114, 115]. With Si technologies, amplifier operation beyond 250 GHz have been reported based on both CMOS [111] and SiGe HBT technologies [109]. For a wideband performance, the distributed amplifier topology was employed for an InP HBT technology, leading to a bandwidth of 40–220 GHz [116]. A significant output power has also been obtained for some of the THz amplifiers reported, with the help of various types of power combining techniques, examples including 650-GHz InP HEMT amplifier with 3-mW output power [117] and 300-GHz InP HBT amplifier with 20 mW [118]. At 225 GHz, a power as high as 350 mW was achieved based on an InP HBT technology [119].

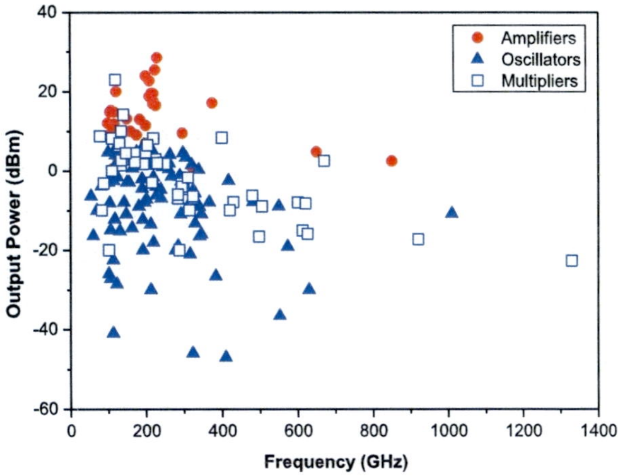

Fig. 2.33 Performance comparison of all three types of transistor-based signal sources considered: oscillators, frequency multipliers, and power amplifiers. The data points include all the harmonic numbers and process technologies (Si CMOS, SiGe HBT, III–V HEMT, III–V HBT)

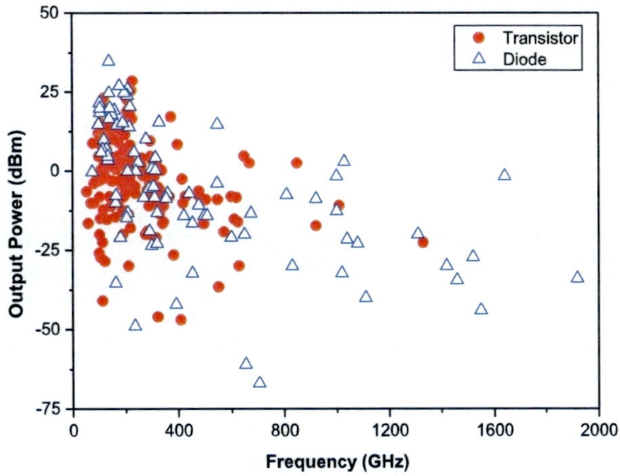

Fig. 2.34 Performance comparison of diode-based and transistor-based signal sources. The data points include all the harmonic numbers, circuit types, and process technologies for transistor circuits (Si CMOS, SiGe HBT, III–V HEMT, III–V HBT)

A comparison is made in Fig. 2.33 for the output power obtained from THz amplifiers and that of other signal sources based on transistor circuits such as oscillators and frequency multipliers. As is obvious from the plot, the output power with the amplifiers surpasses those of other circuits, which supports the idea of employing power amplifiers as a power booster as is typically the case for lower-frequency operations. Figure 2.34 provides a further extended comparison, in which the output power of transistor circuit-based signal sources and that of diode-based sources. It indicates a similar level of output power and frequency range for both types of signal sources, but the transistor-based approach still benefits from the essential advantage of being based on standard semiconductor technologies and able to be integrated with other parts of electronics. Finally, a grand global comparison between vacuum devices and solid-state devices, including both diodes and transistors, is made in Fig. 2.35. It is clear that the vacuum devices provide much greater

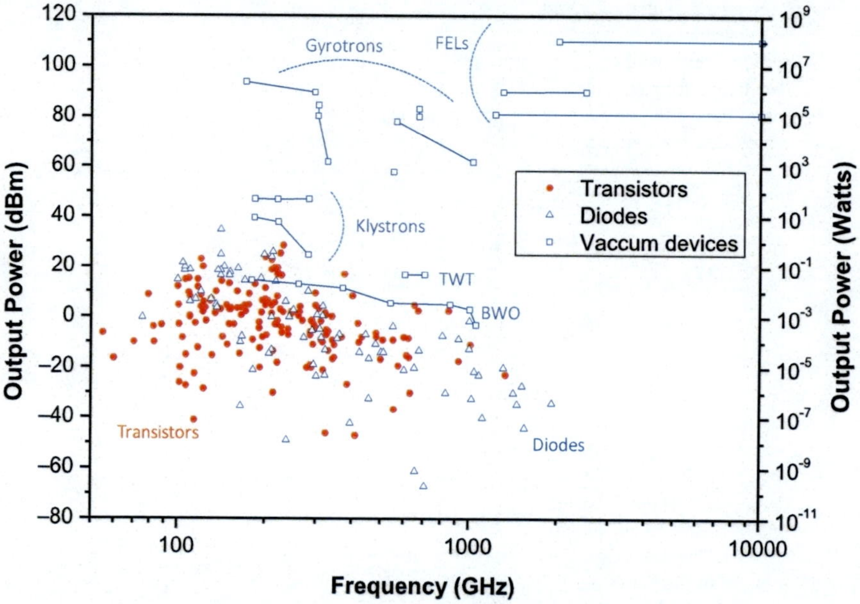

Fig. 2.35 Performance comparison of vacuum devices and solid-state signal sources, including both diode- and transistor-based ones. The data points include all the harmonic numbers, circuit types, and process technologies (Si CMOS, SiGe HBT, III–V HEMT, III–V HBT) for transistor circuits. The data points for vacuum devices are from [7]

output power than what is available with solid-state signal sources, while the compact form factor and lower DC power dissipation are still attractive features for solid-state sources. It should be stated that these two types of signal sources are not competing for performance, but rather they have their own optimal application fields.

2.5.2 Power Combining Techniques

There has been a strong thrust to develop various types of power combiners aiming at achieving output power larger than the saturation power of individual (power) *amplifiers*. At the same time, another drive to develop power combiners has been efforts to achieve output power larger than the power of individual *oscillators* that operate at frequencies beyond the operation frequency of amplifiers. From the implementation point of view, there is little difference between the two needs, and the discussion in this subsection may apply to both cases. Also, as most of the power combiners are reciprocal, they can be used as dividers, too. This property is very useful for amplifier power combining since the input power is initially *divided* into multiple parallel amplifiers and then *combined*, for which symmetry is highly desired. In this section, the discussion will be mostly made from the power combiners' viewpoint, but it can be applied to power dividers as well.

There have been a large number of power-combining techniques developed for microwave and millimeter ranges [120, 121], most of which can still be adopted to the THz band. The techniques can be grouped into many categories, based on various criteria. There are *two-way* (binary) combining and *N-way* combining, depending on how many input channels are combined in a single step. When preferred, multiple two-way combiners can be further combined to realize a large number of combining, in which case two configurations can be considered: corporate (tree) type and chain (serial) type. In the *corporate* type combining, two channels are added in a repeated fashion, whereas, in the *chain* type, the power from each channel is sequentially added to the main path of the outgoing power. The combining techniques can also be grouped into resonant type vs. non-resonant type. In the *resonant* type, the power combining takes place in a resonant cavity, where the input ports are aligned with the resonant modes of the cavity to maximize the combining efficiency, while the *non-resonant* type does not involve such a cavity structure. Yet another way of grouping is to divide them into *planar* power combining and *spatial* power combining, depending on where the combining physically happens. Discussion below will follow this categorization.

2.5.2.1 Planar Power Combining

Planar power combining literally refers to the techniques in which powers are combined on a planar structure, typically on the surface of a semiconductor chip (widely called "on-chip power combining") or other types of substrate. There are several different implementation approaches for planar combining [122–125], which will be reviewed below.

Direct shunt power combining is the simplest way of combining powers, which can be applied to either two-way or *N*-way combining cases (Fig. 2.36a). Since the current from each channel is combined, it is also called direct current combining. Although easily implemented, it suffers from a poor isolation between channels. Also, the matching will be increasingly difficult as the number of combined channels increases. This is because the parallel combination of the channels would require a larger output impedance per each channel to match the fixed load impedance, which contrasts with the low output impedance preferred for power amplifiers to boost the output power.

Another widely employed planar-type power combiner, particularly for Si-based high-frequency circuits, is *transformer-based power combining* (Fig. 2.36b). A conventional transformer is composed of two interleaved solenoid coils, which would require a three-dimensional structure. In this sense, realizing a planar combiner based on transformers would be a challenge. However, as the planar on-chip inductors became a standard component for RF circuits with extensive development, planar on-chip transformers also became readily available and widely adopted [126]. Transformer-based power combiners overall show good isolation and

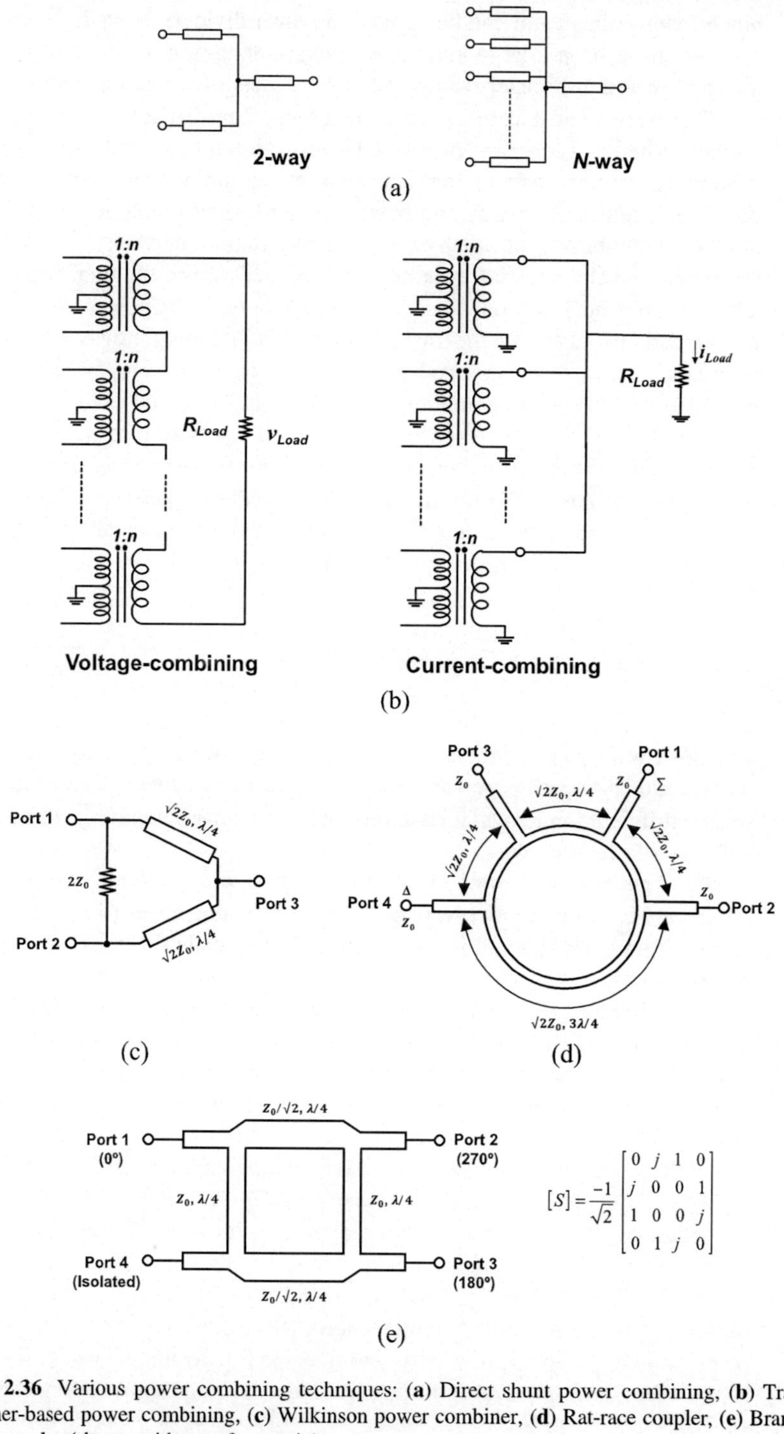

Fig. 2.36 Various power combining techniques: (**a**) Direct shunt power combining, (**b**) Transformer-based power combining, (**c**) Wilkinson power combiner, (**d**) Rat-race coupler, (**e**) Branch-line coupler (shown with transfer matrix)

wideband characteristics. They also benefit from the inherent properties of transformers, such as the impedance transformation and transition between single-ended and differential nodes. There are two possible options for transformer-based power combiners: voltage combiners and current combiners. The transformer-based *voltage combiners* add the voltage from each channel through a series connection of the transformers of the input channels. In this case, the load impedance can be effectively divided over the number of channels (assuming 1:1 turn ratio) for matching, leading to a small required output impedance for each channel, a preferred situation for power amplifiers. Of course, such impedance division can be further controlled by employing different transformer turn ratio. For the transformer-based *current combiners*, current from each channel is added through a parallel connection of the transformers. This configuration may suffer from a similar issue encountered with the direct shunt power combining as discussed above, a large output impedance required per each channel. However, with the presence of transformers, this problem can be elegantly mitigated by adopting a turn ratio larger than 1 (1:N turn ratio). Further, the current-based combining benefits from the symmetry in the channels, a highly preferred property to avoid amplitude and phase mismatch between the channels [122]. Yet another type of transformer-based combiner is the distributed active transformer (DAT) [127], which provides an alternative option for power combining with high efficiency.

Wilkinson power combiner is also a widely adopted planar-type combiner. It can be realized as both two-way and N-way configurations, although the two-way version is by far more widely used. A two-way Wilkinson power combiner is composed of two merged $\lambda/4$ transmission lines with an isolation resistor inserted between the two lines (Fig. 2.36c). The matching is achieved when the characteristic impedance of the $\lambda/4$ lines is $\sqrt{2}Z_0$ and the resistance of the bridging resistor is $2Z_0$, with its input and output terminated with Z_0. The advantages of Wilkinson power combiner include the excellent isolation and ensured matching for all the channels, the major reasons for its popularity. However, as the matching is achieved at a single frequency, it is inherently a narrow band combiner. Also, its large area occupied by the combiner is a downside for on-chip applications, as $\lambda/4$ line is the main component constituting the entire structure. This dimension issue, though, is expected to be gradually relaxed with increasing frequency, a favorable trend for THz applications. For example, a $\lambda/4$ transmission line stretches over 380 μm at 100 GHz when it is immersed in silicon dioxide. This may be bulky, but the number reduces down to 126 μm at 300 GHz and further down to 38 μm at 1 THz. We can manage this range of dimension in most applications.

Other more traditional planar microwave structures can also be used as power combiners, such as rat-race couplers and branch-line couplers. A *rat-race coupler*, as shown in Fig. 2.36d, is composed of two input ports (Ports 2 and 3) and two output ports (Ports 1 and 4). When two input signals with 180° phase difference enter the coupler, the sum and difference can be taken at Port 1 and Port 4, respectively. This is an ideal case for combining two *differential* signals if the output is taken at the sum port. Also, this configuration provides an opportunity for the coupler to serve as a balun, since the input is differential while the output is single-ended. A *branch-line coupler* (also called *quadrature hybrid coupler*) is composed of two input ports (Ports 2 and 3) and one output port (Port 1), the remaining port (Port 4) serving as the isolation port (Fig. 2.36e). The sum of two inputs with 90° phase difference will appear at the output node, leading to power combining. Of course, the opposite operation is possible, making the coupler to be employed as a power divider (Port 1 as input and Ports 2 and 3 as output). In this case, the branch-line coupler will serve as a quadrature signal generator, a valued component for systems that need quadrature signals, such as quadrature modulation communication systems.

Both of these couplers were traditionally built with microstrip lines for off-chip planar circuits. However, increased operation frequency has enabled the on-chip integration of these structures, benefiting from the smaller size with higher frequency. It is still under intense debate whether it is the best way to realize the bulky passive components (including antenna) on the chip, as its "real-estate" value is tremendous. The high area cost largely stems from the complex frontend-of-the-line (FEOL) processes intended for the construction of active devices, which are not much utilized for most (if not all) passive devices that only require backend-of-the-line (BEOL) elements such as metal lines and dielectric layers. It is noted that there are options for size reduction with these couplers. For rat-race couplers, a meander-line structure can be employed with a significantly reduced area [128]. This approach also lifts the main obstacle for rat-race couplers to be realized in commercial Si processes, say, the design rule that prohibits circular shapes. For branch-line couplers, the capacitive open-stub loading technique can be employed for size reduction, which effectively shrinks the coupler dimension with the help of a series of open stubs placed along the line edges [129].

While the techniques described so far can be considered rather classical, there have been high-frequency signal sources operating at the THz band based on novel techniques. One notable approach is the *distributed active radiator* (DAR) [130, 131], which is briefly reviewed here. It is basically four cross-coupled differential pairs mutually injection-locked through a pair of concentric signal loops, which separately link the drains of each pair of the core transistors (see Fig. 2.37). The two loops are ingeniously cross-coupled, so that the second harmonic signals traveling along the two loops are in-phase, while the fundamental signals traveling along the same are out-of-phase. Hence, the two second harmonics add up and the two fundamentals cancel each other as they radiate from the pair of loops, which serve as the antenna. This allows an effective suppression of the fundamental signal without resorting to a filter, which is typically lossy, as is practiced in many conventional approaches. Additionally, an aperture opening was made for the ground plane just beneath the loops, while retaining only radial paths between differential pair source nodes and outer ground plane. This configuration prevents possible radiation by the second harmonic return current on the ground that might cancel out the main radiation because of the opposite direction of circulation if exists. This approach, implemented in a Si CMOS technology as a 2 × 2 array, resulted in a total output power of 80 μW around 290 GHz.

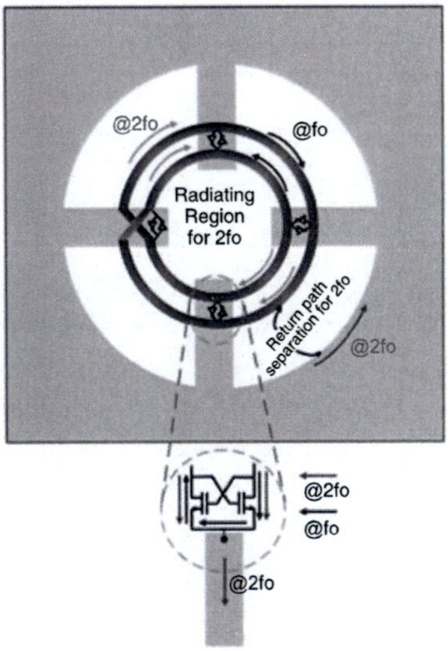

Fig. 2.37 Conceptual view of the distributed active radiator (DAR). (© 2011 IEEE [130])

2.5.2.2 Spatial Power Combining

The planar power combining techniques discussed so far combine powers in the form of voltage or current on a planar surface. In contrast, the spatial power combining techniques to be discussed below combine powers in the form of electromagnetic waves in a guided or open space [132]. There are various advantages expected with the spatial power combining as follows. Planar combiners usually combine powers from multiple components (or channels) arranged in one-dimension as the power propagates on the surface along the signal lines. On the other hand, spatial combining works on a two-dimension array of components as the power often (not always) propagates perpendicular to the surface after radiation. This configuration allows spatial power combining to apply on a massive number of components, counting up to tens or even hundreds. In this way, much higher output power can be achieved than with planar combining. Also, as the powers from the array components directly combine without any branched corporate structure that involves accumulated loss from each stage (for corporate structures, widely employed for multi-stage direct shunt power combining, the total loss is $L \cdot \log_2 N$, where N is the total number of devices combined and L is the loss per stage), a higher combining efficiency can be expected with spatial combining. Additionally, a high directivity can be obtained for the radiated power when the phase is well aligned over the array components, a preferred property for many practical applications.

This effect is better understood in terms of EIRP (equivalent isotropic radiation power) of an array made of N elements, each element with output power of P_{out} and antenna gain of G_{ant}:

$$\mathrm{EIRP}_{\mathrm{array}} = N^2 P_{\mathrm{out}} G_{\mathrm{ant}}. \tag{2.19}$$

It is noted that EIRP rapidly increases with the number of elements by a factor N^2, which results from the fact that both total radiated power and directivity increase by a factor of N. The definition of the antenna gain and a formal treatment of EIRP can be found in Sect. 4.2. Finally, spatial power combining can be extended to beam steering capability by properly incorporating phase shift components, which will be detailed in the next section. On the downside, the bulky size of spatial power combining structure may be an issue for some applications, as is imposed by the antenna size and the minimum required spacing between components of around $\lambda/2$. In a sense, however, this limitation may be an opportunity for THz applications, with reduced dimension for increasing frequency.

There are various ways to implement the spatial power combining. They can be largely divided into two different cases, depending on whether the power combining occurs in the open space or inside metallic waveguide enclosures as depicted in Fig. 2.38 (adapted from [132]). For the open space case, the input signal is incident on the array of amplifying components directly or through a lens without any confining structure. The output is radiated in a similar way. This approach is also called the quasi-optic power combining. Although Fig. 2.38 assumes there exists an incoming input power, there are cases where signals are generated from an on-chip array and do not require input signal. In these cases, the array components will be signal sources instead of amplifiers. Although the combining in the open space does not require a complex packaging structure, it may suffer from focusing error, diffraction loss, and, more importantly, poor heat sink. These issues can be addressed by confining the power combining inside a metallic enclosure. In an example shown in Fig. 2.38b, the input signal is fed through a waveguide and then combined inside a metallic structure, which has a larger cross section than the waveguide to accommodate the two-dimensional array of components. After combined, the output power exits through the waveguide on the opposite side. The widely used two-way or N-way waveguide power combiners, which do not need a separate structure for power combining unlike the one described above, also belong to this category of combiners. In fact, there exist many different configurations when it comes to the details in actual realization [119, 133].

The THz application of the spatial combining has so far limited to arrays with a relatively small number of components [94, 134]. However, a massive spatial combining may appear with the advantage of small size required in the high-frequency range.

Fig. 2.38 Spatial power combining techniques: (a) Open space, (b) Waveguide-enclosed. (© 2002 IEEE [132])

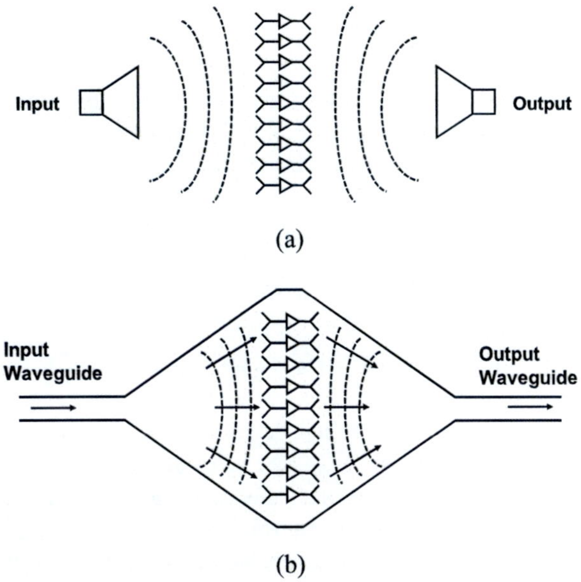

2.6 Beamforming Techniques

As discussed so far, THz sources persistently suffer from the low output power level, which would hinder them from many interesting THz applications. Once radiated, it will be further suppressed by the free-space path loss (FSPL), which increases with the square of frequency, as well as the large attenuation in Earth atmosphere near the THz band. The two methods discussed in the previous section, amplification and power combining, help to enhance the power level as we observed. Yet another way to boost the THz power is the beamforming. In this case, the boost is directional and applies to a specific direction of radiation with the total radiated power unchanged. This is different from the two techniques described above and hence discussed separately in this section.

Beamforming can be understood as a special case of the spatial power combining, which effectively increases the power delivered to a certain direction by improving the directivity in a controlled way. This definition, based on the viewpoint from the signal generation, can be reciprocally applied to the signal detection for increased directional sensitivity as well, but the discussion in this section will be made from the signal generation perspective. We have already observed in the previous section that a spatial power combining from an array of N elements with aligned phase will result in N-factor increase in total power and N^2-factor increase in the power delivered to a certain direction compared to a single-element case. Hence, the "formation of a beam" can be achieved with a simple spatial combining with an N-element array in principle. However, the "beamforming" with a more conventional meaning implies an additional important feature: the ability to *control the direction* of (or steer) the beam by manipulating the phase delay for each element of the array. Typically implemented by a phased array, beamforming allows fast electrical beam steering from a stationary fixture, without the need for bulky rotating mechanical scanners. This surely is a more elegant and cost-effective solution. Below, the readers will be introduced to the basic general principles of beamforming first and then to the recently reported THz phase arrays.

2.6.1 *Basics of Beamforming*

Beamforming can be understood with the help of Fig. 2.39 depicting various cases of signal radiation. We can start with the single-element case, shown in Fig. 2.39a. Any antenna has a certain degree of directivity although not very high. This is assumed in the figure, showing a beam pattern gently pushed toward the direction of radiation. Next, consider the case of an array of composed of *N* elements without any individual phase control, leading to all elements radiating the signals at the same phase. In this case, as is indicated in Fig. 2.39b, the beam pattern becomes sharper as the directivity is increased by a factor of *N*. Hence, the beam direction will receive much stronger power from the array than the single-element case. However, the beam direction is fixed perpendicular to the array plane, and there is no control over for the beam direction yet. Now consider the case shown in Fig. 2.39c, where a phase control component, or phase shifter, is placed on every signal path so that an extra phase can be added to each element. If the extra phase is designed so that it progressively increases along the array of elements from one end to the other, the direction of the beam pattern will not be perpendicular to the array front anymore, but it will show some angle off the center. If the progressive extra phase increase is more rapid, the angle will become larger. In this way, the beam direction can be

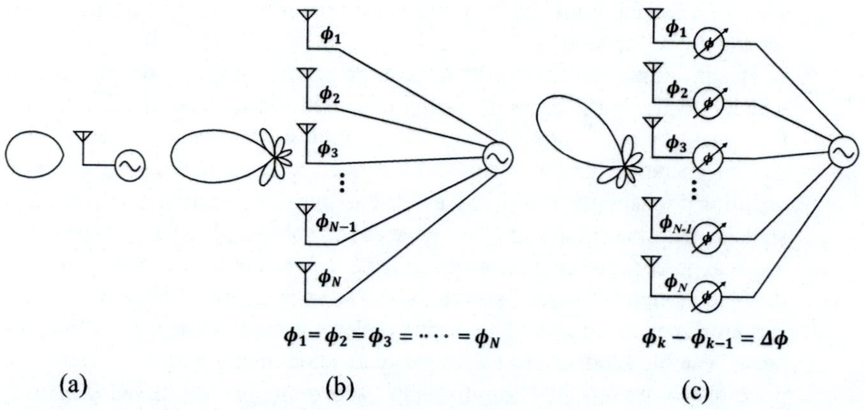

Fig. 2.39 Beam patterns from various source configurations: **(a)** Single source, **(b)** *N*-element array without phase control, **(c)** *N*-element array with phase control

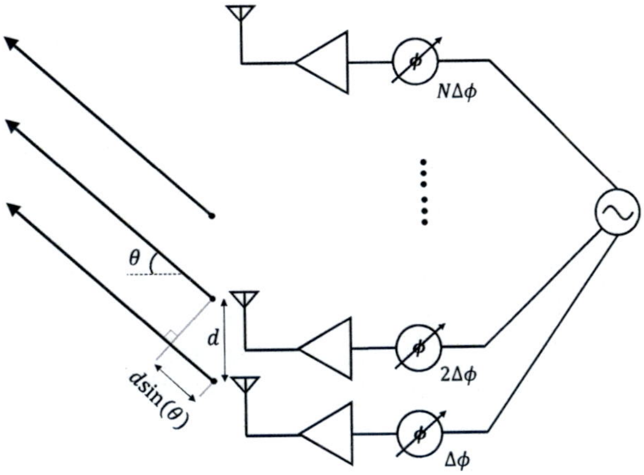

Fig. 2.40 Conceptual schematic of an *N*-element phased array

controlled by precisely adjusting the phase shifts in the array elements. Such a system is called the phased array.

An analytical description can be made with Fig. 2.40, which provides a more detailed look into the phased array. In this *N*-element array, the initial distance between the antenna and the single shared oscillator is assumed to be identical for all the elements. Now suppose phase shift is applied to each channel along the path, which equals $k\Delta\phi$, where k is the element index ranging from 1 (bottom) to N (top) and $\Delta\phi$ is the unit phase difference between adjacent channels. If $\Delta\phi$ is zero, the beam direction will be still perpendicular to the array front as mentioned earlier, and thus the angle between the normal axis and the beam direction, defined as θ, will also be zero. If $\Delta\phi$ is not zero and has a finite value, then the radiation normal to the array front will become smaller due to the destructive interference between the radiations since the phase is not aligned over the elements anymore. Instead, a new phase alignment will be established for a radiation that shows an angle θ with the normal direction, which will be determined by $\Delta\phi$. This is because, for the radiation at this angle, the phase added to each element ($k\Delta\phi$) exactly balances out the extra phase in the free space along the path for each channel. Hence, a strong beam will be formed in this direction due to a constructive interference. In a more quantitative term, the relation between θ and $\Delta\phi$ can be obtained as follows:

$$\frac{d\sin\theta}{\lambda} = \frac{\Delta\phi}{2\pi}, \tag{2.20}$$

where d is the physical distance between the adjacent channels (more precisely, the distance between adjacent antennas) and λ is the free-space wavelength of the beam. Each parameter can be explicitly expressed in terms of the other:

$$\Delta\phi = \frac{2\pi d}{\lambda}\sin\theta \quad \text{or} \quad \theta = \sin^{-1}\left(\Delta\phi\frac{\lambda}{2\pi d}\right). \tag{2.21}$$

When the preferred beam direction (θ) is decided, the required unit phase difference ($\Delta\phi$) can be calculated for a given wavelength and inter-channel distance, or vice versa. Although the analysis is made for one-dimensional array cases based on Fig. 2.40 for a beam steering along a single axis, the analysis can be easily extended to two-dimensional array cases that would allow a beam steering across a biaxial plane. A similar analysis can be made for beamforming in the receiver side, which will not be repeated here.

The expressions above include the wavelength as a parameter, and one may wonder phased arrays are inherently a narrowband system since the given relations are valid only for a given wavelength (or frequency). However, Eq. (2.20) can be equivalently expressed in terms of time shift $\Delta\tau$ instead of phase shift $\Delta\phi$, as is shown below:

$$\frac{d \sin\theta}{c} = \Delta\tau. \tag{2.22}$$

In this expression, the dependence on the wavelength disappeared, implying that the relation is frequency-independent and thus is wideband. With this viewpoint, the *time delay*, not the phase delay, imposed on each channel is balanced out with the path difference in the free space after radiation. Hence, a wideband phased array can be realized if a time delay component is employed in the channels, which is typically achieved by controlling the signal path length. If explained in terms of the phase, for a given time delay created with the time delay component, the corresponding phase delay *linearly* increases with increasing frequency. At the same time, the phase corresponding to the path difference of $d \sin\theta$ also shows a linear increase with frequency. Therefore, the balance between the two phases is retained independent of the frequency, leading to a wideband behavior. In fact, a system configured in this way may be better named timed-array instead of phased-array.

The properties of beamforming depend on the details of the phased-array design. A larger N results in a stronger beam forming, increasing the power along the direction of the main lobe, but a narrower beam pattern. It also leads to a larger number of side lobes. A larger d improves the spatial resolution as the radiation angle is more fine-tuned with a given $\Delta\phi$ as indicated by Eq. (2.21). However, too large d may cause the free-space path difference between adjacent channels ($d \sin\theta$) to become larger than λ. If this happens, there exist multiple angles that satisfy Eq. (2.20) because $\Delta\phi$ is effectively equivalent to $\Delta\phi + 2\pi$. This results in multiple main lobes, an unfavored situation. Hence, to avoid multiple main lobes for the range $0 < \theta < 90°$, d should be smaller than λ. If we consider the full angle range, or $-90° < \theta < 90°$, the condition becomes more stringent, or $d < \lambda/2$ to avoid multiple main lobes for this whole span of θ allowed in the ideal case. However, in reality, the practical range of interest for θ will be narrower than this. In this case, the upper limit of d will be slightly larger than $\lambda/2$, which justifies the popular choice of $d = \lambda/2$ with some margin.

The main advantage of beamforming is the ability to control the beam direction without any mechanical rotation by means of electric beam steering. Another advantage can be discussed in terms of the power efficiency of the transmitter system. We already saw that there is a factor of N^2 increase in the power in the direction of beam radiation for N-element array compared to the single element. The array will consume N times as large DC power as the single-element case. Hence, there is a net advantage of a factor of N in the power efficiency for the intended direction. This is equivalent to saying that we can deliver a certain amount of power to the target point with only $1/N$ of the power consumption for a single-element case. A similar argument can be made for a beamforming in the receiver side with an N-element receiver array. From the desired direction, one can receive N^2 times larger power than the single-element case. The system noise will also be increased with the array, but only by a factor of N if the noise is uncorrelated over the channels. This will lead to SNR improvement by a factor of N in the receiver.

2.6.2 THz Phase Shifters and Phased Arrays

There have been various types of phase shifters developed for high-frequency applications, some of which have also been applied for THz band operation. Popular among those are switched-line type, reflection type, and vector sum type shifters, as depicted in Fig. 2.41, which are briefly reviewed below.

Switched-line phase shifters are composed of multiple phase shifters cascaded, each of which has different shifting phase and can be turned on and off by an embedded switch (Fig. 2.41a). Widely adopted for the individual phase shifters are LC-based low-pass and high-pass filters, either T-type or Π-type, which provides a phase shift that is dictated by the value of the included reactive components as well as the frequency. The detailed relation between the desired phase shift and the required reactive component values can be found in [135] for both low- and high-pass filters with T- and Π-type configurations. The switches are digitally controlled by a set of bits externally applied. For an n-bit phase shifter, n phase shifters are cascaded, each with a shifting phase of $360°/2^m$ where m is an integer satisfying $1 \leq m \leq n$. This results in an overall phase resolution of $360°/2^n$.

Reflection-type phase shifters are based on a branch-line coupler, in which the output is extracted from the isolation port with the input signal applied into the regular input port (Fig. 2.41b). The through and coupled ports are terminated with variable reactive components, for which a varactor or a higher-order LC network is typically used. The applied input signal is reflected by the reactive termination through the coupler and appear at the output port with a shifted phase, the amount of which depends on the reactance of the termination jX. Hence, variable phase shift can be achieved by controlling the terminal reactance. A reflection-type phase shifter operating at 218–268 GHz range has been reported based on GaAs mHEMT technology [136]. This work, which employs a variable gain amplifier and a pair of couplers with reactive termination based on HEMT-based varactors, showed a phase control range up to 247° with an average insertion loss of 1.6 dB.

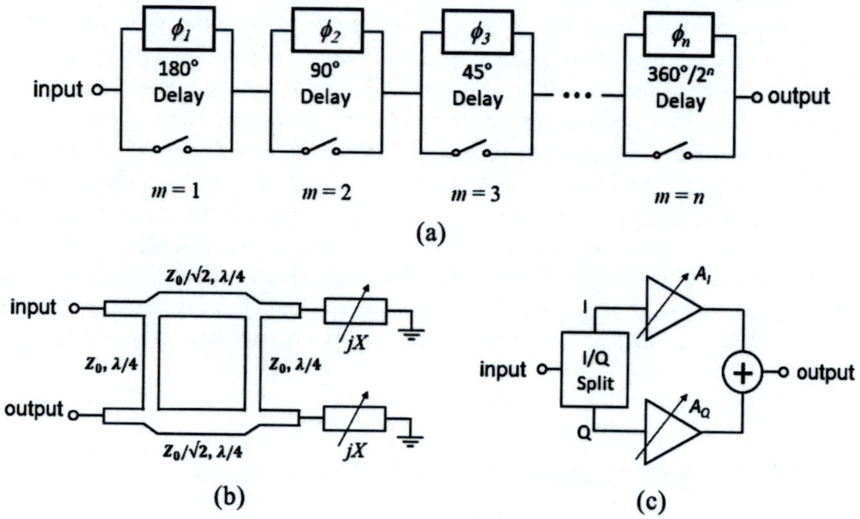

Fig. 2.41 Various phase shifter schemes: (**a**) Switched-line phase shifter (n-bit), (**b**) Reflection-type phase shifter, (**c**) Vector sum phase shifter

The two types of phase shifters discussed above are both passive phase shifters, which benefit from the high linearity and the virtually zero power dissipation. However, being passive, they usually show rather a large insertion loss. For applications where small loss is desired, active phase shifters can be employed. A popular active phase shifter is the *vector sum phase shifter*. As shown in Fig. 2.41c, it is composed of a pair of variable gain amplifiers connected in parallel, with a quadrature signal applied to each amplifier through a quadrature generator. By adjusting the individual gain of amplifiers, the phase of the sum of the two output signals can be controlled. There has been a report on a vector sum phase shifter operating in the range of 220–320 GHz [137]. Built on an InP HBT technology, this wideband phase shifter employed a single Gilbert cell for each amplifier and exhibited continuous phase shift up to 360° with a phase error and insertion loss smaller than 10.2° and 16 dB, respectively. In another vector sum phase shifter, based on a pair of high-gain WR3.4 InP HBT amplifiers, a gain up to 10 dB was obtained although the phase control range was limited up to 90° with a narrow band operation near 300 GHz [138].

Phased arrays require a higher level of integration than phase shifters, involving more challenges, especially at a high-frequency range such as the THz band. While it took some time, integrated THz phased arrays have become reality, leading to exciting recent reports on integrated THz sources based on phased arrays. The 2 × 2 290-GHz source array based on DAR as described earlier [130], was extended to facilitate phase control over each element, enabling a 2-D beamforming based on 4 × 4 DARs [131]. The required phase shift was achieved with a vector sum shifter operating at the frequency of 47 GHz. The signal frequency was subsequently boosted up to near 140 GHz by an injection-locked frequency tripler, then up to

280 GHz through the DARs that internally extract the second harmonic from the incoming input signal. It is noted that the phase shift is obtained at a much lower frequency than the THz output frequency, which relaxes the challenge for shifting phases that grows more stringent for higher frequencies. The phased array, built on a 45-nm SOI CMOS technology, exhibited an EIRP of 9.4 dBm and a beam steering range of over 80° in both axes with 820 mW of DC power dissipation.

A THz phased array based on a novel phase shifting technique was also reported [139]. In this 2 × 2 phased array operating at 318–370 GHz range, fourth harmonic signal is extracted from four unit cells, each composed of two differential pairs made of SiGe HBTs. For beamforming, a new technique that controls base-emitter voltage (V_{BE}) of the transistors is employed. When a slightly larger V_{BE} is applied to a unit cell than that of an adjacent cell, the induced larger swing triggers a traveling wave toward the adjacent cell, which effectively causes a phase difference between the two cells for the generated signal. This rather intrinsic phase shift results in a beamforming, which can be controlled over two axes due to the nature of the 2D array. With this phase array, beam steering angles of 128° and 53° for *E*- and *H*-plane, respectively, were obtained along with a peak radiation power of −6.8 dBm. The phase shift is in fact obtained at the fundamental frequency of around 80–90 GHz, 1/4 of the output frequency, relaxing the challenges as was the case for the previous circuit discussed.

References

1. A.V. Gaponov-Grekhov, V.L. Granatstein, *Applications of High-Power Microwaves* (Artech House Publishers, Boston, 1994)
2. R. Barker, J. Booske, N. Luhmann, G.S. Nusinovich, *Modern Microwave and Millimeter-Wave Power Electronics* (Wiley, New York, 2005)
3. J. Benford, J.A. Swegle, E. Schamiloglu, *High Power Microwaves* (CRC press, Boca Raton, FL, 2007)
4. S.E. Tsimring, *Electron Beams and Microwave Vacuum Electronics* (John Wiley & Sons, Hoboken, NJ, 2006)
5. R.H. Varian, S.F. Varian, A high frequency oscillator and amplifier. J. Appl. Phys. **10**(5), 321–327 (1939). https://doi.org/10.1063/1.1707311
6. J.W. Gewartowski, H.A. Watson, *Principles of Electron Tubes: Including Grid-Controlled Tubes, Microwave Tubes, and Gas Tubes* (Van Nostrand, Princeton, NJ, 1965)
7. J.H. Booske et al., Vacuum electronic high power terahertz sources. IEEE Trans. Terahertz Sci. Technol. **1**(1), 54–75 (2011). https://doi.org/10.1109/TTHZ.2011.2151610
8. R.K. Parker, R.H. Abrams, B.G. Danly, B. Levush, Vacuum electronics. IEEE Trans. Microw. Theory Tech. **50**(3), 835–845 (2002). https://doi.org/10.1109/22.989967
9. R. Dobbs et al., Design and fabrication of terahertz Extended Interaction Klystrons, in *35th International Conference on Infrared, Millimeter, and Terahertz Waves*, 5–10 Sept 2010 (2010), pp. 1–3. https://doi.org/10.1109/ICIMW.2010.5613040
10. R. Kompfner, The invention of traveling wave tubes. IEEE Trans. Electron Devices **23**(7), 730–738 (1976)
11. A. Gilmour, *Principles of Traveling Wave Tubes* (Artech House, Boston, 1994)

12. K.E. Kreischer, J.C. Tucek, D.A. Gallagher, R.E. Mihailovich, Operation of a compact, 0.65 THz source, in *33rd International Conference on Infrared, Millimeter and Terahertz Waves*, (IEEE, New York, 2008), pp. 1–2
13. A.J. Theiss, C.J. Meadows, R. Freeman, R.B. True, J.M. Martin, K.L. Montgomery, High-average-power W-band TWT development. IEEE Trans. Plasma Sci. **38**(6), 1239–1243 (2010). https://doi.org/10.1109/TPS.2010.2041794
14. P. Guenard, O. Doehler, B. Epsztein, R. Warnecke, New UHF tubes with wide electronic tuning range. C. R. Acad. Sci. (Paris) **235**, 236 (1952)
15. R. Kompfner, N. Williams, Backward-wave tubes. Proc. IRE **41**(11), 1602–1611 (1953)
16. A. Negirev, A. Fedorov, Wide-range miniature millimeter-range backward-wave tubes. Radio Commun. Technol. **4**(4), 43–46 (1999)
17. C.Y. Levin, A.Y. Kirichenko, A. Borodkin, S. Churilova, A.Y. Usikov, The clinotron, in *22nd European Microwave Conference*, vol. 1, (IEEE, New York, 1992), pp. 603–607
18. F. Rusin, G. Bogomolov, Orotron—An electronic oscillator with an open resonator and reflecting grating. Proc. IEEE **57**(4), 720–722 (1969)
19. G. Nusinovich, M. Thumm, M.I. Petelin, The Gyrotron at 50: Historical overview. J. Infrared Millimeter Terahertz Waves **35**(4), 325–381 (2014). https://doi.org/10.1007/s10762-014-0050-7
20. V.L. Bratman et al., Gyrotron development for high power THz technologies at IAP RAS. J. Infrared Millimeter Terahertz Waves **33**(7), 715–723 (2012). https://doi.org/10.1007/s10762-012-9898-6
21. K.L. Felch et al., Characteristics and applications of fast-wave gyrodevices. Proc. IEEE **87**(5), 752–781 (1999). https://doi.org/10.1109/5.757254
22. D.A.G. Deacon, L.R. Elias, J.M.J. Madey, G.J. Ramian, H.A. Schwettman, T.I. Smith, First operation of a free-electron laser. Phys. Rev. Lett. **38**(16), 892–894 (1977). https://doi.org/10.1103/PhysRevLett.38.892
23. H. Motz, Applications of the radiation from fast electron beams. J. Appl. Phys. **22**(5), 527–535 (1951)
24. Y.-S. Lee, *Principles of Terahertz Science and Technology* (Springer, New York, 2009)
25. J. Madey, H. Schwettman, W. Fairbank, A free electron laser. IEEE Trans. Nucl. Sci. **20**(3), 980–983 (1973)
26. O. Shevchenko et al., The Novosibirsk free electron laser–Unique source of terahertz and infrared coherent radiation. Phys. Procedia **84**, 13–18 (2016)
27. A.W. Hull, The effect of a uniform magnetic field on the motion of electrons between coaxial cylinders. Phys. Rev. **18**(1), 31–57 (1921). https://doi.org/10.1103/PhysRev.18.31
28. H. Boot, J. Randall, The cavity magnetron. J. Inst. Electric. Eng. IIIA Radiolocation **93**(5), 928–938 (1946)
29. T.K. Ishii, *Practical Microwave Electron Devices* (Academic Press, San Diego, CA, 1990)
30. N. Avtomonov, V. Naumenko, D.M. Vavriv, K. Schünemann, A. Suvorov, V. Markov, Toward terahertz magnetrons: 210-GHz spatial-harmonic magnetron with cold cathode. IEEE Trans. Electron Devices **59**(12), 3608 (2012)
31. J.-I. Kim, S.-G. Jeon, G. Park, Terahertz vacuum electronics, in *Handbook of Terahertz Technologies: Devices and Applications*, ed. by H.-J. Song, T. Nagatsuma, (CRC Press, Boca Raton, FL, 2015)
32. J.B. Gunn, Microwave oscillations of current in III–V semiconductors. Solid State Commun. **1** (4), 88–91 (1963)
33. H. Kroemer, Theory of the Gunn effect. Proc. IEEE **52**(12), 1736–1736 (1964)
34. B. Ridley, T. Watkins, The possibility of negative resistance effects in semiconductors. Proc. Phys. Soc. **78**(2), 293 (1961)
35. C. Hilsum, Transferred electron amplifiers and oscillators. Proc. IRE **50**(2), 185–189 (1962)
36. S.Y. Liao, *Microwave Solid-State Devices* (Prentice-Hall, Englewood Cliffs, NJ, 1985)
37. H. Eisele, 480 GHz oscillator with an InP Gunn device. Electron. Lett. **46**(6), 422–423 (2010). https://doi.org/10.1049/el.2010.3362

38. W.T. Read Jr., A proposed high-frequency, negative-resistance diode. Bell Syst. Tech. J. **37** (2), 401–446 (1958)
39. R.L. Johnston, B.C. DeLoach, B.G. Cohen, A silicon diode microwave oscillator. Bell Syst. Tech. J. **44**(2), 369–372 (1965). https://doi.org/10.1002/j.1538-7305.1965.tb01667.x
40. C. Lee, R. Batdorf, W. Wiegmann, G. Kaminsky, The read diode—An avalanching, transit-time, negative-resistance oscillator. Appl. Phys. Lett. **6**(5), 89–91 (1965)
41. S.M. Sze, *Physics of Semiconductor Devices*, 2nd edn. (John Wiley & Sons, New York, 1981)
42. J. Nishizawa, K. Motoya, Y. Okuno, GaAs TUNNETT diodes. IEEE Trans. Microw. Theory Tech. **26**(12), 1029–1035 (1978). https://doi.org/10.1109/tmtt.1978.1129540
43. M. Ino, T. Ishibashi, M. Ohmori, C.W. oscillation with p+-p-n+ silicon IMPATT diodes in 200 GHz and 300 GHz bands. Electron. Lett. **12**(6), 148–149 (1976). https://doi.org/10.1049/el:19760115
44. J. Nishizawa, P. Plotka, T. Kurabayashi, H. Makabe, 706-GHz GaAs CW fundamental-mode TUNNETT diodes fabricated with molecular layer epitaxy. Phys. Status Solidi C **5**(9), 2802–2804 (2008)
45. M. Mukherjee, N. Mazumder, S.K. Roy, Photosensitivity analysis of gallium nitride and silicon carbide terahertz IMPATT oscillators: Comparison of theoretical reliability and study on experimental feasibility. IEEE Trans. Device Mater. Reliab. **8**(3), 608–620 (2008). https://doi.org/10.1109/TDMR.2008.2002358
46. L.L. Chang, L. Esaki, R. Tsu, Resonant tunneling in semiconductor double barriers. Appl. Phys. Lett. **24**(12), 593–595 (1974). https://doi.org/10.1063/1.1655067
47. R. Tsu, L. Esaki, Tunneling in a finite superlattice. Appl. Phys. Lett. **22**(11), 562–564 (1973)
48. T. Maekawa, H. Kanaya, S. Suzuki, M. Asada, Oscillation up to 1.92 THz in resonant tunneling diode by reduced conduction loss. Appl. Phys. Express **9**(2), 024101 (2016)
49. E.H. Colpitts, Oscillation generator, USA Patent 1,624,537 Patent Appl. 1,624,537 (1927)
50. B. Razavi, R. Behzad, *RF Microelectronics* (Prentice Hall, Upper Saddle River, NJ, 1998)
51. J.K. Clapp, An inductance-capacitance oscillator of unusual frequency stability. Proc. IRE **36** (3), 356–358 (1948)
52. P. Andreani, W. Xiaoyan, L. Vandi, A. Fard, A study of phase noise in colpitts and LC-tank CMOS oscillators. IEEE J. Solid State Circuits **40**(5), 1107–1118 (2005). https://doi.org/10.1109/JSSC.2005.845991
53. V. Marotta, G. Macera, M.P. Kennedy, E. Napoli, Comparative analysis of differential colpitts and cross-coupled VCOs in 180 nm Si-Ge HBT technology, in *2016 IEEE International Symposium on Circuits and Systems (ISCAS)*, 22–25 May 2016 (2016), pp. 1650–1653. https://doi.org/10.1109/ISCAS.2016.7538883
54. L. Li, P. Reynaert, M. Steyaert, A colpitts LC VCO with Miller-capacitance gm enhancing and phase noise reduction techniques, in *2011 Proceedings of the ESSCIRC (ESSCIRC)*, 12–16 Sept 2011 (2011), pp. 491–494. https://doi.org/10.1109/ESSCIRC.2011.6045014
55. Z. Hu, M. Kaynak, R. Han, High-power radiation at 1 THz in silicon: A fully scalable array using a multi-functional radiating mesh structure. IEEE J. Solid State Circuits **53**(5), 1313–1327 (2018). https://doi.org/10.1109/JSSC.2017.2786682
56. B. Heinemann et al., SiGe HBT technology with fT/fmax of 300GHz/500GHz and 2.0 ps CML gate delay, in *2010 International Electron Devices Meeting*, 6–8 Dec 2010 (2010), pp. 30.5.1–30.5.4, https://doi.org/10.1109/IEDM.2010.5703452
57. K. Lundien, R. Mattauch, J. Archer, R. Malik, Hyperabrupt junction varactor diodes for millimeter-wavelength harmonic generators. IEEE Trans. Microw. Theory Tech. **31**(2), 235–238 (1983)
58. P. Andreani, S. Mattisson, On the use of MOS varactors in RF VCOs. IEEE J. Solid State Circuits **35**(6), 905–910 (2000)
59. T. Soorapanth, C.P. Yue, D.K. Shaeffer, T.H. Lee, S.S. Wong, Analysis and optimization of accumulation-mode varactor for RF ICs, in *1998 Symposium on VLSI Circuits*, Citeseer (1998), pp. 32–33

60. Y. Oh, S. Kim, S. Lee, J.-S. Rieh, The island-gate varactor—A high-Q MOS varactor for millimeter-wave applications. IEEE Microw. Wirel. Compon. Lett. **19**(4), 215–217 (2009)
61. Y. Oh, J.-S. Rieh, A comprehensive study of high-Q island-gate varactors (IGVs) for CMOS millimeter-wave applications. IEEE Trans. Microw. Theory Tech. **59**(6), 1520–1528 (2011). https://doi.org/10.1109/tmtt.2011.2117437
62. N. Kim, Y. Oh, J.-S. Rieh, A 47 GHz LC cross-coupled VCO employing high-Q island-gate varactor for phase noise reduction. IEEE Microw. Wirel. Compon. Lett. **20**(2), 94–96 (2010)
63. G. Cusmai, M. Repossi, G. Albasini, A. Mazzanti, F. Svelto, A magnetically tuned quadrature oscillator. IEEE J. Solid State Circuits **42**(12), 2870–2877 (2007). https://doi.org/10.1109/JSSC.2007.908727
64. P. Chiang, O. Momeni, P. Heydari, A 200-GHz inductively tuned VCO with −7-dBm output power in 130-nm SiGe BiCMOS. IEEE Trans. Microw. Theory Tech. **61**(10), 3666–3673 (2013). https://doi.org/10.1109/TMTT.2013.2279779
65. K. Kwok, J.R. Long, J.J. Pekarik, A 23-to-29 GHz differentially tuned varactorless VCO in 0.13 μm CMOS, in *2007 IEEE International Solid-State Circuits Conference*, 11–15 Feb 2007 (2007), pp. 194–596. https://doi.org/10.1109/ISSCC.2007.373360
66. D.D. Kim, J. Kim, C. Cho, A 94GHz locking hysteresis-assisted and tunable CML static divider in 65nm SOI CMOS, in *2008 IEEE International Solid-State Circuits Conference*, 3–7 Feb 2008 (2008), pp. 460–628. https://doi.org/10.1109/ISSCC.2008.4523256
67. R. Miller, Fractional-frequency generators utilizing regenerative modulation. Proc. IRE **27**(7), 446–457 (1939)
68. J. Al-Eryani et al., A 47–217 GHz dynamic frequency divider in SiGe technology, in *Bipolar/BiCMOS Circuits and Technology Meeting*, (IEEE, New York, 2015), pp. 125–128
69. M. Seo, J. Hacker, M. Urteaga, A. Skalare, M. Rodwell, A 529 GHz dynamic frequency divider in 130 nm InP HBT process. IEICE Electron. Express **12**(3), 20141118 (2015)
70. H.R. Rategh, T.H. Lee, Superharmonic injection-locked frequency dividers. IEEE J. Solid State Circuits **34**(6), 813–821 (1999). https://doi.org/10.1109/4.766815
71. H. Seo, S. Seo, S. Jeon, J.-S. Rieh, A Q-band injection-locked frequency divider with inductive feedback for a locking range enhancement. IEEE Microw. Wirel. Compon. Lett. **21**(6), 317–319 (2011)
72. H. Tong, S. Cheng, A.L. Karsilayan, J.S. Martinez, An injection-locked frequency divider with multiple highly nonlinear injection stages and large division ratios. IEEE Trans. Circuits Syst. II Express Briefs **54**(4), 313–317 (2007). https://doi.org/10.1109/TCSII.2007.895071
73. P.-H. Feng, S.-I. Liu, A 300 GHz divide-by-2 ILFD using frequency boosting technique. IEEE Microw. Wirel. Compon. Lett. **23**(11), 599–601 (2013)
74. B. Razavi, A study of injection locking and pulling in oscillators. IEEE J. Solid State Circuits **39**(9), 1415–1424 (2004)
75. N. Kim, K. Song, J. Yun, J. Yoo, J.-S. Rieh, Two 122-GHz phase-locked loops in 65-nm CMOS technology. IEEE Trans. Microw. Theory Tech. **64**(8), 2623–2630 (2016). https://doi.org/10.1109/TMTT.2016.2581816
76. Y. Zhao et al., A 0.56 THz phase-locked frequency synthesizer in 65 nm CMOS technology. IEEE J. Solid State Circuits **51**(12), 3005–3019 (2016)
77. J. Yoo, D. Kim, J. Kim, K. Song, J.-S. Rieh, A 283-GHz fully integrated phase-locked loop based on 65-nm CMOS. IEEE Trans. Terahertz Sci. Technol. **8**(6), 784–792 (2018)
78. S.-C. Yen, T.-H. Chu, An Nth-harmonic oscillator using an N-push coupled oscillator array with voltage-clamping circuits, in *2003 IEEE MTT-S International Microwave Symposium Digest*, vol. 3, (IEEE, New York, 2003), pp. 2169–2172
79. U.L. Rohde, A.K. Poddar, J. Schoepf, R. Rebel, P. Patel, Low noise low cost ultra wideband N-push VCO, in *IEEE MTT-S International Microwave Symposium Digest* (2005), pp. 1171–1174. https://doi.org/10.1109/MWSYM.2005.1516884
80. R. Wanner, R. Lachner, G. Olbrich, A monolithically integrated 190-GHz SiGe push-push oscillator. IEEE Microw. Wirel. Compon. Lett. **15**(12), 862–864 (2005)
81. E. Seok et al., A 410GHz CMOS push-push oscillator with an on-chip patch antenna, in *IEEE International Solid-State Circuits Conference (ISSCC)*, (IEEE, New York, 2008), pp. 472–629

82. T. Yu-Lung, W. Huei, Triple-push oscillator approach: Theory and experiments. IEEE J. Solid State Circuits **36**(10), 1472–1479 (2001). https://doi.org/10.1109/4.953475
83. J. Grzyb, Y. Zhao, U. Pfeiffer, A 288-GHz lens-integrated balanced triple-push source in a 65-nm CMOS technology. IEEE J. Solid State Circuits **48**(7), 1751–1761 (2013)
84. D. Shim, D. Koukis, D.J. Arenas, D.B. Tanner, K. Kenneth, 553-GHz signal generation in CMOS using a quadruple-push oscillator, in *2011 Symposium on VLSI Circuits*, (IEEE, New York, 2011), pp. 154–155
85. J. Yun, D. Yoon, S. Jung, M. Kaynak, B. Tillack, J.-S. Rieh, Two 320 GHz signal sources based on SiGe HBT technology. IEEE Microw. Wirel. Compon. Lett. **25**(3), 178–180 (2015). https://doi.org/10.1109/LMWC.2015.2391011
86. O. Momeni, E. Afshari, High power terahertz and millimeter-wave oscillator design: A systematic approach. IEEE J. Solid State Circuits **46**(3), 583–597 (2011)
87. J. Yoo, N. Kim, J. Yun, M.G. Seo, J.-S. Rieh, A CMOS triple-push 280-GHz VCO integrated with 1/16,384 divider chain, in *2015 IEEE International Symposium on Radio-Frequency Integration Technology (RFIT)* (2015), pp. 58–60. https://doi.org/10.1109/RFIT.2015.7377886
88. A.V. Raisanen, Frequency multipliers for millimeter and submillimeter wavelengths. Proc. IEEE **80**(11), 1842–1852 (1992)
89. G. Chattopadhyay et al., An all-solid-state broad-band frequency multiplier chain at 1500 GHz. IEEE Trans. Microw. Theory Tech. **52**(5), 1538–1547 (2004)
90. A. Maestrini et al., Schottky diode-based terahertz frequency multipliers and mixers. C. R. Phys. **11**(7–8), 480–495 (2010)
91. V. Milanovic, M. Gaitan, J.C. Marshall, M.E. Zaghloul, CMOS foundry implementation of Schottky diodes for RF detection. IEEE Trans. Electron Devices **43**(12), 2210–2214 (1996). https://doi.org/10.1109/16.544393
92. C. Mao, C.S. Nallani, S. Sankaran, E. Seok, K.O. Kenneth, 125-GHz diode frequency doubler in 0.13-um CMOS. IEEE J. Solid State Circuits **44**(5), 1531–1538 (2009)
93. U.R. Pfeiffer, C. Mishra, R.M. Rassel, S. Pinkett, S.K. Reynolds, Schottky barrier diode circuits in silicon for future millimeter-wave and terahertz applications. IEEE Trans. Microw. Theory Tech. **56**(2), 364–371 (2008)
94. E. Ojefors, J. Grzyb, Y. Zhao, B. Heinemann, B. Tillack, U.R. Pfeiffer, A 820GHz SiGe chipset for terahertz active imaging applications, in *IEEE International Solid-State Circuits Conference* (2011), pp. 224–226
95. C.M. Grötsch, H. Mabler, A. Leuther, I. Kallfass, An active multiplier-by-six S-MMIC for 500 GHz, in *2018 43rd International Conference on Infrared, Millimeter, and Terahertz Waves (IRMMW-THz)*, (IEEE, New York, 2018), pp. 1–2
96. O. Momeni, E. Afshari, A broadband mm-wave and terahertz traveling-wave frequency multiplier on CMOS. IEEE J. Solid State Circuits **46**(12), 2966–2976 (2011)
97. N. Mazor, E. Socher, A SiGe distributed millimeter-wave frequency Tripler. IEEE Microw. Wirel. Compon. Lett. **24**(12), 893–895 (2014). https://doi.org/10.1109/LMWC.2014.2361625
98. A. Moussessian et al., A terahertz grid frequency doubler. IEEE Trans. Microw. Theory Tech. **46**(11), 1976–1981 (1998). https://doi.org/10.1109/22.734524
99. E. Monaco, M. Pozzoni, F. Svelto, A. Mazzanti, Injection-locked CMOS frequency doublers for μ-wave and mm-wave applications. IEEE J. Solid State Circuits **45**(8), 1565–1574 (2010)
100. Z. Ahmad, M. Lee, K.O. Kenneth, 1.4 THz, −13dBm-EIRP frequency multiplier chain using symmetric-and asymmetric-CV varactors in 65nm CMOS, in *2016 IEEE International Solid-State Circuits Conference (ISSCC)*, (IEEE, New York, 2016), pp. 350–351
101. A. Zamora et al., A high efficiency 670 GHz x36 InP HEMT multiplier chain, in *2017 IEEE MTT-S International Microwave Symposium (IMS)*, (IEEE, New York, 2017), pp. 977–979
102. Z. Chen, P. Heydari, An 85–95.2 GHz transformer-based injection-locked frequency tripler in 65nm CMOS, in *2010 IEEE MTT-S International Microwave Symposium*, (IEEE, New York, 2010), pp. 776–779

103. N. Mazor, E. Socher, X-band to W-band frequency multiplier in 65 nm CMOS process. IEEE Microw. Wirel. Compon. Lett. **22**(8), 424–426 (2012)
104. V. Jain et al., 1.0 THz f_{max} InP DHBTs in a refractory emitter and self-aligned base process for reduced base access resistance, in *Device Research Conference*, 20–22 June 2011 (2011), pp. 271–272. https://doi.org/10.1109/drc.2011.5994528
105. X. Mei et al., First demonstration of amplification at 1 THz using 25-nm InP high electron mobility transistor process. IEEE Electron Device Lett. **36**(4), 327–329 (2015). https://doi.org/10.1109/LED.2015.2407193
106. H. Lee et al., Intel 22nm FinFET (22FFL) process technology for RF and mm wave applications and circuit design optimization for FinFET technology, in *2018 IEEE International Electron Devices Meeting (IEDM)* (2018), pp. 14.1.1–14.1.4. https://doi.org/10.1109/IEDM.2018.8614490
107. B. Heinemann et al., SiGe HBT with fT/fmax of 505 GHz/720 GHz, in *2016 IEEE International Electron Devices Meeting (IEDM)* (2016), pp. 3.1.1–3.1.4. https://doi.org/10.1109/IEDM.2016.7838335
108. Y. Mao, K. Schmalz, J. Borngräber, J.C. Scheytt, A 245 GHz CB LNA in SiGe, in *2011 6th European Microwave Integrated Circuit Conference*, 10–11 Oct 2011 (2011), pp. 224–227
109. D. Yoon, M. Seo, K. Song, M. Kaynak, B. Tillack, J. Rieh, 260-GHz differential amplifier in SiGe heterojunction bipolar transistor technology. Electron. Lett. **53**(3), 194–196 (2017). https://doi.org/10.1049/el.2016.3882
110. O. Momeni, E. Afshari, A high gain 107 GHz amplifier in 130 nm CMOS, in *2011 IEEE Custom Integrated Circuits Conference (CICC)* (2011), pp. 1–4. https://doi.org/10.1109/CICC.2011.6055322
111. O. Momeni, A 260GHz amplifier with 9.2dB gain and −3.9dBm saturated power in 65nm CMOS, in *2013 IEEE International Solid-State Circuits Conference* (2013), pp. 140–141. https://doi.org/10.1109/ISSCC.2013.6487672
112. D. Park, D.R. Utomo, J. Hong, S. Lee, A 230–260GHz wideband amplifier in 65nm CMOS based on dual-peak Gmax-core, in *2017 Symposium on VLSI Circuits* (2017), pp. C300–C301. https://doi.org/10.23919/VLSIC.2017.8008516
113. D. Yoon, N. Kim, U. Pfeiffer, B. Heinemann, J. Rieh, A wide band 215–255 GHz CB differential amplifier in a 0.25-μm SiGe HBT technology, in *2013 Asia-Pacific Microwave Conference Proceedings (APMC)* (2013), pp. 351–353. https://doi.org/10.1109/APMC.2013.6695143
114. V. Radisic et al., InP HBT transferred substrate amplifiers operating to 600 GHz, in *2015 IEEE MTT-S International Microwave Symposium* (2015) pp. 1–3. https://doi.org/10.1109/MWSYM.2015.7166750
115. J. Hacker, M. Urteaga, M. Seo, A. Skalare, R. Lin, InP HBT amplifier MMICs operating to 0.67 THz, in *2013 IEEE MTT-S International Microwave Symposium Digest (MTT)* (2013), pp. 1–3. https://doi.org/10.1109/MWSYM.2013.6697518
116. S. Yoon, I. Lee, M. Urteaga, M. Kim, S. Jeon, A fully-integrated 40–222 GHz InP HBT distributed amplifier. IEEE Microw. Wirel. Compon. Lett. **24**(7), 460–462 (2014). https://doi.org/10.1109/LMWC.2014.2316223
117. V. Radisic, K.M. Leong, X. Mei, S. Sarkozy, W. Yoshida, W. Deal, Power amplification at 0.65 THz using InP HEMTs. IEEE Trans. Microw. Theory Tech. **60**(3), 724–729 (2012)
118. J. Kim, S. Jeon, M. Kim, M. Urteaga, J. Jeong, H-band power amplifier integrated circuits using 250-nm InP HBT technology. IEEE Trans. Terahertz Sci. Technol. **5**(2), 215–222 (2015). https://doi.org/10.1109/TTHZ.2014.2387259
119. H. Kazemi et al., 350mW G-band medium power amplifier fabricated through a new method of 3D-copper additive manufacturing, in *2015 IEEE MTT-S International Microwave Symposium* (2015), pp. 1–3. https://doi.org/10.1109/MWSYM.2015.7167037
120. K.J. Russell, Microwave power combining techniques. IEEE Trans. Microw. Theory Tech. **27**(5), 472–478 (1979). https://doi.org/10.1109/TMTT.1979.1129651

121. K. Chang, C. Sun, Millimeter-wave power-combining techniques. IEEE Trans. Microw. Theory Tech. **31**(2), 91–107 (1983). https://doi.org/10.1109/TMTT.1983.1131443
122. Q.J. Gu, Z. Xu, M.F. Chang, Two-way current-combining W-band power amplifier in 65-nm CMOS. IEEE Trans. Microw. Theory Tech. **60**(5), 1365–1374 (2012). https://doi.org/10.1109/TMTT.2012.2187536
123. H. Wang, J. Lin, Z. Tsai, Review of millimeter-wave CMOS power amplifiers, in *2018 IEEE International Symposium on Radio-Frequency Integration Technology (RFIT)* (2018), pp. 1–3. https://doi.org/10.1109/RFIT.2018.8524034
124. T.-W. Huang, J.-H. Tsai, J.-F. Yeh, On-chip power-combining techniques for mm-wave silicon power amplifiers, in *mm-Wave Silicon Power Amplifiers and Transmitters*, ed. by H. Hashemi, S. Raman, (Cambridge University Press, Cambridge, UK, 2016), pp. 257–301
125. J.-W. Lai, A. Valdes-Garcia, mm-Wave power-combining architectures: Hybrid combining, in *RF and mm-Wave Power Generation in Silicon*, ed. by H. Wang, K. Sengupta, (Elsevier, Amsterdam, 2016), pp. 335–360
126. J. Long, Monolithic transformers for silicon RF IC design. IEEE J. Solid State Circuits **35**(9), 1368–1382 (2000)
127. I. Aoki, S.D. Kee, D.B. Rutledge, A. Hajimiri, Distributed active transformer-a new power-combining and impedance-transformation technique. IEEE Trans. Microw. Theory Tech. **50** (1), 316–331 (2002). https://doi.org/10.1109/22.981284
128. H.-J. Song, J.-Y. Kim, K. Ajito, N. Kukutsu, M. Yaita, 50-Gb/s direct conversion QPSK modulator and demodulator MMICs for terahertz communications at 300 GHz. IEEE Trans. Microw. Theory Tech. **62**(3), 600–609 (2014)
129. D.-H. Kim, J.-S. Rieh, CMOS 138 GHz low-power active mixer with branch-line coupler. Electron. Lett. **48**(10), 554–555 (2012)
130. K. Sengupta, A. Hajimiri, Distributed active radiation for THz signal generation, in *2011 IEEE International Solid-State Circuits Conference* (2011), pp. 288–289. https://doi.org/10.1109/ISSCC.2011.5746322
131. K. Sengupta, A. Hajimiri, A 0.28 THz power-generation and beam-steering array in CMOS based on distributed active radiators. IEEE J. Solid State Circuits **47**(12), 3013–3031 (2012). https://doi.org/10.1109/JSSC.2012.2217831
132. M.P. DeLisio, R.A. York, Quasi-optical and spatial power combining. IEEE Trans. Microw. Theory Tech. **50**(3), 929–936 (2002). https://doi.org/10.1109/22.989975
133. D. Gritters, K. Brown, E. Ko, 200–260GHz solid state amplifier with 700mW of output power, in *2015 IEEE MTT-S International Microwave Symposium*, 17–22 May 2015 (2015), pp. 1–3. https://doi.org/10.1109/MWSYM.2015.7166808
134. Z. Ahmad, M. Lee, O.K. Kenneth, 1.4THz, −13dBm-EIRP frequency multiplier chain using symmetric- and asymmetric-CV varactors in 65nm CMOS, in *2016 IEEE International Solid-State Circuits Conference (ISSCC)* (2016), pp. 350–351. https://doi.org/10.1109/ISSCC.2016.7418051
135. B.-W. Min, G. Rebeiz, Single-ended and differential Ka-band BiCMOS phased array front-ends. IEEE J. Solid State Circuits **43**(10), 2239–2250 (2008)
136. D. Muller et al., A WR3-band reflective-type phase shifter MMIC with integrated amplifier for error- and loss compensation, in *2017 12th European Microwave Integrated Circuits Conference (EuMIC)* (2017), pp. 1–4. https://doi.org/10.23919/EuMIC.2017.8230645
137. Y. Kim, S. Kim, I. Lee, M. Urteaga, S. Jeon, A 220–320-GHz vector-sum phase shifter using single Gilbert-cell structure with Lossy output matching. IEEE Trans. Microw. Theory Tech. **63**(1), 256–265 (2015). https://doi.org/10.1109/TMTT.2014.2376515
138. H.G. Yu, K.J. Lee, M. Kim, 300 GHz vector-sum phase shifter using InP DHBT amplifiers. Electron. Lett. **49**(4), 263–264 (2013). https://doi.org/10.1049/el.2012.4109
139. H. Jalili, O. Momeni, A 318-to-370GHz standing-wave 2D phased array in 0.13μm BiCMOS, in *2017 IEEE International Solid-State Circuits Conference (ISSCC)* (2017), pp. 310–311. https://doi.org/10.1109/ISSCC.2017.7870385

Chapter 3
THz Detectors and Related Topics

This chapter will provide an overview of various methods of detecting electromagnetic waves, with a focus on THz wave detection. The detection can be made exploiting several different mechanisms. First, various thermal effects caused by THz radiation can be used for detection. Called thermal detectors, this type of detectors include bolometers, pyroelectric detectors, thermopiles, and Golay cells. Second, the rectification of AC signals, often after conversion from electromagnetic waves by antenna, into DC voltage can be used for detection, as in the case for the square-law detectors. Third, there is the heterodyne technique, in which signal is down-converted by a mixer and then detected with a conventional low-frequency AC signal detector. This relaxes the operation frequency requirement for the detector at the back-end of the receiver chain. Historically, the first technique has been widely deployed over infrared detection, while the other two techniques prevailed over the microwave and lower frequency detection. THz wave detection, therefore, can be realized based on these approaches with a proper extension of the frequency range, downward or upward. Yet another detection method that has been dominating the optical band is detection with photodetectors. However, this method does not fit for THz detection since the photon energy corresponding to the THz spectrum is too small (a few meV) for practical detection, and hence will not be treated in this chapter. The rectification and heterodyne techniques can be implemented by either diodes or transistor circuits and will be discussed separately after describing the thermal detector in this chapter. Array detectors will be treated as an additional section.

3.1 Thermal Detectors

The illumination of materials with electromagnetic waves generates heat and causes a temperature increase inside the material. This temperature change induces various secondary thermal effects, which can be exploited for the detection of the THz

© Springer Nature Switzerland AG 2021
J.-S. Rieh, *Introduction to Terahertz Electronics*,
https://doi.org/10.1007/978-3-030-51842-4_3

waves. For example, a temperature increase results in resistance change, which is utilized by bolometers for detection. Temperature variation can also induce charges across a capacitor if a certain type of material is used as the dielectric, which is behind the operation of pyroelectric detectors. Unbalanced temperature across some materials leads to a voltage difference, which is gauged in thermopiles for detection. Increased temperature causes an expansion of the gas volume, which is cleverly combined with an optical technique in Golay cells. An overview of the mechanisms for these operations will be provided in this section. Before moving on to this section, a brief introduction to the parameters used to characterize the performance of detectors will be provided, which will be relevant not only to this section but to the entire chapter.

The two main performance parameters for detectors are the responsivity and the noise equivalent power (NEP). The responsivity $\mathfrak{R}$ is the ratio of the output DC voltage over the input power, and can be expressed as:

$$\mathfrak{R} = \frac{V_{\mathrm{DC,out}}}{P_{\mathrm{in}}}\ [\mathrm{V/W}]. \tag{3.1}$$

As Eq. (3.1) indicates, the responsivity represents the effectiveness of the conversion of the input power into the output voltage. NEP is defined as the input power that leads to the output signal-to-noise ratio (SNR) of unity for a bandwidth of 1 Hz. The average noise voltage of a system with a bandwidth of 1 Hz is often represented by $\overline{v}_{\mathrm{n}}$, and thus the condition for SNR = 1 with 1-Hz bandwidth can be expressed as:

$$\mathrm{SNR}_{\mathrm{out}} = \frac{V_{\mathrm{DC,out}}}{\overline{v}_{\mathrm{n}}} = \frac{\mathfrak{R}P_{\mathrm{in}}}{\overline{v}_{\mathrm{n}}} = 1 \quad \text{if} \quad P_{\mathrm{in}} = \mathrm{NEP}. \tag{3.2}$$

Hence,

$$\mathrm{NEP} = \frac{\overline{v}_{\mathrm{n}}}{\mathfrak{R}}\ \left[\mathrm{W/Hz^{1/2}}\right]. \tag{3.3}$$

NEP can be interpreted as a measure of the sensitivity of the detector, or the minimum detectible input power. It should be remembered again, though, that NEP values assume a bandwidth of 1 Hz, meaning that they are normalized over the bandwidth. As the noise level increases with the system bandwidth, in most cases, the actual minimum detectible power level by the system is larger than the NEP value. For example, NEP of 1 pW/Hz$^{1/2}$ does not necessarily mean that the minimum detectible power is 1 pW. It is very likely that the maximum detectible power will be larger than 1 pW as the system bandwidth is usually larger than 1 Hz. One may have noticed from the unit of NEP that the normalization is made over the square root of the bandwidth instead of the bandwidth itself. This is because the square of the noise voltage (v_{n}^2), a quantity given in the unit of [V^2/Hz], is normalized over the true bandwidth. NEP is proportional to average noise voltage ($\overline{v}_{\mathrm{n}}$) or square root of v_{n}^2, and thus its normalization is made over the square root of the bandwidth.

Let us consider the example case of NEP = 1 pW/Hz$^{1/2}$ (= 10^{-12} W/Hz$^{1/2}$) again. If the bandwidth of the system is 10 kHz (= 10^4 Hz), for example, the minimum detectible power is 100 pW (= 10^{-10} W), as limited by the detection rate. It is useful to remember that in some literatures NEP is expressed in the unit of [W] instead of [W/Hz$^{1/2}$] by omitting bandwidth normalization, but this is rare and not a general practice. It is also noted that Eqs. (3.1)–(3.3) assume the detector output is given in voltage, but the output may be current for some detectors such as photodiodes. In those cases, the responsivity is given as $I_{\mathrm{DC,out}}/P_{\mathrm{in}}$ in the unit of [A/W] and NEP as $\overline{i_\mathrm{n}}/\mathfrak{R}$ in the unit of [W/Hz$^{1/2}$] (unchanged), where $\overline{i_\mathrm{n}}$ is the average noise current in the unit of [A/Hz$^{1/2}$]. While a balanced performance assessment of detectors would show both responsivity and NEP, it is fair to say that NEP is more intrinsic to the detector. This is because the responsivity can be enhanced by placing an amplifier following the detector, while NEP cannot be improved with such a simple add-on scheme. Hence, in this chapter, NEP will be treated as the primary parameter in comparing detectors.

As an additional note, the sensitivity of detectors is sometimes presented in terms of another parameter, the specific detectivity D^* (reads D-star) [1]. D^* is defined as the inverse of NEP normalized over the area, or

$$D^* = \frac{\sqrt{A}}{\mathrm{NEP}}\left[\mathrm{cm\,Hz}^{1/2}/\mathrm{W}\right]. \tag{3.4}$$

D^* may provide a fair comparison over detectors with different active areas if NEP is proportional to the square root of the area, which is often the case for the background limited cases. The readers may find some literatures define D^* as $\sqrt{AB}/\mathrm{NEP}$ with bandwidth B included, but this is based on the unnormalized NEP in [W] as mentioned above. The *specific detectivity* D^* is often referred to as the *detectivity* for simplicity, but there is a separate definition for *detectivity* D, which is simply the inverse of NEP in [Hz$^{1/2}$/W]. Still, D^* is in far more popular use than D as many find its area normalization useful.

3.1.1 Bolometers

Bolometers generally refer to a group of detectors that make use of the resistance change due to the temperature rise caused by radiation. The invention of the bolometer is credited to Langley, who devised a detector for astronomical observation in 1878 and named it "bolometer." It was not until 1960s, however, that bolometers began to be widely adopted with the emergence of the carbon resistance bolometer [2], before which Golay cell was a popular choice. Since then, bolometers have been widely selected for various applications in infrared and neighboring bands, which was fueled by the introduction of germanium thermometer [3], superconducting thermometer [4], and silicon monolithic thermometer [5] for bolometers. As bolometers can be realized in many different ways, there are

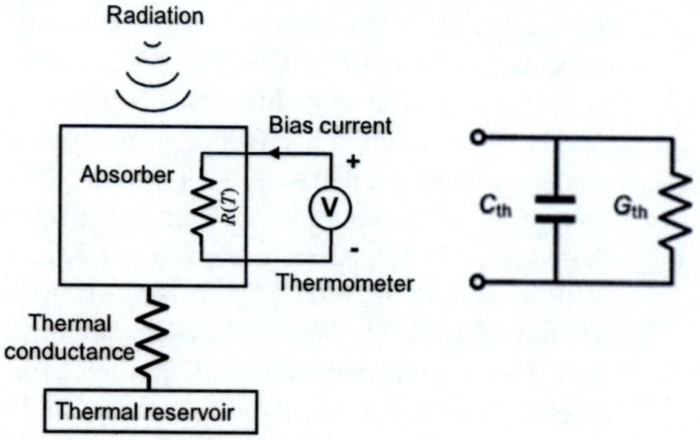

Fig. 3.1 (**a**) Conceptual sketch of a typical bolometer. (**b**) Equivalent thermal circuit

numerous variants depending on the details of the implementation, leading to different structures for each case. In this section, the common feature of bolometers will be briefly described first, followed by an overview of various bolometer types.

The conceptual sketch of a typical bolometer is given in Fig. 3.1a. The key components of a bolometer are radiation absorber and thermometer, which are connected to a thermal reservoir via a thermal link. Sometimes a single part serves as both radiation absorber and thermometer. Electromagnetic waves incident on the absorber generate heat and increase the temperature of the material composing the absorber. With the thermometer attached to the absorber, the resistance change in the absorber due to the temperature rise is converted to a better measurable quantity, such as voltage, from which the incident power is evaluated. In this process, there exist a finite time delay in the temperature change and a finite conductance for the generated heat to be drained to the thermal reservoir. Hence, the absorber and the thermal link can be represented by a thermal capacitor C_{th} and a thermal conductance G_{th}, respectively. Based on this view, a bolometer can be modeled by a thermal *RC* network as shown in Fig. 3.1b, an analogy to an electrical *RC* network. The incident power absorbed by the bolometer, *P*, can be expressed by the following heat transfer equation:

$$P = G_{\mathrm{th}} \Delta T(t) + C_{\mathrm{th}} \frac{d\Delta T(t)}{dt}, \tag{3.5}$$

where $\Delta T(t)$ is the temperature rise induced by the illumination. (If the power absorption from the ambient temperature is also to be considered, *P* will additionally include the ambient term as well.) It is apparent from this equation that the thermal response time will be given as $C_{\mathrm{th}}/G_{\mathrm{th}}$.

In most applications, a fast response is desired for bolometers, which would require a small C_{th} and a large G_{th}. Small C_{th} demands the absorber (and the supporting substrate) to have a low heat capacity, which in turn requires a small volume and a low specific heat. The condition for a small volume, however, contradicts the need for a large absorption area that will improve the bolometer sensitivity. Hence, we need a smart structural optimization based on this tradeoff. A large G_{th} can be obtained with a short and wide thermal link made of a material with a high thermal conductivity. Low heat capacity will also be preferred for the thermal link to suppress the parasitic thermal capacitance. The thermometer physically attached to the absorber also needs to have a low heat capacity, along with a proper temperature dependence for the electrical resistance. For the details on bolometer detection principles, the readers are referred to a comprehensive overview by Richards [6].

While there are various types of bolometers, they can be grouped into chip bolometers and composite bolometers, depending on their structure details. For the former, there is no separation between absorber and thermometer, and they are realized by one single part, leading to a compact structure. For the latter, absorber and thermometer are separate, which allows the independent optimization of the two parts, relaxing the tradeoff between the heat capacity and the absorption area as mentioned above. For chip bolometers, semiconductor materials are widely adopted such as Ge and Si, while Si is now a more preferred option as it shows a lower specific heat leading to a small heat capacity. It also benefits from a better compatibility with standard micro-fabrication processes, which allows monolithic approach and easy extension to array bolometers. For composite bolometers, various materials can be employed for different parts, such as bismuth (Bi), gold (Au), aluminum (Al) for absorbers, and sapphire, diamond, and Si for the supporting substrate, with a separate thermometer.

Another type of bolometers is superconductor bolometers [4]. Since the majority of bolometers operate at cryogenic temperatures, the adoption of superconducting material for bolometers is a natural path in a sense, because most superconductors show the state transition at cryogenic temperatures. If the operation temperature of a superconductor bolometer is set just below the transition temperature, a small increase in the temperature due to radiation will trigger a transition from the superconducting state to the normal state, leading to a sudden resistivity surge that can be used for radiation detection. The hot electron bolometer, which is a widely adopted superconductor bolometer, will be discussed in detail in a separate subsection.

Antenna-coupled bolometers are also worthwhile to mention. Bolometers typically do not need an antenna, as opposed to the case of rectifying detectors that compete with bolometers over some frequency ranges. However, on-chip antennas can be still employed for bolometers, which will effectively increase the absorption cross section, leading to a lowered NEP [7]. Besides, a control over directivity, polarization, and frequency (if needed), can be attained with antennas. However, all these merits will be attained only at the sacrifice of area, which is a critical factor especially for array applications. The matching with antenna is also an issue for antenna-coupled bolometers, for which absorption materials with a high resistivity are preferred for impedance matching. Typically coupled with antennas are microbolometers [8], which has a surface area dimension much smaller than the operation wavelength, as the name indicates. The small dimension of the absorption area, which is preserved by placing the coupled antenna separate from the absorption area, allows a fast response, while a low NEP and a high responsivity can be obtained with the help of the coupled antenna. There are reports on microbolometers without antennas as well, suitable for cases that require high density for array applications [9]. One may note that most of microbolometers are intended for room temperature operation, unlike the mainstream bolometers, although some cryogenic microbolometers have also been reported [10].

Bolometers operating at cryogenic temperatures, mostly near liquid helium (He) temperature (4.2 K) or below, generally exhibit NEP in the range of 10^{-17} to 10^{-12} W/Hz$^{1/2}$ depending on the bolometer type and the actual temperature. On the extreme side, there have been efforts to meet the required NEP of in the order of 10^{-20} W/Hz$^{1/2}$ for spaceborne infrared telescopes being planned such as SPICA or SAFIR, leading to some early results exhibiting NEP lower than 10^{-19} W/Hz$^{1/2}$ [11]. NEP of room temperature bolometers typically falls on the range of 10^{-11} to 10^{-10} W/Hz$^{1/2}$, while lower NEP has also been reported that reaches below 10^{-12} W/Hz$^{1/2}$ at room temperature [12]. There is an interesting observation that NEP of THz bolometers has been reduced by half every 2 years in average over the past 70 years [13]. It is noteworthy to comment that the noise level of bolometers does not show a strong dependence on the frequency. In fact, this wideband response is an inherent property of any thermal detectors, including those to be discussed below, as the radiation at any spectrum causes a temperature rise of the material it illuminates. The response time of bolometers typically ranges over 1–100 ms, while it can be reduced down below 1 μs depending on the structure and temperature [14].

3.1.2 Pyroelectric Detectors

Pyroelectricity refers to a phenomenon observed in pyroelectric materials, in which a voltage difference is induced across the material when there is a temporal temperature variation. Pyroelectricity is believed to have been discovered as early as in ancient Greek era, followed by rather contemporary observations including that of Brewster, who first named it *pyroelectricity* in the early nineteenth century. It was not until the mid-twentieth century, however, that it was widely employed for electromagnetic radiation detection [15, 16] with a mechanism described below.

When there is a temperature change, the permanent electric dipole moments that exist inside pyroelectric materials are momentarily aligned, inducing charges at the material surface. Hence, a transient potential difference is established across the pyroelectric crystal. This potential difference vanishes, however, if the temperature becomes stabilized, because a generated internal depolarization field neutralize the effect and the established dipole moment alignment is relaxed. Therefore, pyroelectric materials respond to the *temporal change* of the temperature. The schematic of a typical pyroelectric detector is shown in Fig. 3.2. It is basically a parallel plate capacitor with a pyroelectric material adopted as the dielectric, whose polar axis is formed normal to the electrode plates. With a temperature change, a voltage difference will be established across the capacitor. This voltage, or the current induced by

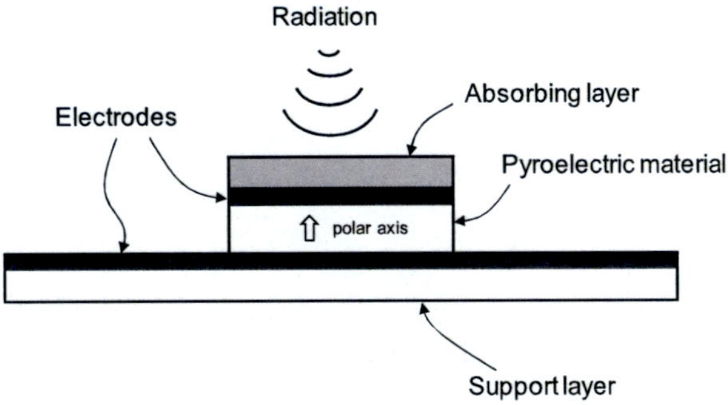

Fig. 3.2 Structure of a typical pyroelectric detector

the voltage along the load impedance (not shown), is the output of the detector, which is usually magnified by an amplifier at the final stage. While pyroelectric detectors are useful for the detection of radiation that changes over time as mentioned above, they can still be employed to detect continuous wave (CW) signal, if the signal is properly modulated by a chopper or other similar modulating schemes.

As for the materials, TGS (triglycine sulfate) was widely used in the early stage of pyroelectric detectors. Later on, various pyroelectric materials have appeared for pyroelectric detectors, including SBN (strontium-barium niobite), $LiTaO_3$ (lithium tantalate), $LiNbO_3$ (lithium niobate), BaTiO (barium titanate), PVF_2 (polyvinylidene fluoride), and so forth. Detailed comparisons between the materials can be found elsewhere [17, 18]. It should be noted that pyroelectric materials are governed by Curie temperature, beyond which the pyroelectric materials lose the unique magnetic property and does not exhibit pyroelectricity anymore. Hence, the detector should be operated below this temperature. In terms of the structure, there is a variance called the compensated pyroelectric detector. As pyroelectric detectors respond to a temperature variation, the ambient temperature variation can be falsely detected. To suppress the effect, a dummy detector, which is screened from the external radiation, is placed in proximity of the true detector. As the ambient temperature variation is detected by both detectors while the radiation is incident only on the true detector, the net effect of the radiation can be obtained by extracting the difference between the two outputs. In fact, this is analogous to the case of the differential pair, which is widely adopted for transistor circuits to suppress the common-mode noise. The principle is basically the same.

Pyroelectric detectors show wideband characteristics like other thermal detectors because of the reason mentioned earlier. Besides, they distinguish themselves from other thermal detectors in that they exhibit a fast response below the thermal relaxation time, since they do not need to wait until the thermal equilibrium is established when there is a temperature change. This leads to the response time easily below μs order. Pyroelectric detectors show fairly good NEP values, typical values ranging around ~10^{-9} W/Hz$^{1/2}$. The actual values vary over a quite wide range depending on the vendors. It is true that this is several orders worse than that of cryogenic bolometers and also falls short of NEPs with typical room-temperature bolometers. However, pyroelectric detectors benefit from the simple structure, often housed into a low-cost transistor-type metal package with a window opening. This makes it an affordable detector still capable of wide-range detection including the THz band [19].

3.1.3 Thermopiles

Another affordable thermal detector operating at room temperature is the thermopile. Thermopiles are one of the traditional device types that has a long history traced back to the early nineteenth century. It is a versatile device used for various cases of thermal detection, one derivative being the radiation detection for infrared and neighboring spectra including the THz band [20].

A thermopile can be understood as a series connection of multiple thermocouples. Hence, its operation principle follows that of thermocouples. The theory of thermocouples starts with Seebeck (reads ze-bek) effect, which can be explained with the help of Fig. 3.3a [21]. When there exists a temperature difference ΔT across a metal rod, a voltage difference ΔV is induced across it. This is because the mean free path of free electrons inside the metal varies over the temperature. In the region with a longer average mean free path, electron density will be lower than that of the region with a shorter average mean free path. Hence, the balance of the electron density will be broken when there is a temperature difference across the metal, and a higher voltage will appear at the region with a lower electron density with respect to the region with a higher electron density. This is called the Seebeck effect, which can be expressed as a simple relation:

$$\Delta V = -S\Delta T, \tag{3.6}$$

where S is the Seebeck coefficient, which itself is a function of the temperature. The minus sign is included here to follow the convention of the Seebeck coefficient, with which the voltage is taken at the cold side with reference to the hot side. For materials with a negative S (e.g., aluminum, platinum), the mean free path is longer at higher temperatures, whereas for materials with a positive S (e.g., copper, gold), the mean free path is shorter at higher temperatures. Note that the Seebeck effect is the inverse process of the Peltier effect, which describes a temperature difference induced across a metal rod when a voltage difference is applied across it.

The Seebeck effect can find various useful applications, one of which is the thermocouple to be discussed here. Consider two metal rods with opposite signs of the Seebeck coefficient, which are connected to form a junction as shown in Fig. 3.3b. If the two sides are placed at different temperatures, voltage differences will be induced for both metal rods, but with opposite polarities. This will lead to a voltage difference established between the two metal tips on the other side of the junction. A similar result will be obtained even for two metals with the same sign of the Seebeck coefficient if the actual values are different, but the induced voltage difference will not be as large as the case with two materials with opposite signs.

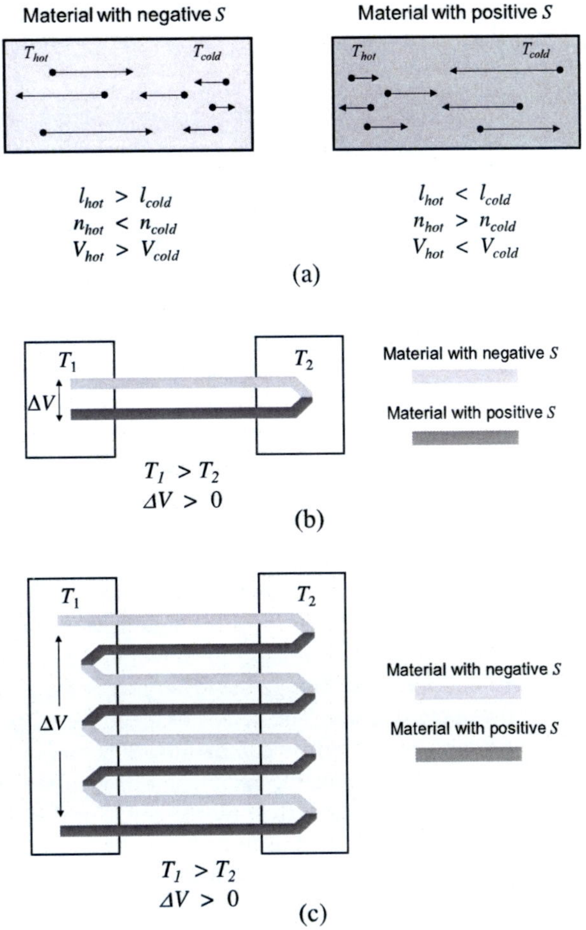

Fig. 3.3 (**a**) Conceptual view of Seebeck effect. For materials with a negative Seebeck coefficient *S*, the mean free path of electrons is larger at higher temperature ($l_{hot} > l_{cold}$). For materials with a positive *S*, the opposite occurs ($l_{hot} < l_{cold}$). As a result, the induced electron concentration (*n*) and voltage (*V*) show the opposite signs for the two materials. (**b**) Structure of a typical thermocouple, which is made of two materials with the opposite signs of *S*. (**c**) Structure of a typical of thermopile, which is basically multiple thermopiles connected in series

With the established voltage taken as the output, this structure can be used as a temperature sensor, which is called the thermocouple. If the temperature difference is induced by electromagnetic wave radiation, it can be employed as a radiation detector. Now, suppose the configuration in which multiple thermocouples are connected in series as depicted in Fig. 3.3c. In this case, the output voltage will be even larger, by a factor of the number of thermocouples cascaded, leading to a larger responsivity by the same factor (assuming the same power absorbed per unit thermocouple). This is the thermopile.

Various materials and structures can be employed for thermopiles, including a modern integrated thermopile based on CMOS technology [20, 22]. A typical NEP of thermopile detectors falls on the order of 10^{-9} to 10^{-8} W/Hz$^{1/2}$, an acceptable value for many practical applications.

3.1.4 Golay Cells

Golay cells are another type of room-temperature uncooled thermal detector that can be used for the detection of a wide range of frequency including the THz band. While all the thermal detectors discussed so far are based on the *solid-state* phenomena that respond to the temperature change, the Golay cell makes use of the expansion of the *gas-phase* material by a temperature rise with the help of an optical method to detect the expansion. It was first invented by Golay in 1947 [23], apparently named after the inventor, and has been used as a very effective radiation detector until today.

The schematic of a typical Golay cell is illustrated in Fig. 3.4 [24]. The main body is composed of a gas cell, an absorbing film, and a flexible mirror. When there is incoming radiation through the cell window incident on the absorption film, the cell temperature increases, resulting in the expansion of the gas confined in the pneumatic gas cell. The gas expansion causes the deformation of the flexible membrane mirror located on the opposite side of the cell window. Hence, if the delicate deformation on the mirror surface can be detected reliably, the temperature rise and thus the radiation power can be estimated. This is performed by an optical system composed of a light source, lenses, a line grid, and a photo detector as shown in the right side of Fig. 3.4. The light from the light source reaches the surface of the flexible mirror through the upper half of the pair of lenses and the line grid, and then is reflected back through the lower half of the grid and the lenses, finally landing on the photo detector. The details of the reflected grid image on the photo detector are sensitive to the surface deformation of the mirror. Hence, by detecting the change in the grid image pattern, the amount of the deformation is determined and so is the radiation power absorbed by the cell. For this, in the Golay's original design, the lower half of the grid pattern was carefully designed so that it is slightly misaligned with the incoming image of the upper-half grid pattern carried by the reflected light beam when the mirror is totally flat. With this tactical scheme, the deformed mirror

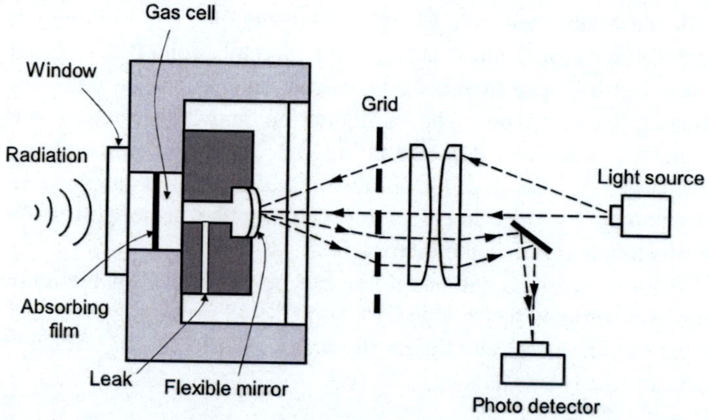

Fig. 3.4 Structure of a typical Golay cell. (Adapted by permission from Springer Nature: [24] © 2012)

surface with the gas expansion, in a concave or convex (in most cases) shape, will either strengthen or weaken the light intensity falling on the photo detector. By reading this change, the absorbed radiation power can be evaluated with a proper calibration.

The desired property for the gas inside the cell is the low heat conductivity, for which xenon (Xe) is typically adopted. The ambient temperature variation can lead to the volume change of the gas cell, which needs to be avoided for a precise radiation detection. A structural solution is to form a small leak between the gas cell and the gas reservoir outside the cell, so that any change in the gas cell volume due to the ambient temperature variation can be compensated by the leak. Only constraint due to this compensation scheme in actual detection is that the detection should be made before the leak takes place. For this reason, the incoming radiation is often modulated by a chopper at a rate faster than the time constant of the leak, typically at around 10–20 Hz with a detection response time of ~20 ms. This also helps to suppress the $1/f$ noise of the detector.

The NEP values available from Golay cells are among the best for room-temperature thermal detectors, ranging over 5×10^{-11} to 10^{-9} W/Hz$^{1/2}$ with a wide spectrum coverage including the THz band. This explains the popular use of this detector long after its invention without much change in the design. However, there do exist modern versions of Golay cells developed for compact implementation. A miniature Golay cell based on planar micromachining technology was reported [25], in which the deformation of a micro gas cell was detected by tunneling displacement sensor instead of the conventional optical method. Built on a Si wafer and intended for a room operation, it showed an NEP of 3×10^{-10} W/Hz$^{1/2}$. There has also been an attempt to adopt a graphene film as the gas cell membrane for micro Golay cell arrays, an interesting crossover of old and new technologies [26].

3.2 Diode Detectors

The term *diode* covers a wide range of devices, and its definition may vary depending on specific application cases. Here, we will assume a rather broader definition for the diode—an electronic device with two terminals. From this point of view, there are various types of diodes that can serve as a THz detector. In this section, three types of diode-based THz detectors will be discussed: SBD (Schottky Barrier Diode), HEB (Hot Electron Bolometer), and SIS (Superconductor–Insulator–Superconductor) mixer. In principle, these detectors operate as a *mixer* based on the heterodyne technique, as opposed to the detectors described in the previous section which are basically direct detectors. To help understand the difference between these two types of detection, a brief comparison will be first made between direct and heterodyne detections before moving on to the description of each type of the diode-based detectors.

3.2.1 *Direct Detection vs. Heterodyne Detection*

As just mentioned, there can be two types of THz signal detection: direct detection and heterodyne detection. Sometimes they are referred to as incoherent detection and coherent detection, respectively, as the latter may preserve the phase information of the RF input signal while the former does not. Conceptual block diagrams for the two cases are illustrated Fig. 3.5. For the direct detection case, detection is made at RF (radio frequency) stage with an RF detector. For the heterodyne case, detection is achieved at IF (intermediate frequency) stage with an IF detector after frequency down-conversion with a mixer driven by an LO (local oscillator). It is true that heterodyne systems can be employed in various receiver configurations, and thus an IF detector does not need to be located at the end as shown in Fig. 3.5b. The comparison with the direct detection, however, will be more straightforward by including an IF detector, which will be assumed here. For both direct and heterodyne cases, the detection sensitivity can be enhanced by employing a front-end amplifier, which is a transistor-based circuit as will be discussed separately in Sect. 3.3.3. The inclusion of an amplifier, however, is optional and will be practiced only when a sufficient gain is available at the incident RF signal frequency, which is not guaranteed for high-frequency regimes such as the THz band.

A comparison between the direct and heterodyne detections can be made in terms of responsivity and noise performances [27]. Let us assume throughout the discussion below that the front-end RF amplifier is not available and thus absent, which is true for lots of THz detection cases. In this case, the responsivities for the direct and heterodyne detections cases will be expressed as:

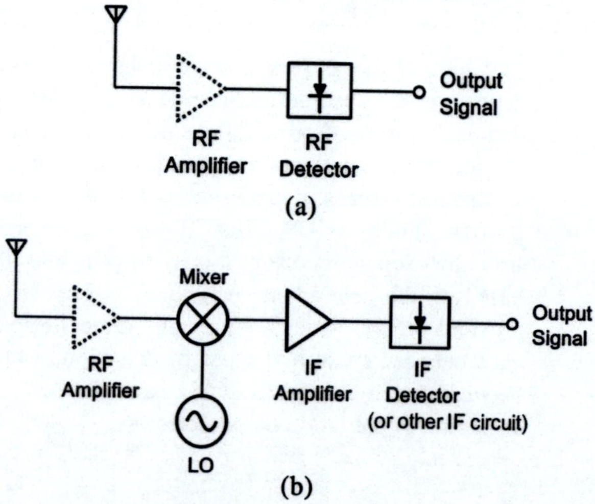

Fig. 3.5 Block diagrams of two cases of detection scheme: (**a**) Direct detection, (**b**) Heterodyne detection

$$\mathfrak{R}_{\text{direct}} = \mathfrak{R}_{\text{RF\_detector}} \tag{3.7}$$

$$\mathfrak{R}_{\text{hetero}} = G_{\text{mixer}} G_{\text{IF\_amp}} \mathfrak{R}_{\text{IF\_detector}}, \tag{3.8}$$

where $\mathfrak{R}_{\text{RF\_detector}}$ and $\mathfrak{R}_{\text{IF\_detector}}$ are the responsivities of the RF detector and the IF detector, and G_{mixer} and $G_{\text{IF\_amp}}$ are the conversion gain of the mixer and the gain of the IF amplifier, respectively. Matched conditions are assumed for all the gain parameters. There are a couple of factors that would favor the heterodyne detection from the responsivity point of view. First, the responsivity of typical detectors tend to show a larger value with lower frequencies, leading to $\mathfrak{R}_{\text{IF\_detector}}$ typically larger than $\mathfrak{R}_{\text{RF\_detector}}$. Second, the responsivity for the heterodyne detection can be boosted with the gain factor, $G_{\text{mixer}} \cdot G_{\text{IF\_amp}}$. The included mixer may exhibit a conversion loss ($G_{\text{mixer}} < 1$) depending on the mixer type and the frequency range. Even for such a case, a sufficiently large $G_{\text{IF\_amp}}$ readily available from the IF amplifier shall override the mixer loss, leading to a significantly large overall gain factor. As a result, the heterodyne detection typically shows a higher responsivity.

The heterodyne detection is favored from the noise perspective as well. To understand this, it will be helpful to exploit the Friis formula [28], which provides a valuable insight on the noise properties of cascaded multi-stage systems such as heterodyne systems. The difficulty here, however, is that the noise level of detectors is conventionally presented in terms of NEP, not the noise factor F or the noise temperature T_n that go along with the Friis formula. Hence, in principle, the Friis formula cannot be directly applied to the noise analysis of detectors based on NEP. The fact that the gain through these types of detectors is nonlinear also adds to the difficulty in utilizing the formula [29]. However, the intuition provided by the Friis formula is so useful, and thus, it can hardly be replaced by other analyses. Hence, here, we will sacrifice rigor in favor of pragmatism to conceptually understand the noise in detection systems by introducing the Friis formula and employing noise factors to represent the detector noise level.

According to the Friis formula, as most readers should be aware, the total noise factor of a cascaded multi-stage system cascaded is given by:

$$F_{\text{total}} = F_1 + \frac{(F_2 - 1)}{G_1} + \frac{(F_3 - 1)}{G_1 G_2} + \frac{(F_4 - 1)}{G_1 G_2 G_3} \cdots, \tag{3.9}$$

where F_i and G_i are the noise factor and gain of the ith stage. The noise factor F is defined as the ratio of the input SNR (signal-to-noise ratio) over the output SNR:

$$F = \frac{\text{SNR}_{\text{in}}}{\text{SNR}_{\text{out}}}. \tag{3.10}$$

Here, the readers are advised not to mix up two parameters, noise factor F (in linear scale) and noise figure NF (in dB scale). The latter is defined as NF = 10log$F$ (dB). There is a close relation between F and T_n, and the two noise parameters can always be converted to the other with simple relations:

Table 3.1 Conversion between T_n, F, and NF at $T_0 = 290$ K

Noise temperature (T_n)	Noise factor (F)	Noise figure (NF) (dB)
10	1.03	0.15
50	1.17	0.69
100	1.34	1.29
500	2.72	4.35
1000	4.45	6.48
5000	18.2	12.6
10,000	35.5	15.5
50,000	173	22.4
100,000	346	25.4

$$F = \frac{T_0 + T_n}{T_0} \quad \text{and} \quad T_n = T_0(F - 1), \tag{3.11}$$

where T_0 is the ambient temperature, for which 290 K is typically assumed. Table 3.1 shows the conversion between T_n and F as well as NF for selected values. The noise temperature T_n of a component can be interpreted as the temperature at which a resistor would generate the same level of thermal noise as the component. The Friis formula can be expressed in terms of the noise temperature as well, which appears more compact:

$$T_{n,\text{total}} = T_{n,1} + \frac{T_{n,2}}{G_1} + \frac{T_{n,3}}{G_1 G_2} + \frac{T_{n,4}}{G_1 G_2 G_3} \cdots, \tag{3.12}$$

where $T_{n,i}$ is the noise temperature of the ith stage.

With these relations fresh in mind, let us return to the two detection cases. For the direct detection, the total noise will be the same as that of the RF detector in the absence of the RF amplifier. Hence, its F is simply given as:

$$F_{\text{direct}} = F_{\text{RF\_detector}}. \tag{3.13}$$

On the other hand, the noise factor of the heterodyne detection case as depicted in Fig. 3.5b will be given as follows:

$$F_{\text{heterodyne}} = F_{\text{mixer}} + \frac{(F_{\text{IF\_amp}} - 1)}{G_{\text{mixer}}} + \frac{(F_{\text{IF\_detector}} - 1)}{G_{\text{mixer}} G_{\text{IF\_amp}}}, \tag{3.14}$$

In most practical cases, the gain of the IF amplifier can be made sufficiently large, so that the noise contribution from the IF detector becomes negligibly small. How about the noise contribution from the IF amplifier itself? It may or may not be suppressed by the mixer depending on whether the mixer gain is larger or smaller than unity (i.e., exhibits gain or loss). Even when the mixer shows a loss instead of gain, however, the noise contribution from the IF amplifier will not be excessive as the noise level of amplifiers is generally much lower than those of mixers and detectors. With this reasoning, the noise comparison of the direct and heterodyne detections is reduced to comparing the relative noise level of the RF detector and the mixer. A direct comparison is hard because the noise level of detectors is typically given by NEP and that of mixers is given by F. However, in a rather indirect way, it can be shown that the required noise performance to obtain a certain level of the output SNR is more relaxed for mixers in heterodyne detection than for an RF detectors in direct detection [30], favoring the heterodyne detection for low-noise operation. This is consistent with the analysis made in [31], where a direct comparison was made between the direct detection and the heterodyne detection in terms of NEP. On the other hand, heterodyne detection systems suffer from their own drawbacks. They include more circuit components, notably a signal source as the LO if integrated, leading to high DC power dissipation and large area occupation. This will be a barrier particularly for array detector applications. Also, they tend to show a narrower bandwidth, as limited by those of the mixer and the IF amplifier. The narrow bandwidth will degrade the overall sensitivity for some applications, such as passive imaging, as is often found in astronomy [32].

Here, although a bit of digression, a few brief comments on the unit of dB (as is used for NF as mentioned above) are made, as they may benefit some readers. As the readers should be aware, a ratio can be conveniently represented by its logarithmic value when it varies over a wide range, such as orders of magnitude. Widely used in engineering (and some science) is its representation in terms of dB (decibel), whose value is obtained as $10\log(x)$, where x is its linear value. The inclusion of the factor 10 is a result of using *decibel* as the unit, where "deci" is a Latin prefix meaning one-tenth. A decibel is a quantity equal to 1/10 of a *bel* (after Alexander Graham Bell). Hence, when a parameter is expressed in terms of decibel (dB), the values are ten times as large as those expressed in terms of bel (B), for example, 10 dB is 1 B, 50 dB is 5 B, and so forth. The unit of B is rarely used despite its simpler definition (i.e., $\log(x)$ when x is the linear value), presumably because the values can be more accurately represented in terms of dB with one more significant digit available before the decimal point. One confusion arises when the dB values are obtained with $20\log(x)$, instead of $10\log(x)$, for certain parameters, such as current, voltage, or field, while power is always given in $10\log(x)$. Why are there these two different ways to obtain dB values? In view of the definition of *deci*-bel, the expression based on $10\log(x)$ is fundamentally correct. The occasional adoption of $20\log(x)$ is for the sake of practical convenience in the conversion between those parameters (current, voltage, field) and corresponding power values. For example, 20-dB voltage gain or 20-dB current gain equals 20-dB power gain for a fixed load impedance with the adopted convention. Finally, dB often comes with a suffix, which provides additional information regarding the nature of the ratio being taken for the logarithmic conversion. Some examples useful for THz researchers include: m for the ratio with reference to 1 mW (dBm), V with reference to 1 V (dBV), c with reference to carrier (dBc), i with reference to isotropic case as in antenna gain (dBi), and so forth. You may invent a new dBx with your own preferred x.

3.2.2 Schottky Barrier Diodes (SBDs)

The Schottky barrier diode (SBD) is basically a metal-junction diode. Its history can be traced back to 1874 when F. Braun first reported a rectifying behavior of a metal-crystal junction [33], which precedes the birth of the *p-n* diode that occurred only in the 1930s. Its much earlier appearance can be attributed to its simpler structure available, a metal tip making a point contact to a crystal. Despite the simplicity, SBDs came to offer various useful properties that are particularly important for high-frequency applications, with its structure gradually evolving into much sophisticated ones. As briefly mentioned earlier in Sect. 2.4.2, it shows a faster switching operation than *p-n* junction diodes owing to the smaller junction charge/recharge time, which is a main characteristic of majority carrier devices. It also benefits from a smaller forward bias voltage drop. These advantages have led to a wide adoption of SBDs for various high-frequency applications, which exploit the nonlinear behavior of SBDs. One is the frequency multiplier (see Sect. 2.4.2), but a more popular adoption has been with detectors. Interestingly, SBDs can be employed for both direct detection and heterodyne detection, the former as a square-law detector and the latter as a mixer. In this subsection, a brief overview of the device characteristics of SBDs will be introduced, followed by the description of various SBD structures. Its use cases as high-frequency detectors will be discussed separately for direct detection and heterodyne detection afterwards.

Let us take a look at the characteristics of SBDs in terms of DC and RF behaviors, starting with the former. An SBD consists of two regions: metal and semiconductor. For the semiconductor side of SBDs, both *p*-type and *n*-type can be employed, while *n*-type is almost exclusively used in favor of its superior speed owing to the higher mobility of electrons. The energy band diagram for an *n*-type SBD is shown in Fig. 3.6. The voltage barrier height ϕ_B, which is determined by the metal work function and the semiconductor doping concentration, is a key parameter that dictates the *I–V* characteristics. The *I–V* relation of SBDs shows an exponential dependence on the bias voltage V_D, as is the case for other junction diodes, which is shown below:

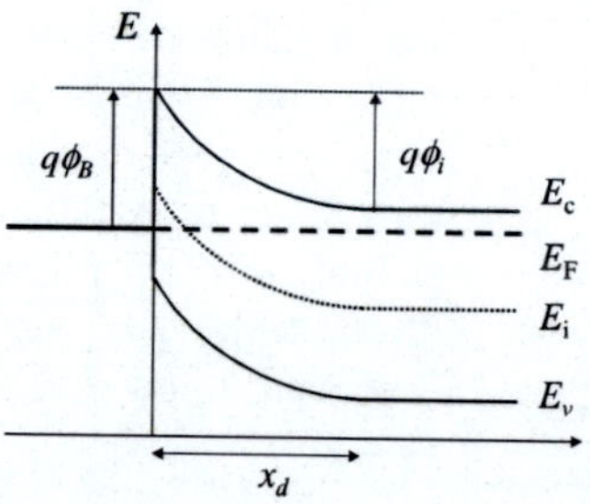

Fig. 3.6 Energy band diagram of a typical SBD on *n*-type Si

$$I_D = I_S\left(\exp\left(\frac{V_D}{V_T}\right) - 1\right), \tag{3.15}$$

where I_S is the saturation current, V_T is the thermal voltage given by kT/q with k for Boltzmann constant, T for temperature, and q for elementary charge. It is useful to remember $V_T \sim 26$ meV at room temperature. The unique behavior of SBDs is included in I_S, which can be explicitly shown as:

$$I_S = AA^* \exp\left(-\frac{\phi_B}{V_T}\right), \tag{3.16}$$

where A is the junction area and A^* is the Richardson constant of the semiconductor material employed for the diode. It is noted that I_S has an exponential dependence on ϕ_B, which governs the thermionic emission of electrons, making it sensitive to the metal type and also the semiconductor doping concentration. The saturation current density of SBDs is greater by a few orders of magnitude than that of *p-n* diodes in typical cases. Such a high current level is the key property of SBDs that is derived from its being a majority carrier device, also resulting in a smaller forward voltage drop.

SBDs are widely known for their high-frequency operation. The RF or high-frequency characteristics of SBDs can be represented by the diode cutoff frequency, which is given by:

$$f_T = \frac{1}{2\pi R_S C_0}, \tag{3.17}$$

where C_0 is the main junction capacitance component of the SBD measured at zero bias (cold condition) and R_S is the total series resistance. Equation (3.17) assumes a simple SBD model consisting of a capacitance and a series resistance, although more detailed SBD device models would involve additional components to account for the contributions from various parts of the device [34]. f_T is basically the inverse of RC time constant and the high f_T values typically found for SBDs are mainly due to a small R_S, which also explains the smaller charging/discharging time as mentioned earlier. The cutoff frequency of SBDs reported so far has reached beyond 10 THz with a GaAs SBD [35], while Si SBDs has provided a record f_T around 4 THz [36]. These values are approximately one order higher than the cutoff frequencies of transistors based on similar material systems. However, the readers should note the difference between the cutoff frequency of diodes and that of transistors. For transistors, the cutoff frequency is defined as the frequency where the gain (current gain for f_T and power gain for f_{max}) reaches unity. Beyond this frequency, therefore, the transistors cease to behave as an active device. For diodes, the cutoff frequency does not point to a frequency of a clear physical significance. Instead, it can be understood as a figure of merit to represent the high-frequency characteristics of diodes, which simply corresponds to the inverse RC time constant.

For the implementation of SBDs, GaAs and Si are the most popular choices for the semiconductor side of the device. GaAs benefits from its higher electron mobility, which would further improve the high-frequency operation of SBDs. While various metals have been employed to match GaAs to form a junction in SBDs, aluminum (Al) has been the most widely adopted owing to its high thermal stability as well as the stable contact formation on the surface of GaAs [37]. Gold (Au) is also among the candidate metals that have been employed for GaAs SBDs. On the other hand, Si-based SBDs are favored from a practical point of view in that they are compatible with standard Si process technologies. Although they are yet to be offered as a standard device from foundry Si processes with a supporting device model, they can be easily realized by mix-and-matching various process options available from the commercial technologies, which are widely practiced for Si SBDs. While Al may serve as the metal region of Si SBDs, silicides are often adopted owing to their great process compatibility. Figure 3.7a depicts the cross section of a Si SBD built on the standard Si CMOS technology (adapted from [38]). An n-well is used as the main body of the SBD to form a Schottky contact, while the ohmic contact to the n-well is formed through the n^+ source/drain diffusion regions. Optionally, p-well could be used, in which case the ohmic contact will be made with p^+ source/drain regions. Cobalt silicide, widely employed to make ohmic contacts to highly doped regions, is directly formed on top of the n-well region (lowly doped) to serve as the metal side of the Schottky junction. The shallow trench isolations (STIs) provide device isolation as well as a separation between the Schottky and ohmic regions.

On a different approach, Si SBDs can be built on BiCMOS technologies as well. In this case, the devices can benefit from process elements that are not offered from the standard CMOS process. For example, with BiCMOS technologies, a highly doped subcollector buried layer and the accompanying reach-through implant can be used to provide contact to the cathode, which is the n-type semiconductor side of the SBD. Also, the availability of the deep trench isolation (DTI) is another merit of BiCMOS technology, which offers a better device isolation. An example of SBDs based on a SiGe BiCMOS technology is shown in Fig. 3.7b (adapted from [39]). It also relies on the cobalt silicide and n-well for the metal and semiconductor regions of the SBD, respectively. For the cathode contact, in this particular example, a dedicated deeper n^+ layer was introduced together with an n-well tailored ion-implanted layer, to support not only SBDs but also p-i-n diodes that were used as a part of the circuit to be built.

With the device characteristics and structure given above, SBDs can be used for THz detection, either as an RF detector for the direct detection or as an RF mixer for the heterodyne detection. For the direct detection, SBDs basically operate as a square-law detector, whose output voltage is proportional to the square of the input voltage. Its square-law behavior can be easily derived from the exponential I–V relation of SBDs mentioned above. With the help of Taylor expansion, the diode current can be expressed as below, assuming the diode is under a forward bias and the third and higher order terms are negligible:

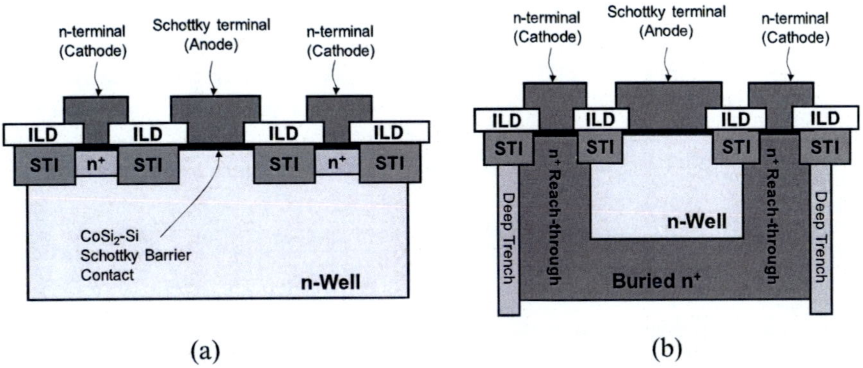

Fig. 3.7 Cross section of SBDs implemented on Si technologies: (**a**) Si CMOS technology (© 2005 IEEE [38]), (**b**) SiGe HBT technology (© 2008 IEEE [39])

$$I_{\mathrm{D}} = I_{\mathrm{S}} \exp\left(\frac{V_{\mathrm{D}}}{V_{\mathrm{T}}}\right) \simeq I_{\mathrm{S}}\left(1 + \frac{V_{\mathrm{D}}}{V_{\mathrm{T}}} + \frac{V_{\mathrm{D}}^2}{2V_{\mathrm{T}}^2}\right). \tag{3.18}$$

With a sinusoidal input voltage of $V_{\mathrm{D}} = V_{\mathrm{in}} \cos \omega t$, the output current can be expressed as:

$$\begin{aligned} I_{\mathrm{D}} &= I_{\mathrm{S}}\left(1 + \frac{V_{\mathrm{in}} \cos \omega t}{V_{\mathrm{T}}} + \frac{V_{\mathrm{in}}^2 \cos^2 \omega t}{2V_{\mathrm{T}}^2}\right) \\ &= I_{\mathrm{S}}\left(1 + \frac{V_{\mathrm{in}} \cos \omega t}{V_{\mathrm{T}}} + \frac{V_{\mathrm{in}}^2 (1 + \cos 2\omega t)}{4V_{\mathrm{T}}^2}\right). \end{aligned} \tag{3.19}$$

If only the DC component is extracted with a proper low-pass filtering, the output voltage induced across the load resistance R_{L} will be given as:

$$V_{\mathrm{out,DC}} = I_{\mathrm{D,DC}} R_{\mathrm{L}} = I_{\mathrm{S}}\left(1 + \frac{V_{\mathrm{in}}^2}{4V_{\mathrm{T}}^2}\right) R_{\mathrm{L}} \simeq \frac{I_{\mathrm{S}} R_{\mathrm{L}}}{4V_{\mathrm{T}}^2} V_{\mathrm{in}}^2. \tag{3.20}$$

Hence, the output voltage is proportional to the square of the input voltage magnitude and thus proportional to the input power. This can also be interpreted as a rectifying behavior of SBDs, as the rectification process generally gives rise to a DC component. The detailed analysis may vary depending on the specific operation case, but the basic operation can be explained in this way. With this response behavior and the high-frequency properties as discussed earlier, SBDs can be used for the direct detection of the incoming THz signal power. It would be worthwhile to mention that an SBD can detect not only the THz radiation incident on an antenna linked to the SBD, but also the THz signals injected by an adjacent electrical device integrated on the same chip. This can be compared to the thermal detectors described earlier that respond only to THz radiation.

For the heterodyne detection, the nonlinear property of SBDs used for detectors can be now exploited for mixing. The approximated response shown in Eq. (3.18) can be applied to the case where two sinusoidal input signals of different frequencies are taken as an input. With $V_D = V_1 \cos \omega_1 t + V_2 \cos \omega_2 t$, it can be easily shown that the output current for the mixed frequency components will be given as:

$$I_{\mathrm{D,IF}} = I_S \left(\frac{V_1 V_2 (\cos(\omega_1 + \omega_2)t + \cos(\omega_1 - \omega_2)t)}{4V_T^2} \right). \tag{3.21}$$

With a proper low-pass filtering, the lower-side band output voltage induced across the load R_L will be:

$$V_{\mathrm{out,IF}} = I_{\mathrm{D,IF}} R_L = \frac{I_S R_L}{4V_T^2} V_1 V_2 \cos(\omega_1 - \omega_2)t. \tag{3.22}$$

Thus, a mixing of the two input signals is achieved with an SBD. Hence, with the help of the well-known high-frequency behavior of SBDs, a down-conversion can be readily carried out for THz-band signals. Once the signal is converted down to the IF spectrum, the resultant IF signal can be detected by a conventional detector that does not particularly require a very high-frequency operation property. This will complete the heterodyne detection of the signals at the THz band.

There have been a great number of reports on SBD-based THz detection with various types of SBDs and for different applications. Most of the reported SBDs operate at room temperature because the operation of SBDs in principle does not require a cryogenic temperature, which is a major advantage of SBDs. There are, though, some reports on cryogenic operation intended for further improved performance as well. It is interesting to notice an obvious trend in the reports on SBDs. Earlier, the reported THz SBDs were mostly based on GaAs and intended for use as a mixer for heterodyne detection. More recent reports on THz SBDs, however, are more likely based on standard Si technologies, targeting at the direct detection. This is an intriguing shift in the points of interest regarding the THz application of SBDs.

As for the direct detection cases, SBD detectors operating in the range of several hundred GHz routinely show NEPs in the range of 10^{-11} W/Hz$^{1/2}$. One SBD detector that operates at 860 GHz showed an NEP of 4.2×10^{-11} W/Hz$^{1/2}$ [40]. NEP of this range is in fact lower than that of most Golay cells. It is in a similar range as the better performances of room temperature bolometers, but it should be noted that these SBDs are based on standard CMOS technology and thus can be on-chip integrated with other circuits. This integration-friendly nature also allows SBDs easily incorporated for array detectors.

To discuss the mixer application of SBDs, it should be noted that the noise performance of SBD mixers is rarely represented by NEP unlike other direct detectors considered so far. This is because a mixer is not the last device in a receiver chain that generates a DC voltage (or current) as the final output of the detection process. In this regard, the noise level of mixers is better expressed in terms of the

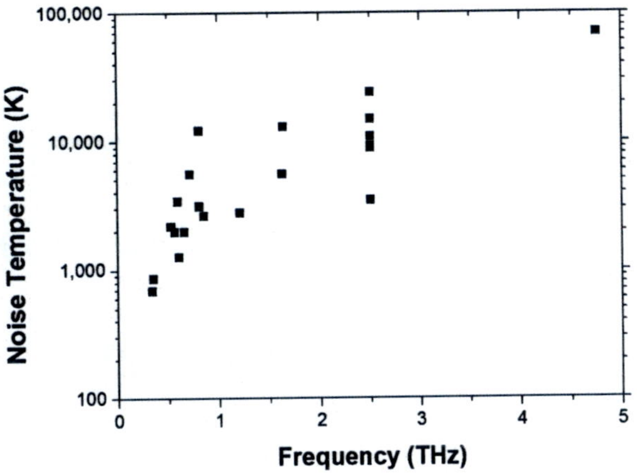

Fig. 3.8 Reported noise temperature of SBDs shown as a function of frequency (room temperature data only)

noise factor F, as the output is still in the form of AC signals. Or, alternatively, they can be quoted in terms of the noise temperature T_n, as preferred by the traditional THz community, which has a simple relation with F as shown earlier in Eq. (3.11) and Table 3.1. In general, not only for SBDs, the noise of devices that have both input and output as AC signals, such as mixers and amplifiers, will be represented by F or T_n, while the noise of devices with AC input and DC output, such as detectors, will be shown in terms of NEP, as mentioned earlier in this section.

The reported noise temperatures of SBD mixers operating at the THz band are shown in Fig. 3.8. The plot includes only room temperature data for a fair comparison, while there are reports on cryogenic temperatures as well. As the plot indicates, they typically exhibit noise temperature in the order of 1000–100,000 K. It increases rapidly with frequency (be aware that the plot is in logarithmic scale), which is a primary distinct feature from the thermal detectors. The operation has reached 4.75 THz with a noise temperature (double-side band, DSB) measured as 70,000 K, which was achieved with a room-temperature GaAs SBD mixer that was installed on an airborne observatory for astronomy researches, including the observation of Orion Nebula as reported in [41].

3.2.3 SIS Mixers

An SIS (Superconductor–Insulator–Superconductor) structure consists of a thin insulator (10–20 Å) sandwiched by two superconductor layers on both sides. The possibility of microwave detection with SIS structures was first proposed in the early 1960s [42], but the actual implementation was made only in the late 1970s, when two reports on SIS-based mm-wave heterodyne detection appeared back-to-back on the same issue of a journal [43, 44]. As a cryogenic device with a strong nonlinearity, the SIS structure can be used as a direct detector as well as a mixer in principle, although a vast majority of SIS structures are used as a mixer for the heterodyne detection. SIS mixers are particularly popular for radio astronomy applications, widely adopted for radio telescopes over the world operating below 1 THz. At ALMA (Atacama Large Millimeter/submillimeter Array), for example, eight bands (84–950 GHz) out of ten adopt SIS mixers for their heterodyne receiver front-end [45].

Before the discussion of the operation of SIS mixers, the readers are first directed to the basic properties of superconductors. When superconductors are cooled down below a critical temperature T_C, a pair of electrons form a Cooper pair. A Cooper pair is a boson with an integer spin, because two half integers of electrons, which are fermions, are combined. (In fact, a Cooper pair will have a zero spin because two electrons with opposite spins will form a pair.) Hence, multiple Cooper pairs can share an energy state, unlike the case of individual electrons that are governed by the Pauli exclusion principle. When the Cooper pair gains enough energy from an external cause, such as non-zero temperature, bias voltage, or external radiation, that exceeds the binding energy, the pair is broken back to individual electrons (to Bogoliubov quasiparticles, to be more accurate, which are often called just *quasiparticles* for simplicity). The binding energy, typically represented by 2Δ, is only a few meV. This explains why the superconductivity is observed only at low temperatures below T_C, because otherwise the thermal energy from the ambient temperature can easily exceed the binding energy and break the Cooper pairs into quasiparticles.

While Cooper pairs are the particles that are responsible for the famous zero-resistance in the superconductor materials, it is the quasiparticles that play the key role in SIS mixers. There used to be early attempts to use Cooper pairs for mixers, but it gave way to quasiparticle-based approaches in favor of the better noise performance. Hence, the operation of SIS mixers can be explained in terms of the quasiparticles [46]. The behavior of quasiparticles can be described by first introducing the concept of the energy bandgap whose value is equal to the binding energy of the Cooper pairs, or 2Δ. Unlike the electrons in metals, in which the energy states are in continuum near the Fermi energy E_F, there is a forbidden band for quasiparticles in superconductors around E_F. In a sense, this energy bandgap concept is similar to that of semiconductors, although its bandgap, 2Δ, is much smaller (in the order of meV) than those of typical semiconductors (in the order of eV). Another notable difference lies in the profile of the density of states (DOS) near the bandgap edge. For semiconductors, DOS for electrons decreases as it approaches the bandgap edge, while DOS for quasiparticles in superconductors increases as it gets closer to the edge.

Now consider the SIS structure described in Fig. 3.9a, which shows the energy bandgap and DOS along the energy axis (vertical axis) at the equilibrium [46]. Note in the plot that the horizontal axis is for DOS magnitude, not the distance as is the case for the typical semiconductor energy band diagram. Still, the central region indicates the insulator layer and the two sides represent the two superconductor layers across the insulator. In this conceptual diagram, it is assumed for simplicity that the same superconductor materials are used on both sides, which is often (but

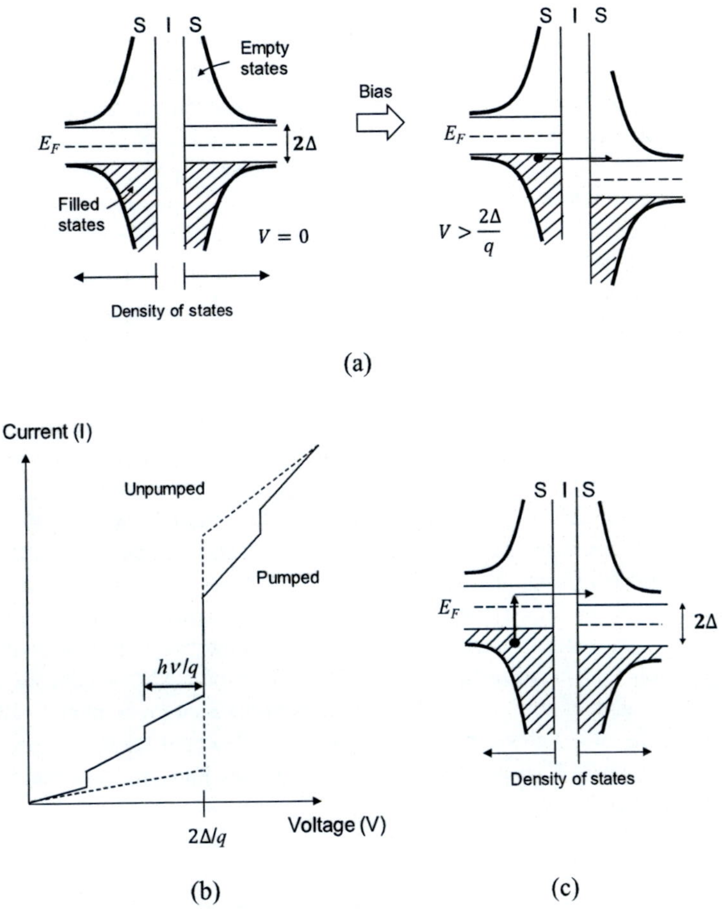

Fig. 3.9 **(a)** Energy band diagram of a typical SIS mixer before (left) and after (right) bias is applied. **(b)** *I–V* relation of SIS mixers with and without LO pumping. **(c)** Photon-assisted tunneling process at the presence of LO pumping. (Adapted by permission from Springer Nature: [24] © 2012)

not always) the case. When a bias voltage is applied across the structure, the energy profiles of the two superconductors is shifted as shown on the right side of Fig. 3.9a. If the bias barely exceeds the bandgap voltage ($2\Delta/q$), the lower band of one side and upper band of the other side of the superconductor layers become slightly overlapped. This will trigger a tunneling of quasiparticles across the thin insulator barrier. This quasiparticle tunneling leads to an abrupt increase in the current across the structure, which is by nature a tunneling current.

The abrupt current increase leads to a strong nonlinearity in the *I–V* curve of the terminal current, which shows a nearly vertical increase near the critical (or turn-on) voltage, as is depicted in Fig. 3.9b. On the other hand, when LO radiation (pumping) is additionally applied, a photon-assisted tunneling will take place, the process of which is depicted in Fig. 3.9c. It will induce a tunneling at voltages lower than $2\Delta/q$ by an amount that equals the photon voltage $h\nu/q$ or its multiples, where h is the Planck constant and ν is the frequency of the photons. In this case, the I–V relation will be modified, now showing a step-like increase of the current, with a slight relaxation in the nonlinearity, which is shown in Fig. 3.9b as a solid line in comparison with the case without LO pumping shown in a dotted line. When the signal reaches an SIS mixer together with LO pumping, the nonlinearity will cause a mixing between the incoming RF signal and LO, leading to a down-conversion of the RF signal. The final detection will be made at IF stage as in the cases with usual heterodyne detection systems.

The operation frequency of SIS mixers is limited by the bandgap energy of the superconductors, since the incident radiation with a larger photon energy (and thus higher frequency) than the bandgap ($h\nu > 2\Delta$) will prevent the normal operation of SIS mixers. This implies that the employment of superconducting materials with a higher bandgap will help to boost the operation frequency. For this reason, NbN (niobium nitride) or NbTiN (niobium titanium nitride), whose 2Δ are larger than 5 meV (corresponding gap frequency of ~1.2 THz), can be used instead of the rather standard Nb (niobium) with 2Δ around 2.8 meV (gap frequency of ~680 GHz) if high-frequency operation is desired. This will increase the fabrication difficulty, but these high-bandgap materials are still often employed for one or both side(s) of the junction for operation frequency extension beyond what is covered with Nb. The critical temperature T_C of these superconductors ranges over 9–18 K for bulk materials. It is known that T_C is significantly lowered for thin films, but liquid He temperature (4 K) is usually low enough to ensure their superconductivity and thus is widely employed.

For the insulator barrier, AlO_x (aluminum oxide) is commonly used, while AlN (aluminum nitride) is also employed in favor of the film quality in some cases. A typical cross-sectional view of an SIS mixer is shown in Fig. 3.10 [13]. It is also noted that SIS mixers are often integrated with a tuning circuit, which will tune out the intrinsic capacitance from the parallel plate structure of the SIS mixers. As for the signal feeding, the RF signal and LO pumping, typically coupled by a beam combiner, are received by an antenna, which can be implemented either on-chip or off-chip. In the case with an on-chip antenna (quasi-optic approach), the chip with the SIS mixer is mounted on an elliptical or hyper-hemispherical lens that focuses the incoming radiation onto the on-chip wideband antenna, such as a log-spiral or log-periodic, integrated with the SIS mixer. In the case with an off-chip antenna

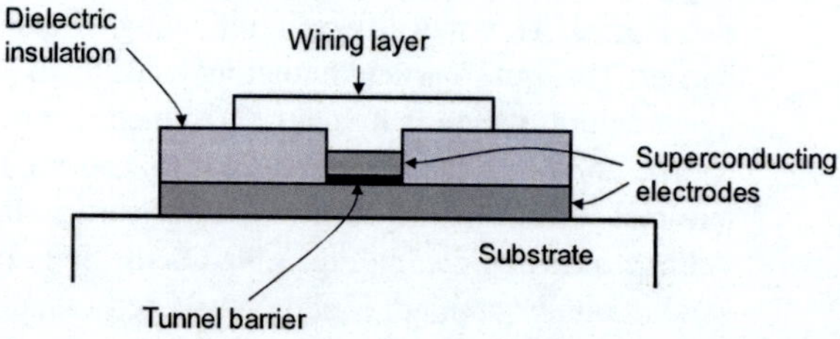

Fig. 3.10 Cross-sectional view of a typical SIS mixer. (Adapted by permission from Springer Nature: [24] © 2012)

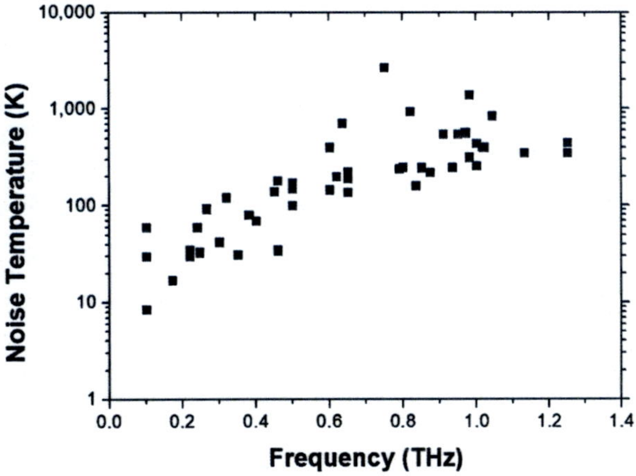

Fig. 3.11 Reported noise temperature of SIS mixers shown as a function of frequency

(waveguide approach), the incoming beam is collected by a separate external antenna, such as a horn antenna, then coupled to the SIS mixer through a waveguide and waveguide transition.

Figure 3.11 shows the reported noise temperature of SIS mixers as a function of frequency. As are widely known for their extremely low noise levels, SIS mixers exhibit noise temperatures mostly below 1000 K up to beyond 1 THz. At the lower frequency side, around 100 GHz, the noise temperature is further reduced even down to ~10 K for some reports [47]. This is close to the quantum limit of $h\nu/k$, which is 4.8 K at 100 GHz. This low noise level explains the popularity of SIS mixers for heterodyne receivers in radio telescopes, for which extremely high sensitivity is critical to capture the faint signals coming from the objects in the space. The upper frequency limit is relatively low, though, compared to other types of diode mixers, which is dictated by the gap frequency as mentioned above. The highest operation frequency of 1.25 THz in the plot was obtained with an SIS mixer based on Nb/AlN/NbTiN junction as developed for Herschel Space Observatory, which showed a noise temperature of 450 K [48].

3.2.4 Superconducting Hot Electron Bolometers (HEBs)

While a general description on bolometers was made in Sect. 3.1.1, superconducting HEBs would merit a separate discussion with a focus on its mixer application for the heterodyne detection. It is particularly favored for THz detection owing to its large operation bandwidth reaching up to several THz, which exceeds that of SIS mixers. The first demonstration of superconductor bolometers as a high-sensitivity detector can be traced back to the early 1940s [49], after which extensive researches have been made on this topic. Superconducting HEBs share a common feature with SIS mixers in that they are both based on superconductors. The popular choices of

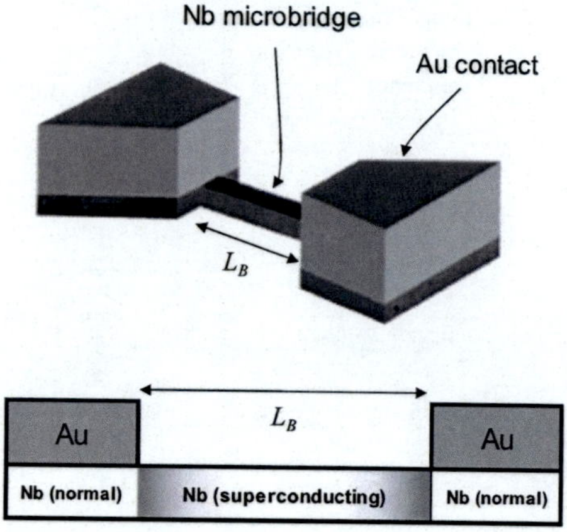

Fig. 3.12 Structure of a typical superconducting HEB. (Adapted from [50], © 1998 American Institute of Physics)

superconducting materials for superconducting HEBs, Nb, NbN, and NbTiN, are also shared with SIS mixers. When it comes to the structure, however, there is a clear difference. While an SIS mixer involves a *junction* made of a thin insulator and two superconductor layers, a superconducting HEB is, being a thermal detector, composed of a *single* superconductor piece, typically in a form of a narrow wire bridge (called microbridge) with contacts formed on both ends. Hence, one can expect that the way of utilizing superconductivity for mixing is much different for the two cases.

The structure of a typical superconducting HEB is shown in Fig. 3.12 [50]. The HEB under mixer operation is cooled down below the critical temperature T_C of the superconductor forming the microbridge. When it is biased with an external DC source, the microbridge will be heated up by joule heating from the DC power dissipation, leading to a temperature rise. Excessive heating, additionally promoted by the radiation of the LO pumping beam, will push the device operation beyond T_C, which triggers a transition from the superconducting state into the normal conducting state. The resultant sudden increase in resistance will induce a nonlinearity in the *I–V* relation of the bolometer, which can be utilized for mixing of the RF signal and the LO pumping. The IF signal obtained with this heterodyne process can be detected by a low-frequency detector. As the operation is related to the transition from the superconducting to the normal conducting states, it is also called the transition-edge sensor [51]. A more detailed model [52] indicates that a hot spot will be created in the middle of the microbridge, in which a high temperature exceeding T_C will be induced. This leads to a local normal-conducting region in the center with the other regions remaining as superconducting. The formation of the regional hot spot will affect the device current level as the total resistance will be dominated by the highly resistive hot spot. Because the electrons are heated up faster than the phonons in the device due to the different heat capacity, the device is called "hot electron" bolometer. The way the RF signal and the LO pumping are coupled into a HEB mixer is in most cases similar to the SIS mixer cases described above. The coupling may rely on a quasi-optic technique for HEB mixers integrated with a wideband on-chip antenna, or waveguide technique that involves a waveguide

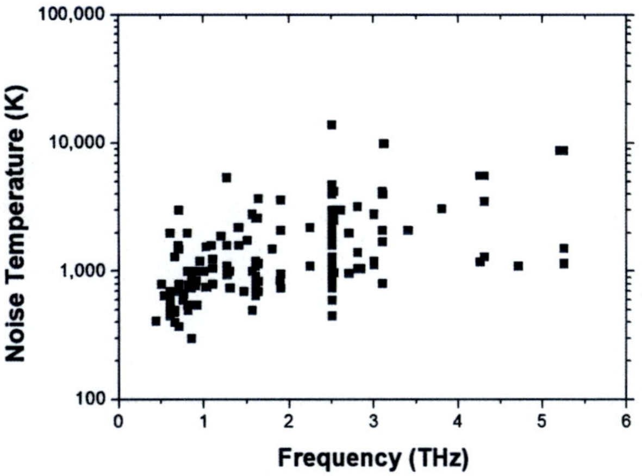

Fig. 3.13 Reported noise temperature of superconducting HEBs shown as a function of frequency

transition with an external antenna. In some cases, though, a direct coupling through an absorption region without antenna is still employed as is the case for conventional bolometers [4].

The reported performance of superconducting HEBs are shown in Fig. 3.13. As indicated by the plot, the operation frequencies of superconducting HEBs are distributed roughly over 0.5–5 THz range. The HEB that marks the highest operation frequency of 5.25 THz in the plot is based on NbN microbridge embedded in a log-spiral antenna built on high-resistivity Si substrate, which shows a noise temperature of 1150 K [53]. Figure 3.14a compares the performances of the three diode detectors discussed so far, SBDs, SIS mixers, and superconducting HEBs, by overlapping the data points from Figs. 3.8, 3.11, and 3.13. The same set of data are repeated in Fig. 3.14b, now presented in terms of NF based on the conversion with Eq. (3.11). As Fig. 3.14 clearly shows, the operation of HEB stretches up to much higher frequencies than that of SIS mixers, but the noise level of HEB mixers is still higher than that of SIS mixers at lower frequencies. For example, near 1 THz, the noise temperatures of superconducting HEBs are reported around 1000 K, while those of SIS mixers are mostly populated in the range of 200–500 K. Near 100 GHz, the noise temperature as low as below 10 K is obtained with SIS mixers, which translates to NF of only 0.15 dB (see Table 3.1 for conversion between noise parameters). The maximum operation frequency ranges are similar for superconducting HEBs and SBDs, while the noise level is much higher for SBDs. This is obviously due to the fact that SBDs mostly operating at room temperature (only room temperature data included in the plots for SBD), while the operation of superconducting HEBs as well as that of SIS mixers are based on cryonic temperatures. Although rarely reported, SBDs operating at cryogenic temperatures show a lower noise than room-temperature SBDs, but larger than typical HEBs.

Fig. 3.14 Comparison of noise performance of SBDs, SIS mixers, and superconducting HEBs based on the data from Figs. 3.8, 3.11, and 3.13. The same set of data are represented by two noise parameters: (**a**) Noise temperature T_n, (**b**) Noise figure NF ($T_0 = 290$ K is assumed)

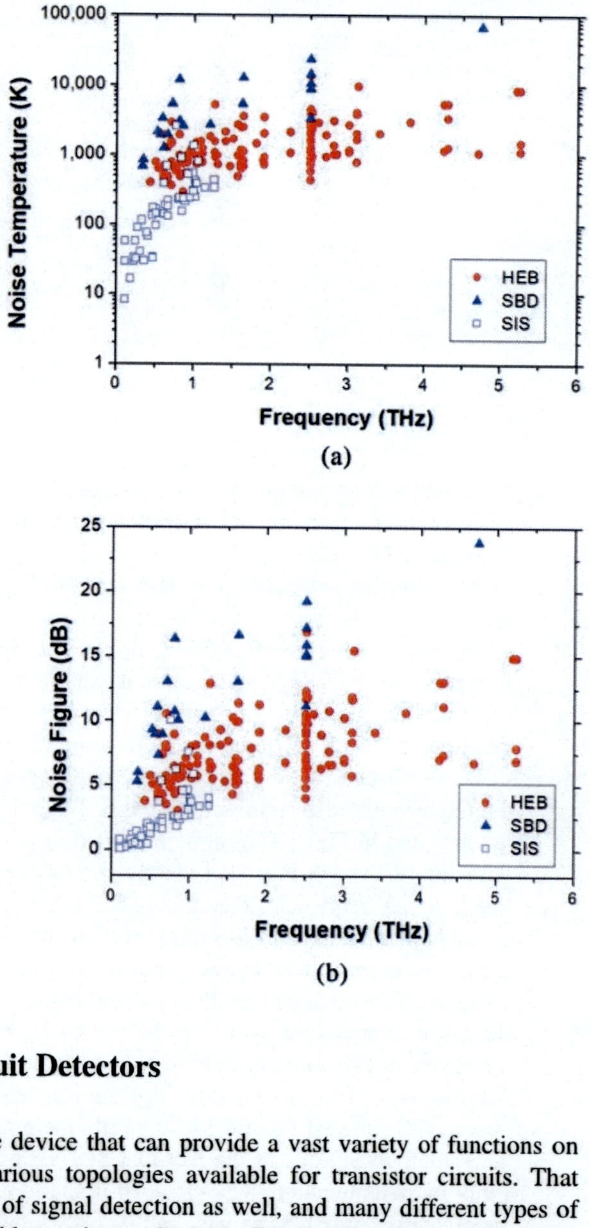

3.3 Transistor Circuit Detectors

The transistor is a versatile device that can provide a vast variety of functions on electronic systems with various topologies available for transistor circuits. That versatility reaches the field of signal detection as well, and many different types of detectors can be realized with transistors, for both direct and heterodyne detection. The availability of gain, which is the most favored aspect of transistors over diodes, may not be a necessity when it comes to signal detection. However, it still helps various aspects related to the detection, and the enhancement of transistor speed, or the frequency limit of the gain availability, contributes to the THz detection based on transistors. In this section, the direct detection with transistors will be first introduced. Then, the key circuit blocks of the heterodyne detection, such as mixers and low noise amplifiers (LNAs), and their implementation by transistor circuits will be discussed. This will be followed by the description of integrated heterodyne detectors and the examples. Although LO is also a core part of heterodyne architecture, it will not be discussed in this section as it can be realized by the transistor-based signal sources described in Sect. 2.3.

3.3.1 Direct Detectors

Direct detectors based on transistors can be largely divided into two groups: active mode and passive mode. In the active-mode detectors, which can be implemented with either FETs or BJTs, the detection is made exploiting the nonlinearity of the transistor operation, typically realized with a pair of transistors in the differential configuration. The topology of a typical active-mode direct detector is shown in Fig. 3.15, for both FET and BJT cases, in the differential configuration (the figure shows the input applied through the gate, but it can be applied through the source as well). It is basically a differential pair with the two drain (or collector) nodes tied together, which serves as the output node. One may notice that it shares almost the same topology as the frequency multiplier with the even harmonics described in Sect. 2.4. In fact, the two operations are very similar in their nature. Only difference is that, for the output, the DC component is taken for direct detectors while the even harmonic component is extracted for frequency multipliers, with a proper filtering network. The distinct feature of the active-mode detection in contrast with the passive-mode detection is that the transistor is fully biased typically with the gate (base) *and* drain (collector) biases. For the passive-mode detection, as shown in

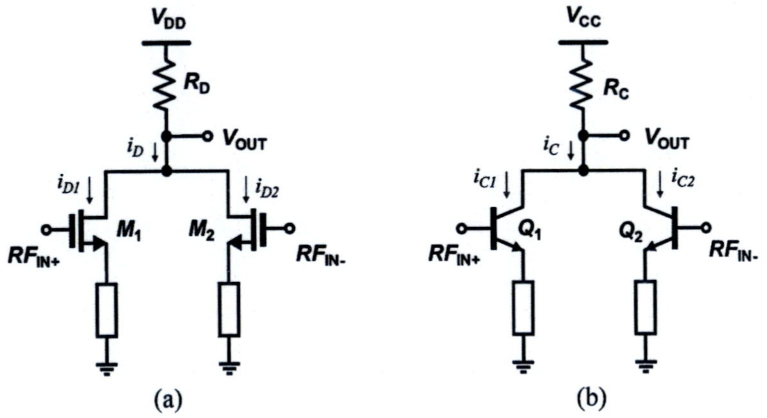

Fig. 3.15 Topology of a typical active-mode direct detector: (**a**) FET-based, (**b**) BJT-based

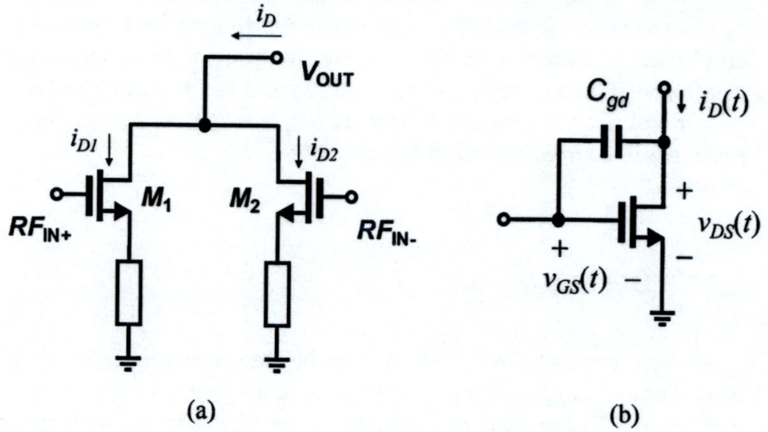

Fig. 3.16 (**a**) Topology of a typical passive-mode direct detector based on FETs. (**b**) Simplified single-ended picture of the passive-mode detector, showing C_{gd} for resistive self-mixing

Fig. 3.16a based on FETs, only gate is biased (not explicitly shown in the figure) without drain bias.

3.3.1.1 Active-Mode Detection

A brief small-signal analysis is made below to help clarify the operation of the active-mode detection with the circuit shown in Fig. 3.15a. Let i_{D1} and i_{D2} be the small-signal drain currents through M_1 and M_2, respectively, and i_D the sum of the two drain currents at the output. At the presence of the transistor nonlinearity,

$$i_{D1} = g_m^{(1)} v_{GS1} + g_m^{(2)} v_{GS1}^2 + g_m^{(3)} v_{GS1}^3 + \cdots \tag{3.23}$$

$$\begin{aligned} i_{D2} &= g_m^{(1)} v_{GS2} + g_m^{(2)} v_{GS2}^2 + g_m^{(3)} v_{GS2}^3 + \cdots \\ &= -g_m^{(1)} v_{GS1} + g_m^{(2)} v_{GS1}^2 - g_m^{(3)} v_{GS1}^3 + \cdots \end{aligned} \tag{3.24}$$

where it was assumed that $v_{GS2} = -v_{GS1}$ with an ideal differential operation. Note that $g_m^{(n)}$ is given as:

$$g_m^{(n)} = \frac{1}{n!} \frac{\partial^n I_D}{\partial V_{GS}^n}\bigg|_{V_{DS}=\text{const}}. \tag{3.25}$$

If we ignore the fourth and higher order terms, i_D is given as:

$$i_{\rm D} = i_{\rm D1} + i_{\rm D2} = 2g_{\rm m}^{(2)} v_{\rm GS1}^2. \tag{3.26}$$

Now assume the input signal is sinusoidal, so that $v_{\rm GS1}(t) = V_{\rm RF}\cos\omega t$. Then,

$$i_{\rm D} = g_{\rm m}^{(2)} V_{\rm RF}^2 (1 + \cos 2\omega t), \tag{3.27}$$

which suggests that only DC and the second harmonic (and even harmonics if higher order terms were included) terms survive at the common node. Odd harmonics will vanish at this node as they are canceled out due to the differential operation. With a proper low-pass filtering, a DC voltage will appear across the load resistance $R_{\rm L}$ (neglect $R_{\rm C}$ or $R_{\rm D}$ for now) as follows:

$$V_{\rm out} = g_{\rm m}^{(2)} R_{\rm L} V_{\rm RF}^2. \tag{3.28}$$

This clearly indicates the circuit serves as a square-law detector. For MOSFETs, the explicit form of $g_{\rm m}^{(2)}$ can be obtained from the drain current expression assuming a long channel device:

$$I_{\rm D} = \frac{1}{2}\mu C_{\rm ox}\frac{W}{L}(V_{\rm GS} - V_{\rm th})^2, \tag{3.29}$$

where W and L represent the gate width and length, respectively, μ is the channel electron mobility, and $C_{\rm ox}$ is the gate oxide capacitance per unit area.
With this current relation, $g_{\rm m}^{(2)}$ is given as:

$$g_{\rm m}^{(2)} = \frac{1}{2}\frac{\partial^2 I_{\rm D}}{\partial V_{\rm GS}^2} = \frac{1}{2}\mu C_{\rm ox}\frac{W}{L}. \tag{3.30}$$

Then the output voltage can be expressed in terms of the device parameters:

$$V_{\rm out} = \frac{\mu C_{\rm ox} R_{\rm L}}{2}\frac{W}{L} V_{\rm RF}^2. \tag{3.31}$$

The square-law relation to the magnitude of the input voltage is obvious from the expression. The dependence on the device parameter is mostly as expected.

The same analysis can be applied to the BJT case shown in Fig. 3.15b. The output voltage can be accordingly obtained by applying the expression for the collector current in place of the drain current. By assuming a one-side step junction with emitter doping much higher than base, which is often the case, the collector current for an *npn* bipolar transistor is given as:

$$I_{\mathrm{C}} = I_{\mathrm{S}} \exp\left(\frac{V_{\mathrm{BE}}}{V_{\mathrm{T}}}\right) = \frac{qAD_{\mathrm{n}}n_i^2}{N_{\mathrm{B}}W_{\mathrm{B}}} \exp\left(\frac{V_{\mathrm{BE}}}{V_{\mathrm{T}}}\right), \tag{3.32}$$

where A is the emitter–base junction area, D_{n} is the electron diffusion constant in the base, n_i is the intrinsic carrier concentration in the base, N_{B} is the base doping concentration (assumed uniform or average value), and W_{B} is the neutral base width. Then, based on Eq. (3.25) with V_{GS} replaced by V_{BE}, $g_{\mathrm{m}}^{(2)}$ can be obtained as:

$$g_{\mathrm{m}}^{(2)} = \frac{I_{\mathrm{C}}}{2V_{\mathrm{T}}^2} = \frac{qAD_{\mathrm{n}}n_i^2}{2N_{\mathrm{B}}W_{\mathrm{B}}V_{\mathrm{T}}^2} \exp\left(\frac{V_{\mathrm{BE}}}{V_{\mathrm{T}}}\right), \tag{3.33}$$

which leads to the output voltage of:

$$V_{\mathrm{out}} = \frac{I_{\mathrm{C}}R_{\mathrm{L}}}{2V_{\mathrm{T}}^2} V_{\mathrm{RF}}^2 = \frac{qAD_{\mathrm{n}}n_i^2 R_{\mathrm{L}}}{2N_{\mathrm{B}}W_{\mathrm{B}}V_{\mathrm{T}}^2} V_{\mathrm{RF}}^2 \exp\left(\frac{V_{\mathrm{BE}}}{V_{\mathrm{T}}}\right). \tag{3.34}$$

Again, the square-law relation is clearly shown. One notable difference from the FET case is that the output voltage is proportional to the DC bias current, or the collector current I_{C} for BJT case.

3.3.1.2 Passive-Mode Detection

The passive-mode detection can be interpreted as the resistive mixer operation of FETs [54] in the self-mixing condition and, therefore, can be realized by FETs only in principle. Before moving into the detector operation of this mode, a brief description of the FET resistive mixer would be helpful. A FET-resistive mixer is basically a FET-based mixer operating in the passive mode with only gate DC bias applied. As no intended drain bias is applied, FET resistive mixers operate in the linear region (or triode region) of the FET operation modes, leading to a linear or "resistive" *I–V* relation of the device. This is in contrast with the active-mode mixers in which DC bias is applied to both gate and drain, mostly likely leading to the device operation in the saturation region. To attain the resistive mixing, LO and RF signals are applied to the gate and the drain nodes, respectively, along with a gate DC bias (with reference to the source), which naturally leads to a drain current (i_{D}) determined by the product of the drain-source voltage (v_{DS}) and the drain-source conductance (g_{DS}) (this will be explicitly shown in the next paragraph). As the drain-source conductance is roughly proportional to the gate voltage in the linear region operation, the output is eventually given as a product of gate and drain voltages, or the product of RF and LO signals, resulting in a mixing process. Note that device nonlinearity is not involved in this mixing process. As the resistive mixing is based on the passive-mode operation of FETs, it generally exhibits a much higher linearity than the conventional active-mode mixing.

To employ the resistive mixing for the passive-mode *detection* with FETs, the RF signal for detection should be applied to both gate and drain nodes. For this to happen, a strong AC coupling between the gate and the drain needs to be established with the RF signal applied to the gate. The coupling can be achieved with the parasitic gate-drain capacitance C_{gd} alone, or it can be further enhanced with an addition of external capacitance across the gate and the drain. Note that this situation is opposite to the normal mixing operation of resistive mixers where the coupling should be suppressed, which may require LO and RF filters for isolation instead of coupling. An analysis can be made to explain the detector operation based on the self-mixing as presented in [55]. Consider the simplified circuit schematic shown in Fig. 3.16b, in which C_{gd} may be purely parasitic or enhanced by extrinsic capacitance. Assume that the RF signal injected to the gate is sinusoidal, so that the total gate voltage is given by $v_{GS}(t) = V_{RF}\cos\omega t + V_g$, where V_g is the DC gate bias. Due to the capacitive coupling, the AC component of the gate voltage will appear at the drain node, leading to $v_{DS}(t) = V_{RF}\cos\omega t$ with no drain DC bias applied. Since the RF signal is small enough to keep the device to stay in the triode region in most practical detector applications, the drain current will be given as a product of the voltage and the conductance across the drain and the source, leading to:

$$i_D(t) = v_{DS}(t)g_{DS}(t) = V_{RF}\cos\omega t \cdot g_{DS}(t). \tag{3.35}$$

Note that $g_{DS}(t)$, the conductance between the drain and the source, is basically the channel conductance, which is approximated in strong inversion as [55]:

$$g_{DS}(t) = \frac{W}{L}\mu C_{ox}\left(v_{GS}(t) - V_{th} - \frac{v_{DS}(t)}{2}\right). \tag{3.36}$$

With $v_{GS}(t)$ and $v_{DS}(t)$ as given earlier substituted to Eq. (3.36), one can obtain the DC component of the drain current as:

$$I_D = \frac{\mu C_{ox}}{4}\frac{W}{L}V_{RF}^2. \tag{3.37}$$

If the output is terminated with a load resistor of R_L, then the output voltage will be simply:

$$V_{out} = \frac{\mu C_{ox} R_L}{4}\frac{W}{L}V_{RF}^2. \tag{3.38}$$

Therefore, again, a square-law detection is achieved with a FET, now operating in the passive-mode based on the resistive self-mixing operation. The analysis above is made for a single FET, but it can be easily modified for the differential configuration as shown in Fig. 3.16a. In this case, the output voltage will be doubled, leading to the same output voltage as the active-mode case obtained earlier (Eq. 3.31). While the analysis so far assumes a quasi-static operation of FETs, it can be extended to non-quasi-static operation that applies to the THz detection, in which the channel is represented by distributed RC components. The result for non-quasi-static operation as presented in [55] indicates that detection is still achieved, more effective with a shorter channel length. For devices with a longer channel than needed for the resistive mixing, the extra channel length will only add capacitance and series resistance, degrading the noise and responsivity of the detector.

3.3.1.3 Plasma-Wave Resonant-Mode Detection

There is yet another mode of transistor-based direct detection, widely known as the plasma wave resonant-mode detection. The principle behind this phenomenon, as suggested by Dyakonov and Shur [56], is briefly introduced here. If the mean free path for electron-electron scattering in the channel of an FET is shorter than those for electron-impurity or electron-phonon scatterings, the total scattering is dominated by electron-electron scattering. In this case, the electrons in the channel will behave as if they are 2D fluid rather than 2D gas. This condition can be achieved with a channel with a high electron mobility and high electron concentration. In this situation, the electrons can be excited to generate waves inside the channel, which is called the plasma wave. With the presence of the plasma wave, the channel can serve as a resonator with a proper set of boundary conditions. Assuming the boundary condition of the source and the drain being AC-wise short and open, respectively, the resonance will be obtained at the fundamental frequency f_0 shown below and its odd-harmonics:

$$f_0 = \frac{s}{4L}, \tag{3.39}$$

where L is the channel length and s is the velocity of the plasma wave in the channel. The velocity s can be expressed as a function of device parameters as:

$$s = \sqrt{\frac{qV_0}{m^*}} = \sqrt{\frac{q^2 n}{m^* C}}, \tag{3.40}$$

where q is the elementary charge, V_0 is the gate overdrive ($V_g - V_{th}$), m^* is the effective mass of electron, n is the channel electron concentration per unit area, and C is the gate capacitance per unit area. The equality on the right-hand side is based on the assumption that the device is in the strong inversion region and $n \approx CV_0/q$ [57]. Since s is given as a function of device parameters as well as the bias condition, its value and thus the resonant frequency can be tuned by the external gate bias. It is noted that a typical value of s is in the order of 10^8 cm/s, larger than the electron saturation velocity in most semiconductors by a factor of an order. This leads to the fundamental resonant frequency in the order of THz, opening the possibility of THz detection. For this resonance to show a sufficiently large Q-factor, the channel length L should satisfy the following condition:

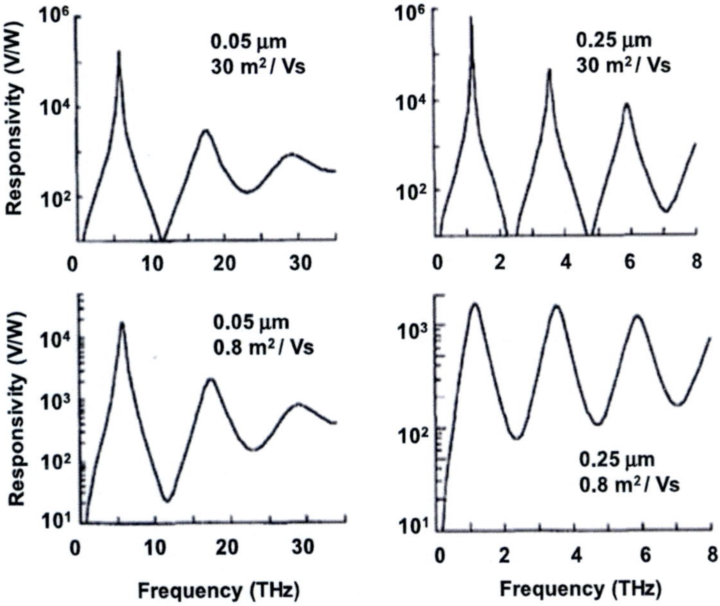

Fig. 3.17 Calculated responsivity profiles for the resonant-mode detection of plasma-wave detector with various gate lengths and mobilities. (© 1996 IEEE [56])

$$L \ll L_{\mathrm{cr}} = \frac{s\mu m^*}{q}, \tag{3.41}$$

where L_{cr} is the critical channel length and μ is the electron mobility. Hence, for the resonant-mode operation of the plasma-wave FET, a sufficiently small channel length is also required. This is an additional condition that adds to the conditions for the existence of the plasma wave, or high mobility and high carrier concentration as mentioned above.

When all these conditions are satisfied, a plasma-wave *resonant-mode* detection can be obtained at the resonant frequencies. The signature of the resonant-mode detection is the periodic occurrence of sharp peaks in the responsivity with very large values at the resonant frequencies (f_0 and its odd harmonics) along the frequency axis. Figure 3.17 shows calculated responsivities with different channel length and mobility conditions that clearly exhibit such peaks, indicating the onset of the plasma-wave resonance [56]. The profiles in Fig. 3.17 implies that a highly sensitive narrow-band THz detection can be achieved if the above-listed conditions for the plasma-wave resonance are satisfied. Those conditions are quite stringent and hard to achieve, limiting the early demonstration of the resonant-mode detection to FETs based on III–V semiconductors [58]. However, the evidence of possible resonant-mode plasma wave detection with Si MOSFETs has been subsequently reported with aggressively scaled gate length [57]. While the resonant-mode detection is narrow-band by nature, less sensitive broadband detection can still be made with a relaxed condition for the

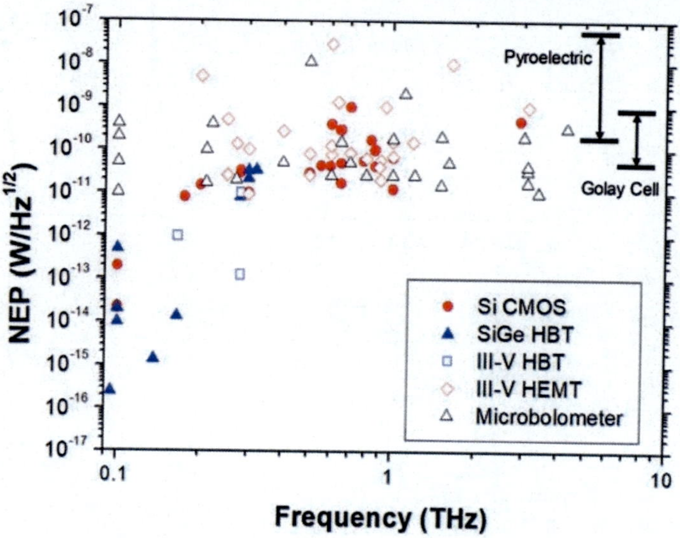

Fig. 3.18 Reported NEP of transistor-based direct detectors with various technologies, shown together with selected thermal detectors for comparison (room temperature data only). For Golay cells and pyroelectric detectors, only the noise range is presented assuming the responses are frequency-independent

resonant-mode detection, which is conveniently called the non-resonant-mode plasma wave detection. In fact, the passive-mode detection with the resistive self-mixing discussed earlier in this section corresponds to this case. In that no drain bias is applied, the plasma-wave detection, both resonant and non-resonant, can be regarded as a passive-mode detection. The application of the FETs with the resonant-mode plasma wave is not limited to the detection, but also expected to be further extended to mixers, frequency multipliers, and even to oscillators [56].

So far, three types of transistor-based direct detection schemes have been described for their operation basics. There have been a growing number of reports on these types of detections for operation frequency near or beyond 100 GHz based on various process technologies, which reaches beyond 3 THz [59]. Figure 3.18 shows the reported performance of transistor-based direct detectors in terms of NEP against the frequency. In the plot, the mode of operation (active or passive) is not distinguished. It also includes the detectors with an LNA at the front-end. Also included in the plot are the NEP performances of selected thermal detectors. Only room-temperature data of the thermal detectors are considered for a fair comparison. For the case of Golay cells and pyroelectric detectors, the responses are assumed frequency-independent so only the noise range is presented in the plot. The frequency coverage is wide also for the microbolometers, but the data in the plot were taken at the measured frequency points as reported by the literature. The NEP values for $f > 200$ GHz are mostly populated in the order of 10^{-11} to 10^{-9} W/Hz$^{1/2}$ for both transistor-based detectors and microbolometers. However, for lower frequencies,

NEP for transistor-based detectors drops rapidly, reaching as low as below 10^{-15} W/Hz$^{1/2}$ around 100 GHz (with the help of an LNA) [60], while microbolometers do not show a significant drop. Hence, it is this region of frequency, roughly below 200 GHz, where transistor-based detectors exhibit a huge advantage over the room-temperature thermal detectors. In fact, the NEP values exhibited by the transistor-base detectors around 100 GHz at room temperature is comparable to some of bolometers operating at cryogenic temperatures. Further, when integrated with a low noise amplifier, the noise level will be further reduced. This will be discussed shortly later in this section.

3.3.2 *Mixers*

Mixers are a circuit component that performs a frequency conversion based on mixing of two input signals. We have reviewed other frequency conversion circuits such as frequency multipliers and frequency dividers in Chap. 2. They are two-port circuits with a single input, and the frequency conversion is limited to integer-multiplication or integer-division of the input signal. Mixers are three-port circuits composed of two input ports and one output port. Hence, they provide a higher flexibility in the frequency conversion than frequency multipliers or frequency dividers, since the output frequency is determined by the difference or sum of the two input frequencies, one of which (typically LO) can be controlled. There can be two frequency conversion types, up-conversion and down-conversion, depending on whether the output frequency is higher or lower, respectively, than the input frequencies. In this section, the discussion will be focused on the down-conversion operation of mixers based on transistors, which provides a key function to heterodyne receivers that employ transistors as building blocks.

3.3.2.1 Mixer Operation Basics

It will be useful to take a brief review of the basic mixer operation principles before moving into the details of various mixer circuit topologies, although we have already partially started discussion on mixers in the previous sections. Mixing can be obtained by either nonlinearity or time-varying response of a system. These two mechanisms will be explained by taking RF and LO as the two inputs and IF as the output, as is the convention for down-conversion mixers.

For the mixers based on a *nonlinearity*, the sum of RF and LO signals, possibly through a power combiner located at the front, enter the mixer and go through the nonlinear process. Assume a two-tone sinusoidal input signal of

$$x(t) = A_0 \cos \omega_0 t + A_1 \cos \omega_1 t, \tag{3.42}$$

and a nonlinear response, taking the third-order nonlinearity as an example,

$$y(t) = \alpha x(t) + \beta x^2(t) + \gamma x^3(t). \tag{3.43}$$

Then, one can easily show that the resultant output will have the following frequency components: ω_0, ω_1, $2\omega_0$, $2\omega_1$, $\omega_0 + \omega_1$, $\omega_0 - \omega_1$, $3\omega_0$, $3\omega_1$, $2\omega_0 + \omega_1$, $2\omega_0 - \omega_1$, $2\omega_1 + \omega_0$, $2\omega_1 - \omega_0$. In general, at the presence of the higher order terms, the following frequency components will be obtained:

$$\omega_{out} = |n\omega_0 \pm m\omega_1|, \tag{3.44}$$

where n and m are integers. The desired frequency component can be selected with a filter at the output. Hence, any nonlinear system can serve as a mixer if two input signals are applied together.

Now consider mixers based on a *time-varying* response. If we assume the system is *linear* although time-variant, the response can be shown as below with an input of $x(t)$:

$$y(t) = \alpha(t)x(t). \tag{3.45}$$

When the time variance of the system, $\alpha(t)$, is periodic with an angular frequency of ω_0, the response can be expressed with the help of Taylor expansion as:

$$y(t) = \sum_{n=-\infty}^{\infty} \alpha_n e^{jn\omega_0 t} x(t), \tag{3.46}$$

where n is an integer. Now consider a single-tone sinusoidal input signal of

$$x(t) = A_1 \cos \omega_1 t. \tag{3.47}$$

Then the response will be

$$y(t) = \sum_{n=-\infty}^{\infty} \alpha_n e^{jn\omega_0 t} A_1 \cos \omega_1 t = \frac{A_1}{2} \sum_{n=-\infty}^{\infty} \alpha_n e^{jn\omega_0 t} \left(e^{\,j\omega_1 t} + e^{-j\omega_1 t}\right). \tag{3.48}$$

Hence, the output will include the following frequency components:

$$\omega_{out} = |n\omega_0 \pm \omega_1|. \tag{3.49}$$

Again, a filtering at the output will provide only the desired frequency component. Hence, a linear time-varying system can serve as a mixer if one of the input signals, typically LO, induces a periodic time-variance in the system response. A *nonlinear* time-varying system can also serve as a mixer, of course, in which case the output will include frequency components of $|n\omega_0 \pm m\omega_1|$ with an additional integer of m. If the response variation over time is extreme so that the system periodically turns on and off, then we can say the mixing is due to switching.

3.3.2.2 Passive Mixers and Active Mixers

Transistors are versatile devices, and they can be assembled in many different ways to form circuits intended for certain functions. Transistor-based mixers are no exception, and they can be realized based on various approaches. Depending on the mode of operation, transistor mixers can be divided into passive mixers and active mixers. For passive mixers, transistors are typically not fully biased and operate in the absence of DC bias current. One example is the resistive mixer briefly mentioned in the previous section [54], which belongs to the case where mixing is based on a time-varying response. Another example is a circuit shown in Fig. 3.19, in which a transistor acts as a switch driven by LO that controls the RF signal path, probably the simplest implementation case of a switching mixer. In this configuration, the transistor does not need the drain DC bias current for switching, and thus involves no DC power dissipation, which is a major advantage of passive mixers. Passive mixers also exhibit an excellent linearity, a signature of passive operation, which stems from the absence of the saturation at high power levels as is generally observed in active circuits. Noise level can be maintained low owing to the reduced $1/f$ noise that often plagues active devices. Let us take a brief digression here. There may be a notion that using transistors for *passive*-mode circuits is a waste of the precious functional resources that can be provided by the transistors employed in the circuit, such as the gain. However, ironically, the transistor is the cheapest device available on a chip when fabricated in a standard semiconductor technology, since it occupies the smallest area. True expense for using transistors comes from the DC power dissipation, rather than from the area, at least for sparsely integrated analog and RF circuits. Hence, there is no particular economic loss in employing transistors for passive circuits.

While passive mixers based on transistors are suited for some occasions, active mixers are still largely favored for most applications whenever they are available. The reason for this favor is the availability of the gain, the hallmark feature of transistors, for this type of mixers. With a conversion gain (larger than unity), the noise from the following stages will be suppressed by the mixer gain, as is indicated by the Friis formula shown in Eq. (3.9). Note that, in passive mixers, the noise from the following stages will be effectively magnified by the conversion loss, as opposed to the active mixer case. Additionally, a conversion gain will contribute to the total gain of the heterodyne detector, enhancing the responsivity of the entire detector. It is still true that a conversion gain larger than unity cannot be always guaranteed for active mixers, especially for very high-frequency operations due to the gain roll-off

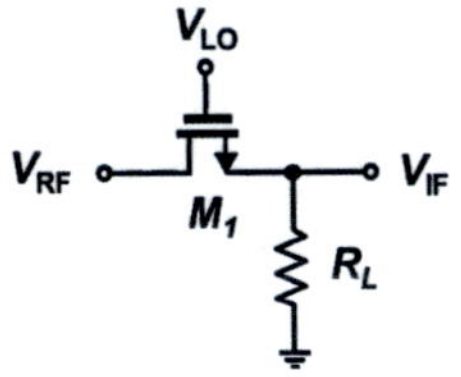

Fig. 3.19 Schematic of a typical passive-mode mixer, in which a transistor behaves as an LO-driven switch

at high frequencies. However, even if an active mixer shows a conversion loss instead of gain, the loss level will be much smaller than that of passive mixers, as the transistor gain will partially compensate for the other losses in the mixers. The availability of the gain also relaxes the requirement for LO power level, which particularly benefits high-frequency applications that usually suffer from the lack of sufficient LO power.

3.3.2.3 Single-Ended Mixers and Differential Mixers

Mixers can be based on either single-ended or differential topologies. Although this is true for both passive and active mixers, the discussion below will be focused on active mixers. Two examples of single-ended transistor mixers are shown in Fig. 3.20. The single-transistor type is shown in Fig. 3.20a, in which RF and LO signals enter the gate of the transistor together, presumably after combined with a power combiner [61]. The LO signal, with a much larger power than the RF signal in

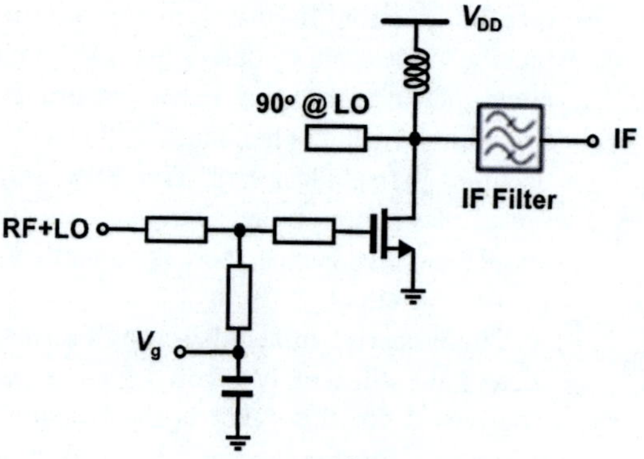

(a)

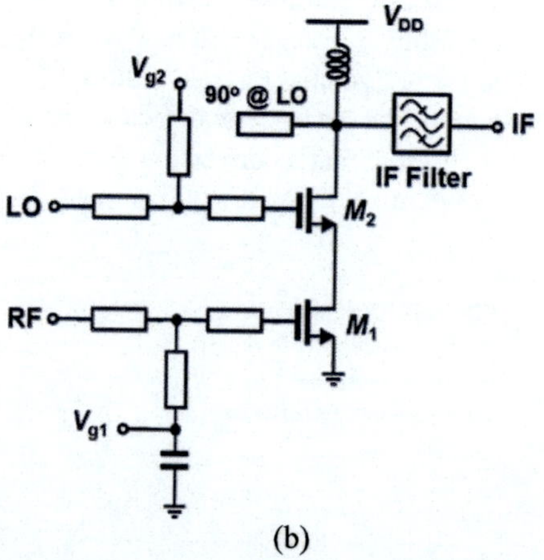

(b)

Fig. 3.20 Example circuits of single-ended mixers: **(a)** LO and RF applied through a shared input (© 2005 IEEE [61]), **(b)** LO and RF applied through two separate inputs

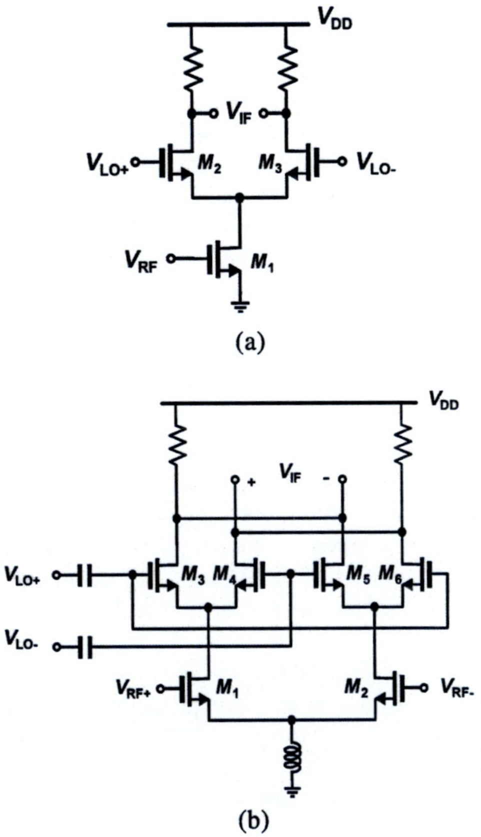

Fig. 3.21 Example circuits of differential mixers: (**a**) Single-balanced, (**b**) Double-balanced

most cases, periodically alters the bias of the transistor up and down. This in turn periodically changes the response of the transistor stage, or the transconductance of the transistor, that works on the RF signal, resulting in mixing between the LO and the RF signals. In this view, one can consider this mixing as resulting from the time-variance of the transistor response as driven by the LO signal. Or, at the same time, it can be interpreted as stemming from the nonlinearity of the transistor stage, which is inherent in transistors operating in the normal operating condition (forward active mode for FETs and saturation mode for BJTs). Dual-transistor mixer shown in Fig. 3.20b have separate inputs for RF and LO. In this scheme, the drain current that carries the RF signal amplified by M_1 is switched by M_2 driven by the LO signal, leading to mixing between the two input signals. Hence, this is another example of switching mixers operating in the active-mode.

Much more popular among the active mixers, especially with Si-based technologies, are differential mixers, examples being shown in Fig. 3.21. Differential mixers, or balanced mixers as sometimes referred to, include one or more differential pair(s), and generally grouped into two categories, single-balanced and double-balanced. For a single-balanced mixer, as depicted in Fig. 3.21a, a single-ended

RF signal enter the tail transistor M_1 at the bottom (transconductance stage), while a differential LO signal is applied through the differential pair M_2/M_3 at the top (switching stage). Similar to the dual-transistor mixer described above, the LO differential pair switches the RF signal riding on the drain current of M_1. The difference is that the differential LO signal alternatively turns on and off, switching transistors M_2 and M_3, leading to a differential IF output signal. For the positive cycle of LO, RF signal is delivered to the positive node of the IF output, while for the negative cycle of LO, RF is delivered to the negative node of IF. For the double-balanced mixer shown in Fig. 3.21b, which is commonly called Gilbert mixer, both RF and LO are applied as differential signals. Hence, the transconductance stage that accepts RF signal is formed by a differential pair (M_1/M_2), and the switching stage driven by LO is composed of two sets of differential pairs (M_3/M_4 and M_5/M_6), each of which controls the drain current of M_1 and M_2. The drain nodes of the two switching differential pairs are cross-coupled to generate a differential IF signal. For the positive and negative cycles of LO, RF signal is delivered to IF output in-phase and out-of-phase, respectively, by a current-commutating action [62, 63]. Double-balanced mixers benefit from the smaller LO-to-IF leakage, because the DC component of the RF signal is absent unlike the single-balanced case, eliminating the mechanism to carry LO to the output. They also show a less even-order distortion. However, their noise level is higher due to more active components included and the voltage budget is tighter with an additional stage stacked (if a tail transistor is employed as a current source) when compared to the single-balanced mixers.

3.3.2.4 Subharmonic Mixers

Up to this point, a rather general overview of RF mixers has been provided. For THz mixers, the biggest challenge stems from the limitation in the power and frequency of the LO signal. This issue can be addressed with the introduction of subharmonic mixers. As Eqs. (3.44) and (3.49) indicate, a mixing can be made between the harmonics of RF and LO signals, with either nonlinear or switching (or time-varying response) mixers. This implies that IF signal frequency does not need to be the difference between the actual frequencies of the LO and the RF signals as in the case of the fundamental mixer. It can be difference between the integer multiples of RF and/or LO signal frequencies, and such a mixer is called the subharmonic mixer. For practical purposes, the case that takes the harmonics only from LO signal is almost exclusively employed, such that $\omega_{IF} = |n\omega_{LO} \pm \omega_{RF}|$. In this case, the requirement for the LO frequency is relaxed by a factor of the harmonic number n, which is a great relief for THz operations. Also, there is a good chance that a higher LO power level will be obtained at the lower frequency. Its own challenge, however, lies in the degradation in the conversion gain and the noise level. An analysis in [27] shows that the conversion gain and the noise can be degraded by several dB when second-order subharmonic mixing is introduced instead of fundamental mixing for a 300 GHz mixer assuming a fixed LO power. Despite the expected performance degradation, however, subharmonic mixing is a widely practiced solution in many THz-band

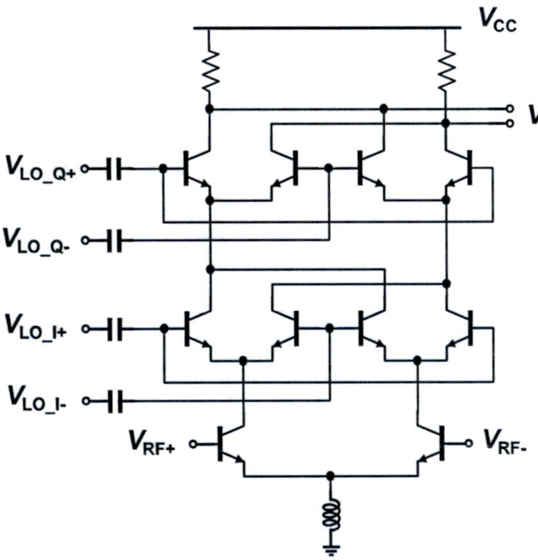

Fig. 3.22 Example circuit of subharmonic mixers based on double-stacked Gilbert cell. (© 2010 IEEE [64])

operations. Improved LO power expected at the lowered frequency may partly compensate for the performance drop.

There can be various approaches to implement subharmonic mixers at high-frequency bands. Figure 3.22 shows one widely used topology that operates at the THz band [64]. In this mixer, one additional stack of switching core is inserted to the standard Gilbert cell, and *I* (in-phase) and *Q* (quadrature) LO signals are applied to the double-stacked switching core, resulting in second-order subharmonic mixing. It requires a separate circuit block to generate the quadrature signals, but has been widely adopted owing to the preferred performance provided including a high conversion gain [65, 66]. Other similar approaches have been reported to achieve a subharmonic mixing based on *I* and *Q* LO signals, with slight variations in the way to apply the quadrature signals to the mixer [67, 68]. There are other techniques toward high-frequency subharmonic mixers, such as those based on active anti-parallel diode pair (APDP) [69] or relying on the natural subharmonic mixing based on the standard Gilbert cell that simplifies the structure [70]. The need for subharmonic mixers for THz operation may gradually diminish, though, with the continuous improvement of THz signal source performance that pushes the upper frequency limit of the fundamental-mode mixers.

3.3.2.5 G_m-Boosting Technique

Another challenge for mixers operating at the THz band is the low conversion gain. Even with the active operation, the lowered transconductance of transistors at raised frequencies affects the availability of sufficient conversion gains for THz mixers. One frequently adopted technique to enhance the conversion gain of active mixers is

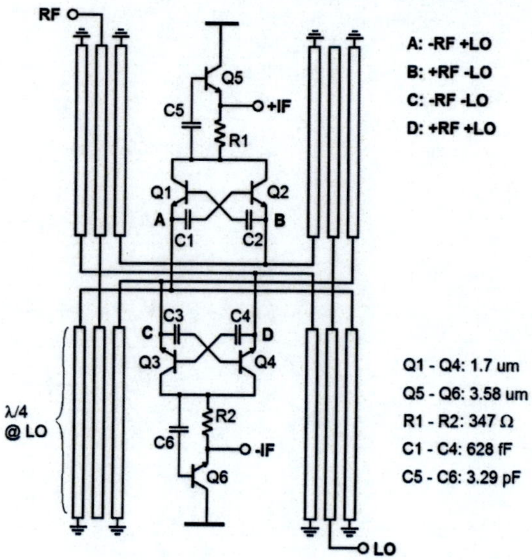

Fig. 3.23 A D-band SiGe HBT mixer with G_m-boosting technique based on differential pair cross-coupling. (© 2012 IEEE [72])

the G_m-boosting technique, which effectively boosts the G_m of the mixer core. There are various ways to boost G_m, one of which is the cross-coupling technique that can be applied to the common-gate or common-base differential configurations. One example is presented in Fig. 3.23, which shows a D-band SiGe HBT mixer with the cross-coupling technique [71, 72]. The cross connection between the base of Q_1 (Q_2) to the emitter of Q_2 (Q_1) in the differential pair enhances the V_{BE} swing of the transistors. This happens because, for the positive cycle applied to Q_1 (Q_2) base, Q_2 (Q_1) base is biased with the negative cycle, pulling down the potential at the Q_1 (Q_2) emitter due to the coupling, which results in increased V_{BE1} (V_{BE2}) for every cycle. The enhanced V_{BE} drive effectively boosts G_m of the differential pair and thus the conversion gain of the mixer. In another view of the same mechanism, it can be interpreted such that the cross-coupling provides a feedback with a gain of $-A$, which results in G_m boost by a factor of $(1 + A)$. The boosted G_m will help to improve the noise performance of the mixers in addition to the conversion gain enhancement. There are other techniques to boost G_m in mixers, such as back-gate connection and current reuse configuration adopted in [73]. It is also noted that the G_m boosting is a general technique that can be widely adopted for various types of high-frequency circuits other than mixers as well, including amplifiers [74] and detectors [27].

There have been quite a lot of mixers reported for operation beyond 100 GHz. Figure 3.24 presents the measured noise figure of mixers shown as a function of frequency based on various technologies (the plot includes the cases where additional circuits, such as a buffer amplifier, are included after the mixer, as they do not affect the noise performance significantly and, in a sense, they can be considered as a part of the mixer circuit). Depending on the mixer configuration, NF as low as ~10 dB has been reported around 200 GHz [75, 76], while it rapidly increases with the frequency leading to ~50 dB near 800 GHz [77].

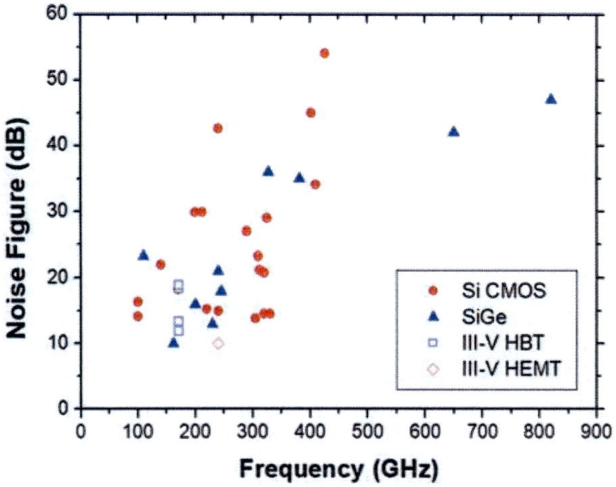

Fig. 3.24 Reported noise figure of THz mixers based on various technologies

3.3.3 Low Noise Amplifiers (LNAs)

As mentioned in the earlier discussion on the amplifying technique in Sect. 2.5.1, ensuring a required gain is a big challenge for THz amplifier design with limited device gain. This is true for THz LNAs as well. The baseline of the THz LNA design will not deviate much from the conventional approach for high-frequency LNAs, such as the input noise matching and the selection of the device size and biases in favor of the noise performance, particularly at the first stage if cascaded. However, such a basic design process becomes partly limited by the insufficient gain when the LNA operates near device f_{max}. For example, the design of the first stage may not be fully optimized for the noise performance if the gain is not sufficient, since a full noise matching would necessarily entail a partial sacrifice of the gain. This is because the noise matching condition does not coincide with gain (power) matching condition in general. If the gain is compromised, it may lead to insufficient suppression of the noise from subsequent stages or even a total loss of the gain at the first stage in the worst case. Hence, a precise balance between noise and gain is needed in the design of THz LNAs to attain the best available noise performance of the entire amplifier. This additional challenge pertinent to the LNA design around the noise is further complicated by the limited accuracy in the device noise model at high frequencies. As discussed earlier in Sect. 2.5.1, there is an accuracy issue for device models around THz frequencies. This issue becomes more stringent when it comes to LNA design since achieving precise *noise model* is particularly difficult. Hence, the designers should be aware that the simulated noise performance at high frequencies may deviate from the reality, and double checking with measurements whenever available is a suggested practice.

3.3.3.1 LNA Configurations

As for the amplifier topology, CE (CS) or cascode configurations have been widely adopted for high-frequency LNAs owing to their favorable noise properties, which trend also applies to THz LNAs. The adoption of CB (CG) for THz LNA is less popular because its noise performance is not as good as those with the other topology options, despite its known advantage of wideband performance that arises from the absence of the Miller capacitance. One aspect that should not be overlooked, however, is that the gain of CB (CG) is larger than that of CE (CS). This fact is particularly relevant to the noise-gain balance just discussed above, since the high gain of CB (CG) should help the suppression of the noise from the following stages. In fact, there is an analysis indicating that a two-stage CB amplifier shows a lower noise and higher gain than those of cascode, apparently owing to the higher gain of CB than CE at the front stage [78].

The issues surrounding the differential vs. single-ended amplifiers for THz operation were discussed in Sect. 2.5.1, and they are valid for LNA cases as well. To repeat, the differential topology will be preferred from the design and performance aspects, such as common-mode noise rejection and the availability of the virtual ground, while the interface with single-ended antennas and test environments may favor the single-ended topology although integrated baluns can always be considered. Particularly for LNAs, their outputs often drive mixers that follow in heterodyne structures, so it is affected by the topology of mixers widely employed. From this aspect, it is not surprising to see that more of Si-based LNAs are designed in differential topology while III–V LNAs are mostly in single-ended scheme, consistent with the trend in mixer topologies. In this regard, it is clear that III–V community is more influenced by the legacy from the hybrid designs based on individually packaged components.

3.3.3.2 Noise Characterization Methods

The noise characterization of LNAs (and receivers) at the THz band is challenging, and a discussion on this topic will benefit the readers. The noise figure (or noise factor), a widely adopted parameter that represents the noise level of two-port networks including LNAs, is usually obtained with a measurement employing a noise figure analyzer and a noise source. A mixer may be optionally included to down-convert the frequency into the range covered by the noise figure analyzer when the operating frequency is higher than the range. However, noise sources operating at the THz band are unfortunately not readily available as a commercial product, currently limited below 200 GHz. Without noise sources available, which are usually compact and convenient for use, one needs to resort to a bulky physical setup that produces noise at multiple temperatures required for the *Y*-factor method, a standard method to extract the noise figure. A brief introduction to the *Y*-factor method and the measurement setup is provided below [79, 80].

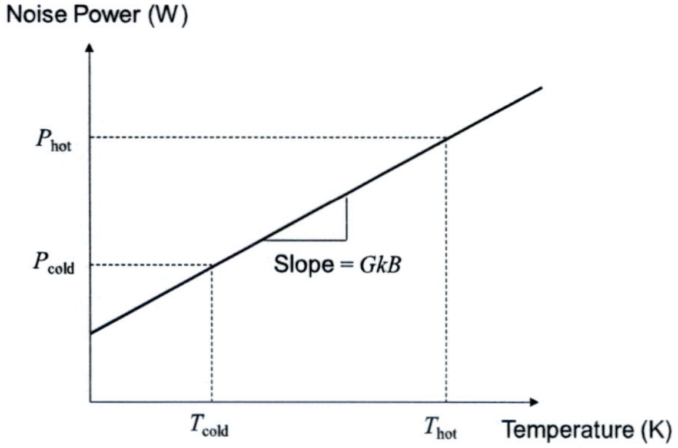

Fig. 3.25 Conceptual diagram showing the linear relation between the noise power and temperature, which is the basis for *Y*-factor method

The starting point of the *Y*-factor method is the expression for the noise power P_n measured at the output of a system to be characterized in the absence of a signal, which is given by:

$$P_n = Gk(T_{ref} + T_{sys})B, \tag{3.50}$$

where T_{ref} is the reference temperature faced by the input of the system, T_{sys} is the noise temperature of the system, G is the system gain, and B is the system bandwidth. k is, of course, Boltzmann constant. According to Eq. (3.50), P_n is given as a linear function of the reference temperature, as is conceptually drawn in Fig. 3.25. Remember that the main object of the *Y*-factor method is to obtain T_{sys}, and this requires the linear function to be determined first. Obviously, this can be achieved if we know the noise power levels at two different reference temperature points such that:

$$P_{hot} = Gk(T_{hot} + T_{sys})B \tag{3.51}$$

$$P_{cold} = Gk(T_{cold} + T_{sys})B, \tag{3.52}$$

where the subscripts *hot* and *cold* represent the two temperatures employed. Now, the *Y*-factor is defined as the ratio between P_{hot} and P_{cold}, thus leading to:

$$Y = \frac{P_{hot}}{P_{cold}} = \frac{T_{hot} + T_{sys}}{T_{cold} + T_{sys}}, \tag{3.53}$$

which can be rearranged for T_{sys}:

$$T_{sys} = \frac{T_{hot} - YT_{cold}}{Y - 1}. \tag{3.54}$$

Based on Eq. (3.54), one can obtain T_{sys} with the knowledge of the *Y*-factor as well as the two temperatures T_{hot} and T_{cold}. A measurement is needed to determine the value of the Y-factor by evaluating the noise power at the system output with the injection of external noise corresponding to the two reference temperatures. With the obtained Y-factor and the temperature values, the noise temperature of the system T_{sys} will be decided. If noise figure is preferred as a noise parameter instead of the noise temperature, a simple conversion can be made with a relation $T_n = T_0(F - 1)$ as mentioned earlier (Eq. 3.11). The elegance of the Y-factor method lies in the fact that no information on system gain G or bandwidth B is needed, because it is the *ratio* of the noise powers that matters, which is independent of G or B, as indicated by Eq. (3.53). In fact, the system gain G is a parameter that can be simultaneously determined from this procedure because the slope of the curve in Fig. 3.25, or GkB, is proportional to the gain. One additional information needed to accurately determine G from the slope, though, is the value of B. To ensure a control over B, a filter of well-defined bandwidth is typically inserted in the measurement setup. As for the reference temperatures, the room temperature (around 300 K) and the liquid nitrogen temperature (77 K) are typically employed for their easy availability, although a larger temperature difference would result in a more reliable measurement with a less fluctuation in the data.

A typical setup for noise measurement with the Y-factor method is illustrated in Fig. 3.26a. A cryogenic container with liquid nitrogen (LN2) and a moving shutter covered with an absorber are placed in front of the DUT (Device Under Test), which is interfaced with a horn antenna at the input and a subharmonic mixer at the output. This is followed by a low-frequency LNA and a band-pass filter. With this configuration, the DUT will face the room-temperature absorber when the shutter is closed (the hot-load state) while it looks into LN2 through a parabolic mirror when the shutter is open (the cold-load state). By measuring the noise power at the output with these two states, the Y-factor can be obtained and thus T_{sys} can be determined. To be more precise, the obtained T_{sys} with this process is not for the DUT alone but for a structure that comprises the DUT as well as the probe and post-DUT components. Hence, the extraction of T_{sys} for the isolated DUT will require an additional calibration measurement that should be made in the absence of the DUT while maintaining the rest based on otherwise the same setup. Also, when the noise power difference between the two states are not large enough, which occurs when the noise level of the device is relatively high, repeated measurements are necessary to extract the reliable averaged value. It is noted that the conventional noise measurement setup with a noise figure analyzer and a noise source is in fact based on the Y-factor method, too, in which turning on and off the noise source emulates the hot and cold states in the Y-factor method, respectively.

There is another method to extract the noise figure of LNAs that merits a brief review here, too. Called the signal generator N-times power method, it makes use of the ratio of the output power of the system with and without the input signal at a fixed temperature, most likely the room temperature [80–82]. The measurement setup for the signal generator N-times power method is depicted in Fig. 3.26b. It is similar to the setup used for the Y-factor method, the main difference being the front cryogenic part replaced with a signal generator. As it is performed at a single fixed temperature, it alleviates the need for a low-temperature measurement and thus can be

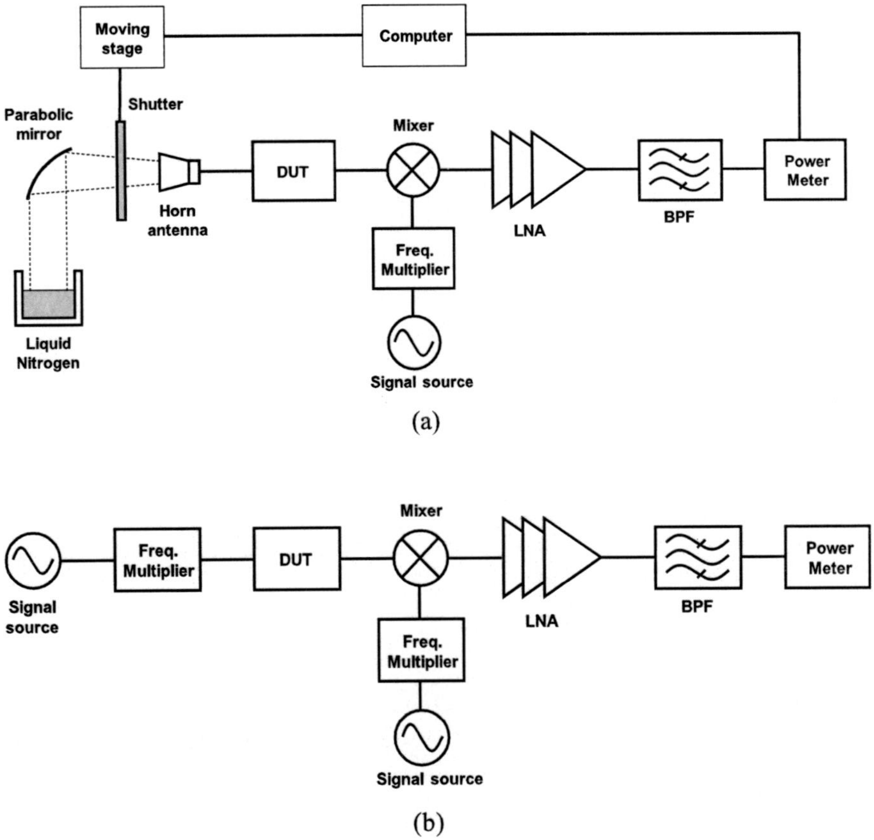

Fig. 3.26 Setups for noise measurement: (**a**) *Y*-factor method with cryogenic temperature, (**b**) Signal generator *N*-times power method

conveniently practiced without a cryogenic setup. On the downside, however, this method requires accurate values of the injected input power and the system bandwidth. The output power of the system with and without the signal injected at the input can be expressed as:

$$P_{\text{on}} = G\left(P_{\text{sig}} + k\left(T_{\text{ref}} + T_{\text{sys}}\right)B\right) \tag{3.55}$$

$$P_{\text{off}} = Gk\left(T_{\text{ref}} + T_{\text{sys}}\right)B, \tag{3.56}$$

where P_{sig} is the input signal power. From these relations, one can easily arrive at the following expression for the system noise temperature:

$$T_{sys} = \frac{P_{sig}}{kB(P_{on}/P_{off} - 1)} - T_{ref}. \quad (3.57)$$

Hence, with the values of input power P_{sig}, two output powers P_{on} and P_{off}, and the system bandwidth B, the noise temperature T_{sys} can be obtained. It is worthwhile to note that Eqs. (3.54) and (3.57) are in fact equivalent if the following relation is valid:

$$T_{hot} = \frac{P_{sig}}{kB} + T_{cold}. \quad (3.58)$$

3.3.3.3 THz LNA Performance

As mentioned earlier, an LNA is an optional component for THz detection systems depending on its availability at the frequency of detector operation. Since the noise level of a multi-stage system is dominated by the component located at the first stage of the chain as Friis formula indicates, the availability of LNAs critically affects the noise level of heterodyne detectors. The absence of LNA exposes the RF detector (in direct detection) or the mixer (in heterodyne detection) at the front of the detection circuit, whose noise levels may be higher by a factor of an order or larger when compared with LNAs. There has been a report that compares two direct detection cases with and without a front-end LNA at 140 GHz [83]. According to the experiment based on detector circuits implemented in a SiGe HBT technology, the detection including an LNA of a 14-dB gain and a 10-dB NF showed an NEP of 80 fW/Hz$^{1/2}$, as compared to 3.3 pW/Hz$^{1/2}$ obtained without the LNA based on otherwise identical circuit, leading to an improvement by a factor of around 40. For detector systems, another effective method to lower the noise level by a factor of orders is cooling the system down to cryogenic temperature, which is commonly adopted for noise reduction in diode-based systems. For background-limited thermal detectors, NEP will decrease with the temperature at a rate of ~$T^{2.5}$ [84], which leads to a reduction by a factor of ~30 and ~50,000 when cooled from 300 K down to 77 K (liquid nitrogen) and 4 K (liquid helium), respectively. Hence, one can argue that an inclusion of an LNA in front of a square-law detector has roughly a similar effect as cooling a thermal detector down to liquid-nitrogen temperature (if not liquid-helium temperature), without requiring a bulky cryogenic setup.

As the operation frequency of amplifiers keeps moving upward, the frequency range that benefits from the inclusion of LNA is being further extended. There is an increasing number of LNAs operating at the THz band, as is indicated by Fig. 3.27 that shows the measured noise figure of amplifiers as a function of frequency based on various technologies. One notable result is an LNA operating at 850 GHz, which is based on NGC (Northrop Grumman Corporation) 25-nm InP HEMT technology [85]. The packaged ten-stage common-source amplifier shows a NF as low as 11.1 dB, which can be converted to a noise temperature of T_n = 4300 K. A

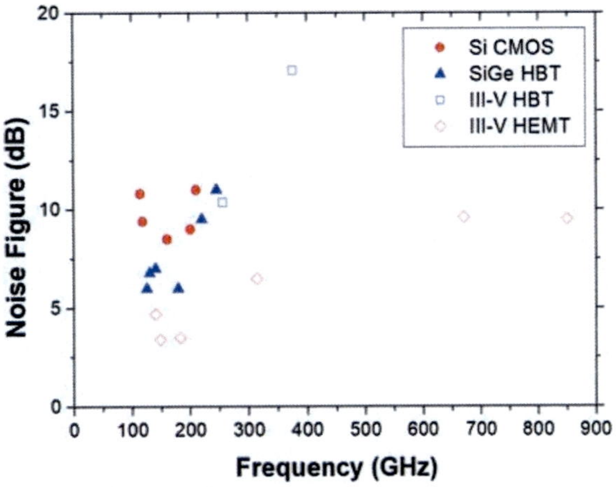

Fig. 3.27 Reported noise figure of THz LNAs based on various technologies

measurement made on unpackaged circuits shows an even lower NF of 9.3 dB, implying a noise penalty close to 2 dB due to the packaging. The associated gain was 13.6 dB, which translates to only 1.36 dB per stage. This barely positive gain obtained from a single stage well indicates the challenge in ensuring a sufficient gain at this high frequency range. As for more practical frequency range of around 200 GHz, a NF of ~3.5 dB was obtained near 180 GHz with NGC 35-nm InP HEMT [86] and with Fraunhofer 50-nm mHEMT technology [87]. With Si technologies, a NF of around 11 dB has been reported for a 210-GHz CMOS LNA [88] as well as a 245-GHz SiGe HBT LNA [89]. Figure 3.28 compares the reported NF of THz LNAs with that of mixers as well as those of diode-based detectors such as SBDs, SIS mixers, and superconducting HEBs. The data points in the plot are taken from Figs. 3.14, 3.24, and 3.27. It is noted that the NF of LNAs and SBDs are based on room-temperature operation, while those of SIS mixers and superconducting HEBs are from cryogenic operation. It clearly shows that the cryogenic detectors exhibit the lowest NF. Also notable is that transistor circuit-based mixers show much larger noise than SBD mixers. In fact, SBD mixers show NF as low as those of LNAs at frequencies that range from a few to several hundreds of GHz [90–92].

3.3.4 *Integrated Heterodyne Detectors*

The biggest advantage of using transistor-based circuits for detection is that all the circuit components needed to build a detector circuit (especially that of heterodyne detector) can be integrated and implemented in a single chip. Also, if desired, they can be further integrated with other part of electronics, including digital parts, to enhance the function of the chip. This integration feasibility is the most outstanding feature of the transistor detector circuits when compared to other approaches based on thermal detectors and diode detectors. The major limitation for the transistor-

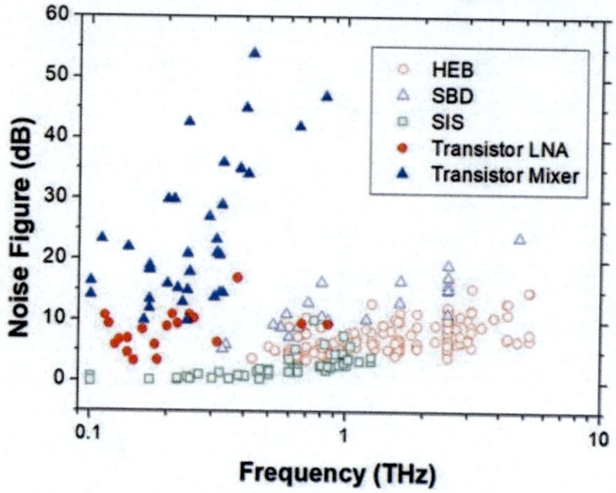

Fig. 3.28 Comparison of reported noise figure of diode-based detectors including SBDs, SIS mixers, superconducting HEBs, and transistor circuits including mixers and LNAs. The data points are taken from Figs. 3.14, 3.24, and 3.27. Data for SIS mixers and superconducting HEBs were obtained at cryogenic temperatures. The others are all room-temperature data

based approach has been the upper frequency limit for active operation (with gain), which is being gradually lifted with the advent of transistor technologies with a device speed much faster than before. The availability of discrete circuit components, though, does not guarantee the successful operation of integrated heterodyne detectors as there exist additional design considerations in stitching them into one piece meticulously. Besides the matching between the circuit blocks, the configuration (differential or single-ended) needs to be aligned, which may require additional balun circuits. The harmonic order of the mixer needs to be decided based on the available power and frequency of the LO. Although rarely happens at the THz band, the signal saturation also requires attention. Satisfying all these requirements is challenging, but a growing number of integrated heterodyne detectors are being reported.

It would be useful to review an example of integrated heterodyne detectors, as presented Fig. 3.29 [27]. In this design, a 300-GHz SiGe HBT heterodyne detector is composed of an on-chip antenna, a mixer, an oscillator as LO, an IF amplifier, and an IF detector, all integrated together on a chip. LNA is not included since an amplifier operating at 300 GHz was not readily available with the SiGe HBT technology used in this work. The generation of the 300-GHz signal for LO, however, was feasible with the help of the push-push technique employed to attain the second harmonic signal based on a differential Colpitts oscillator. Owing to this 300-GHz LO available on-chip, the mixer was designed to operate at the fundamental mode, which would benefit from the lower noise and higher conversion gain compared to the cases with subharmonic mixers. Since the oscillator output is single-ended as is typical with the push-push configuration, the mixer adopted a single-balanced topology that accepts differential RF and single-ended LO inputs. The two-stage

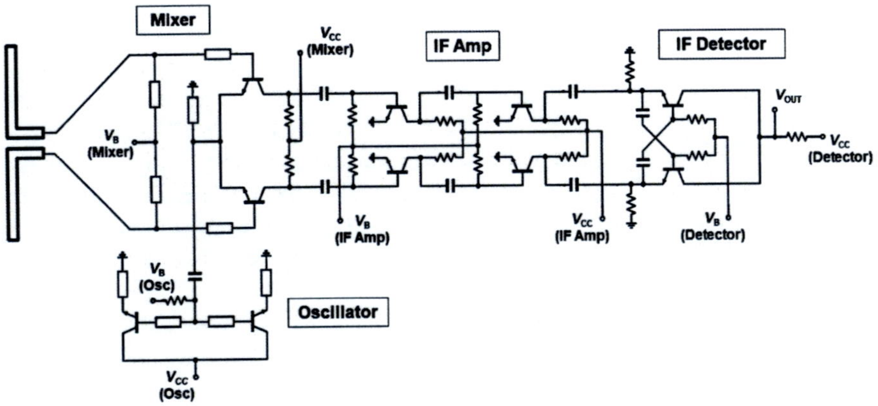

Fig. 3.29 Schematic of the 300-GHz SiGe HBT heterodyne detector circuit

IF amplifier is based on differential CE configuration, which is followed by the IF detector that adopts a G_m-boosting cross-coupling technique. The front-end of this heterodyne imager is interfaced with an on-chip dipole antenna. One favored aspect of increased frequency for circuit design is the smaller size of passive devices, including the antenna, owing to the reduced wavelength λ. The characteristic dimension of dipole antennas, $\lambda/2$, is about 500 μm at 300 GHz in the air and shrinks down to around 250 μm on silicon dioxide or 150 μm on silicon, acceptable sizes for most high-frequency circuits. Details on antennas will be described in Chap. 4. The heterodyne detector exhibited a maximum responsivity of 322 kV/W and a minimum NEP of 3.9 pW/Hz$^{1/2}$ at 300 GHz. Compared with a direct detector implemented on the same chip for comparison, which showed a maximum responsivity of 6121 V/W and a minimum NEP (noise equivalent power) of 21.2 pW/Hz$^{1/2}$, the heterodyne detector exhibits an improvement by a factor of 53 and 5.4 for responsivity and NEP, respectively. The price for this performance improvement, however, is the larger DC power dissipation (21 mW vs. 0.6 mW) and the larger area (610 × 610 μm^2 vs. 460 × 410 μm^2) as well as the reduced bandwidth (6 GHz vs. 60 GHz). The THz images obtained with this circuit will be introduced in Sect. 6.2.

There have been other THz heterodyne circuits reported based on various approaches. On the high-frequency front, a heterodyne receiver operating at as high as 820 GHz has been reported [77]. Based on a SiGe HBT technology, this heterodyne receiver employs a fifth-order subharmonic mixer driven by a 164-GHz LO signal generated by an on-chip amplifier-multiplier chain. Notably, the receiver is configured as an array with four channels, each driven by a 0-dBm LO pumping power and interfaced with a differential patch antenna at the front. The circuit exhibited a peak conversion gain of −22 dB and a noise figure of 47 dB around 820 GHz. Note that this receiver does not include an IF detector at the back-end, and thus the down-converted IF signal is taken as the output, in which case the noise level is represented by the noise figure instead of NEP. On a different path, an example including an LNA can be found in [79]. The 300-GHz heterodyne receiver, based on a III–V HBT technology, integrates a five-stage differential common-base amplifier with a double-balanced fundamental mixer driven by an on-chip 300-GHz fundamental-mode oscillator for LO signal generation. The availability of the 300-GHz amplification and 300-GHz fundamental-mode signal generation can be attributed to the superior performance of III–V technologies over Si-based technologies. The receiver showed a peak conversion gain of 26 dB and a noise figure of 12.0–16.3 dB around the operation frequency of 300 GHz.

3.4 Array Detectors

3.4.1 Array vs. Single-Pixel Detectors

In the previous sections, we have reviewed various detectors that can be used for THz wave detection. Those detectors can be used as a stand-alone individual detector, or employed as part of a group of multiple detectors operating collectively. The former case is widely referred to as a single-pixel detector, a term apparently originated from imaging applications, while the latter is called an array detector or a detector array. The array detectors are a highly popular option for imaging applications, while spectroscopy and communication systems can benefit from them, too.

3.4.1.1 Advantages of Array Detectors

Let us briefly review the merits of array detectors. Essentially, array detectors can improve the effectiveness in the detection process over the single-pixel counterpart, by converting the serial process in the time domain into the parallel process in the spatial domain in detection. For example, in imaging applications, single-pixel detectors will need a scan of the object (or, alternatively, scan of the detector) to acquire 2D images, but array detectors will remove the scan step or at least reduce the number of scans, depending on the array size (pixel count). For spectroscopy, array detectors (usually 1D array in this case) will allow the simultaneous detection of a wide spectrum spread over a distance, alleviating the need for mechanical scanning of the detector that would be needed with single-pixel detectors. In fact, communication systems will also benefit from the array detectors. MIMO (multiple-input and multiple-output) systems [93] include multiple receiver antennas, together with multiple transmitter antennas, for improved channel capacity of the data link, which can be considered as one kind of array detectors. The discussion below, though, will assume array detectors largely intended for imaging applications. It should be mentioned at this point that another advantage of the array detector is the capability of steering the sensing direction (reverse of beam forming) by controlling the phase of each pixel, or by phased array. This topic, however, will not be treated in this section, as it was covered in Sect. 2.6 as a separate topic.

3.4.1.2 Challenges for Array Detectors

Despite the apparent advantages briefly mentioned above, there are various issues in realizing detector arrays. First, there is an issue in the power level received by each pixel. For an array detector, the total power incident on the detector is spread over the pixels, leading to a significant reduction in the received power per pixel and degraded SNR, especially when the pixel count is large. Simply put, when there are N pixels in the array, each pixel will receive $1/N$ of the total incident power. This contrasts with the case with a single-pixel detector, where nearly the entire incident power can be focused on the single detector in the ideal case. Additionally, in most practical cases, the signal detected by each pixel is sequentially extracted from the array. Although the situation may depend on the detection scheme details, it is likely that the detector in each pixel will not be collecting the incident power for the entire exposure time but only for a fractional time allocated per each pixel. Again, with a pixel count of N, the allocated time per each pixel will be $1/N$ of the total exposure time in the ideal case. In reality, it will be even smaller than this due to the overhead such as the time required for transition between pixels. This will be more significant with a larger video bandwidth (higher frame rate) if the array is employed for real-time imaging, as the time allowed for the entire array scan will be smaller. As a result, both spatial and temporal distribution of the power over the pixels will lead to a significantly degraded SNR for each pixel, severely limiting the receiver performance and image quality.

Second, there is an issue of nonuniformity. When an array is made of discrete components, it typically suffers from serious nonuniformity due to the performance variance over the components. If the array is made of a single chip with monolithic approach, the situation will be much better with a tighter performance distribution, but it cannot avoid the variation over the pixels. For example, Fig. 3.30 shows the measured distribution of NEP of a Si CMOS 7 × 7 array implemented on a single-

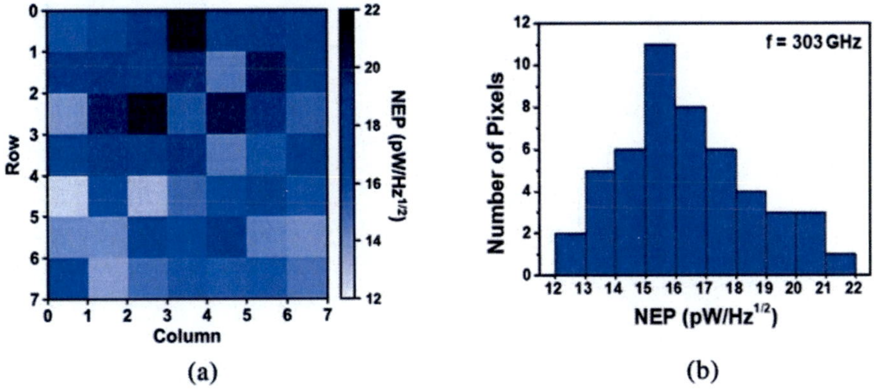

Fig. 3.30 Nonuniformity in measured NEP over a single-chip Si CMOS 7 × 7 array: (**a**) NEP distribution over the array chip, (**b**) Distribution of the NEP values. (Reprinted by permission from Springer Nature: [94] © 2020)

chip, which still shows a significant variation [94]. This expected nonuniformity mandates a calibration process to be carried out after the image acquisition, which can mitigate the issue to a certain extent. Yet another nonuniformity issue comes from the variance in the THz beam intensity across the beam cross section. As will be detailed in Sect. 4.1, THz beams will have a finite lateral dimension (even at focal points) and will show a Gaussian profile in its intensity along the transverse direction. Hence, when the beam from a THz source arrives at a 2D array detector, the incident power will show variation over the pixels, showing the peak at the center and gradually decreasing toward the edge. This nonuniform power distribution will cause distortion in the acquired images and thus also require a calibration process for correction after data acquisition.

Third, the effect of the readout circuit needs to be minimized. Array detectors necessarily include a readout circuit to sequentially select a pixel at a time and transfer the data to the output. The readout circuit will include switches for row/column selection and physical signal paths between each pixel and the output node. Although indispensable for array operation, the readout circuit will affect the performance of the array detector. It will cause loss and add noise as the signal from the selected pixel travels down to the output node. Hence, an additional design care should be taken, and an amplifier, on-chip or off-chip, may be needed to compensate for the losses. It is also noted that the loss caused by the readout circuit will be different over the pixels depending on the location of each pixel. This is another cause of nonuniformity in array detectors. For example, the schematic of a Si CMOS 7×7 array shown in Fig. 3.31 indicates that the signal from Pixel 0 should travel a

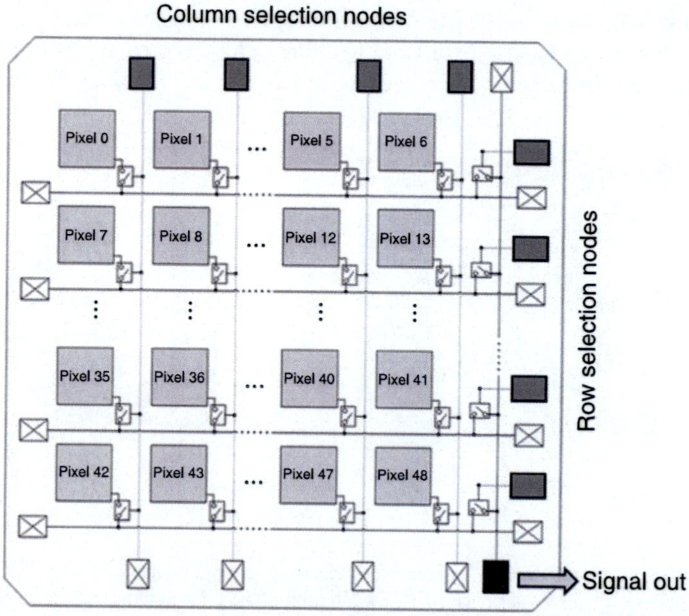

Fig. 3.31 Chip-level schematic of a Si CMOS 7×7 array (4×4 mm^2). (© 2020 K. Song [95])

much longer path than that of Pixel 48 to reach the output node, suffering from a larger loss [94, 95]. Fortunately, for this type of nonuniformity, the variation is not random but rather systematic, which can be calibrated and canceled out with a relative ease.

3.4.1.3 Array Detector Schemes

For the advantages and challenges of the array detectors discussed above, the direct detection scheme was implicitly assumed. While it is true that array detectors are mostly adopted with the direct detection scheme, arrays can be applied to heterodyne systems as well. The above-mentioned advantages and issues of arrays are valid for the arrays with the heterodyne scheme as well, plus there will be additional issues pertinent to heterodyne arrays. Compared with the direct detector, the unit detector in the heterodyne arrays will occupy a large area, as it contains more components, leading to a larger pixel size. This also means a higher power dissipation per pixel, which may accumulate to a massive level for the entire array. To avoid excessive power dissipation, it is highly desired to turn on only the pixel that is selected for signal detection at a given time. In fact, this is true also for the direct detection arrays if power-hungry components are included in each pixel. A symmetric placement of the pixels will also be hard for heterodyne arrays, especially for 2D arrays, because of the elongated shape often found for heterodyne detector layout. This would make 1D arrays a better candidate for heterodyne arrays. LO implementation is another main issue for heterodyne arrays. The central distribution scheme, which is based on a single LO source located in the center of the array, will suffer from the low LO power per pixel (or channel) as the power from the LO is shared by all the pixels. Also, if the phase information is to be retained, which is the key advantage of the heterodyne detection over the direct detection, the distance from the LO to the pixel needs to be identical for all the pixels. This requirement is challenging particularly for THz applications due to the small wavelength, as a small distance delta will be significant in terms of the electrical length. An alternative LO scheme for heterodyne arrays is the localized LOs, in which each pixel has its own LO source. In this case, a reasonable LO power can be obtained for each pixel, but the phase synchronization over the pixels will need a precise phase locking between adjacent oscillators. Also, there may arise an issue of LO power uniformity over the pixels, which is difficult to achieve when individual LO drives each pixel.

As for the array dimension, they can be either 1D or 2D. With 1D arrays, usually in a linear configuration, the scan step cannot be completely eliminated to acquire 2D images, since a scan along the direction perpendicular to the array axis is still needed. Nonetheless, the total scan time can be reduced by a factor of the pixel count along the array. In fact, 1D arrays are widely deployed for practical applications, including the body scan at the airport security points as well as the product scan over a moving conveyer belt for quality inspection. 2D arrays provide possibility of acquiring 2D images without any scan step, in which case the 2D array can be considered as a camera. In this case, 2D images can be obtained with a single exposure and real-time imaging becomes also available, if the pixel count is large enough to satisfy the required image resolution. If the pixel count is not large enough, scan may be still needed to meet the resolution target, but the total scan time will be significantly reduced compared with the imaging with single-pixel detectors [94].

A couple of comments can be made related to 2D arrays for imaging. The 2D detector arrays used for imaging applications are often referred to as the focal plane array (FPA) [13, 55, 96]. The name, focal plane array, tends to imply that the array is supposed to be located at the focal plane of the imaging system. However, this is not always true. Technically, the focal plane is the plane that contains the focal point on it, but not necessarily the plane on which images are focused. The focused images will be obtained on the "image plane" instead, which is in general different from the focal plane. Hence, to acquire high-quality images, the array should be located at the imaging plane. There is a special case, though, when the focal plane will coincide with the image plane. This happens when the object is located at the infinity, as in the case with telescopes for astronomy including radio telescopes. It is no surprise that the term, focal plane array, originated from the astronomy, in which all the objects are virtually at the infinity. We will come back to this issue again in Sect. 6.2. Another comment can be made on 2D array configuration as follows. In typical 2D arrays, pixels are packed inside the given array dimension, arranged in rows and columns. Sometimes, however, images of a similar quality can be obtained with detectors sparsely scattered over the given area, not necessarily being packed, with the help of a signal-processing technique called the compressed sensing [97]. In this case, a large number of detectors can be saved, which will benefit especially the arrays based on individual components rather than monolithic arrays. The compressed sensing can also be used with single-pixel detectors, in which case raster scan can be possibly eliminated [98].

3.4.2 *Array Detector Examples*

Based on CMOS technology, a single-chip 1k-pixel array detector was developed with a target detection frequency range of 0.7–1.1 THz [99, 100]. It is composed of an array of 32 × 32 detectors, each occupying 80 × 80 μm^2, integrated into a chip area of 2.9 × 2.9 mm^2. The unit pixel consists of an antenna-integrated direct detector in a differential topology, the common drain node of which forms a second differential pair with an antenna-free dummy reference detector for offset compensation purpose. The pixels can be reset with a global reset circuit that drives the gate of the detector transistors. The rows and columns of the pixels are selected by 5-bit addresses that are decoded by on-chip decoders, and the signal from the selected pixel is extracted at the output node with optional amplification by an on-chip OP-amp. The block diagram of the array chip is shown in Fig. 3.32a. The maximum

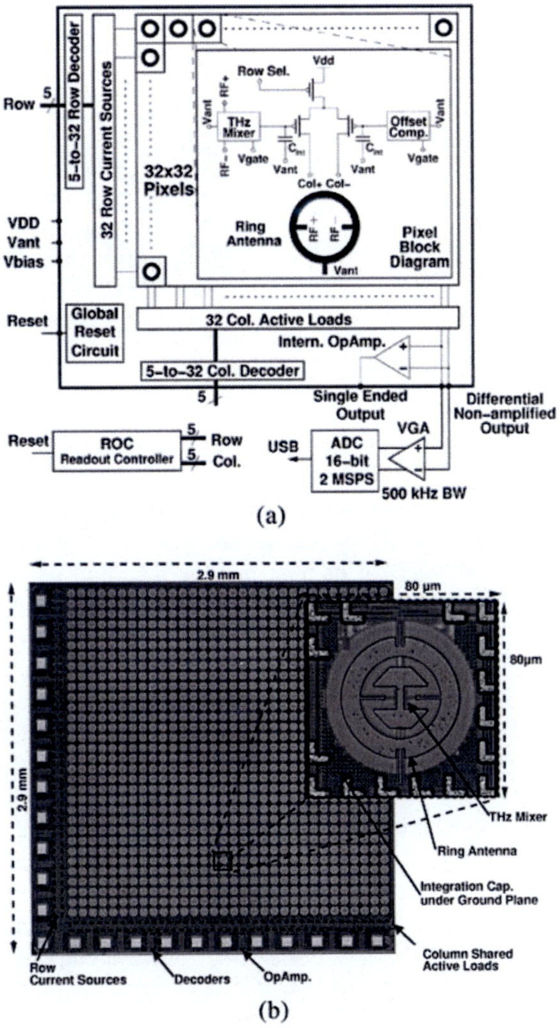

Fig. 3.32 One kilo-pixel THz Si CMOS array detector: (**a**) Chip block diagram, (**b**) Photo of the fabricated array chip. (© 2012 IEEE [100])

responsivity and minimum NEP of the unit pixel detector at the optimum bias point were measured to be 140 kV/W and 100 pW/Hz$^{1/2}$, respectively, at a chopping speed of 5 kHz. The completed chip is mounted on a hyper-hemispherical Si lens for an efficient backside detection. The photo of the fabricated array detector is shown in Fig. 3.32b. Note that this array detector functions as a THz camera, capable of acquiring still images in a single shot and real-time images, which will be discussed in Sect. 6.2.4.

For the detector chip described above operating near 1 THz, pixels counting up to ~1000 were able to be contained within a reasonable chip size. For lower frequencies, however, the allowed spacing between adjacent pixels is larger as dictated by

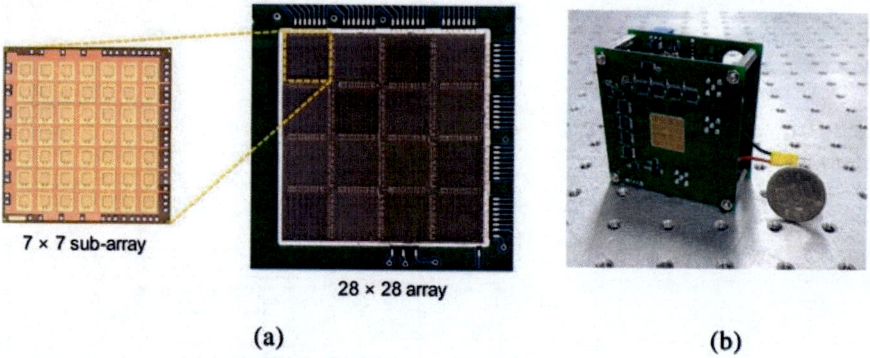

Fig. 3.33 (**a**) Photo of the CMOS multi-chip 2-D array with 28 × 28 pixels (16 × 16 mm^2). (**b**) The multi-chip array mounted on a package module

the maximum resolution of diffraction-limited optical systems (~$\lambda/2$) and the antenna dimension. Hence, the number of pixels allowed in a single chip decreases with decreasing frequency. A Si CMOS 300-GHz 2D detector array reported in [94] contains 7 × 7 detectors, each with a size of 500 × 500 μm^2. The pixel count of 49 is limited by the chip size of 4 × 4 mm^2, a typical reticle size of MPW (multi project wafer) chips. Detector arrays with this order of pixel count may help to reduce the raster scan time, but may not eliminate the entire raster scan, which would be needed for single-shot 2D images or real-time imaging. One plausible solution to circumvent the pixel count limit imposed by the single-chip size is to adopt the multi-chip array [95]. In this modular approach, multiple chips that carry their own sub-arrays are assembled to constitute a larger array. This approach relaxes the limit on the maximum size of the entire array and also allows flexibility in selecting the array size. Figure 3.33a shows one of such examples, a detector array of 28 × 28 pixels consisting of 16 chips stitched together, each carrying a 7 × 7 detector sub-array operating at 300 GHz. The 16 chips arranged in a 4 × 4 configuration are electrically connected with bonding wires, which allow detected signals and readout signals to pass through the chip interfaces. One practical issue for this approach is that the peripheral area of each chip is consumed for bonding pads that are needed for inter-chip connection, creating a dead zone in the images along the individual chip edges as pixels are missing on those areas. A solution is that the lost portion of the image is recovered by interpolating the data on the adjacent regular pixels. Including these virtual pixels, which have the same physical dimension as the regular pixels on purpose by design, the total effective pixel count becomes 31 × 31 = 961 for the multi-chip array. The photos of the completed array and its packaged module are shown in Fig. 3.33b.

While both the arrays described above are based on Si CMOS detectors, 2D THz array detector with bolometers were also reported [9, 101]. The unit cell is a room-temperature microbolometer made of vanadium oxide with a pixel pitch of 23.5 μm. Figure 3.34 presents the cross section of the bolometer structure, which shows the bolometer layer formed on a diaphragm, thermally isolated from the Si substrate that carries the integrated read-out circuits [9]. The array consists of 320 × 240 bolometer pixels. A partial photo view of the array is given in Fig. 3.34b. With a very wide operation bandwidth of 1.5–100 THz, the bolometer exhibited NEP of around 0.1–0.2 pW/Hz$^{1/2}$ within the band. With the large pixel count, real-time images were successfully obtained by employing quantum cascade laser (QCL) as a THz source.

The array examples introduced so far are all microscopic, built on tiny chips based on microelectronics. There are macroscopic large-scale THz array detectors as well. One good example is Atacama Large Millimeter/submillimeter Array (ALMA) [102], as briefly mentioned earlier, which consists of 66 radio telescopes with dish reflector sizes of 12 or 7 m in diameter. Built on Chajnantor plateau in Chile at 5000-m altitude for optimal observation condition, they operate in a wide frequency range of 35–950 GHz, which are sub-divided into ten bands. The antennas are mobile and can move around over a baseline ranging up to 16 km for a flexible arrangement of the array configuration. The operation of ALMA is in fact based on the principle of interferometry, which provides a resolution of an aperture size that is equal to the maximum separation between antenna elements. Accordingly, ALMA operates with an effective aperture size of over a 10-km range. Based on the same principle, radio telescopes scattered over the Earth can be linked for co-operation, in which case the Earth effectively becomes a single enormous antenna for space observation. This is the basis for Very-Long-Baseline Interferometry (VLBI), which was utilized for the Event Horizon Telescope (EHT) (see Fig. 3.35b) that has successfully taken the very first (shadow) image of a black hole at the center of galaxy M87 at a wavelength of 1.3 mm ($f = 230$ GHz) [103].

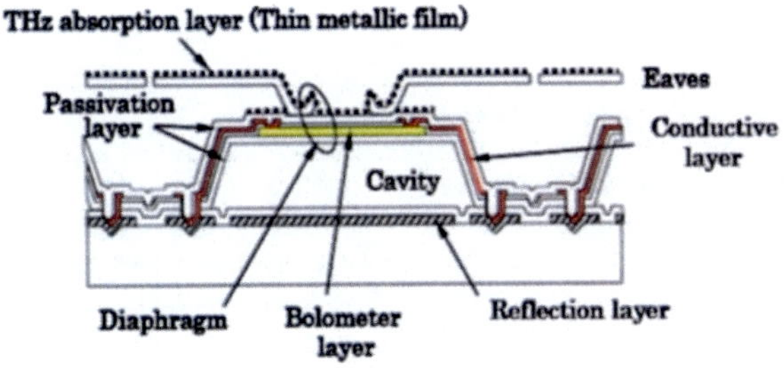

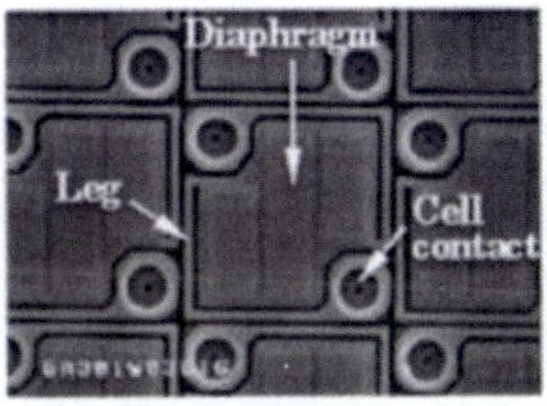

Fig. 3.34 Microbolometer-based 320 × 240 array detector: (**a**) Unit bolometer structure, (**b**) Partial image of the fabricated array. (Reproduced with permission from [9], © 2008 SPIE)

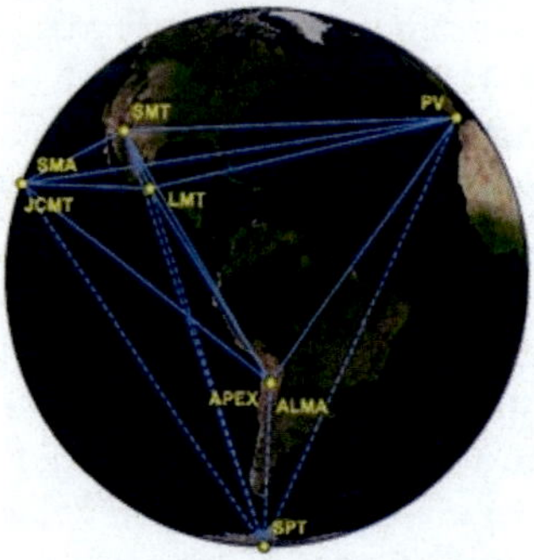

(a)

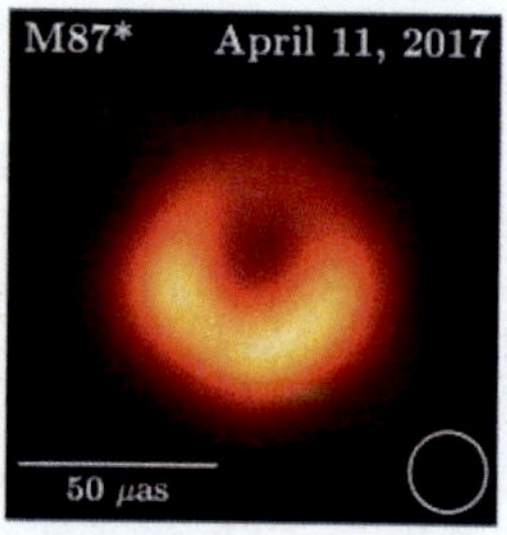

(b)

Fig. 3.35 (**a**) Locations of the eight stations of the Event Horizon Telescope (EHT), which renders the Earth an enormous virtual antenna. (**b**) The first (shadow) image of the black hole taken with the EHT at 230 GHz. (Published in [103] under CC BY 3.0 license)

References

1. R.C. Jones, Phenomenological description of the response and detecting ability of radiation detectors. Proc. IRE **47**(9), 1495–1502 (1959)
2. W.S. Boyle, K.F. Rodgers, Performance characteristics of a new low-temperature bolometer. J. Opt. Soc. Am. **49**(1), 66–69 (1959). https://doi.org/10.1364/JOSA.49.000066
3. F. Low, Low-temperature germanium bolometer. J. Opt. Soc. Am. **51**(11), 1300–1304 (1961)
4. J. Clarke, G. Hoffer, P. Richards, N.H. Yeh, Superconductive bolometers for submillimeter wavelengths. J. Appl. Phys. **48**(12), 4865–4879 (1977). https://doi.org/10.1063/1.323612
5. P.M. Downey et al., Monolithic silicon bolometers. Appl. Opt. **23**(6), 910–914 (1984). https://doi.org/10.1364/AO.23.000910
6. P. Richards, Bolometers for infrared and millimeter waves. J. Appl. Phys. **76**(1), 1–24 (1994)
7. S.E. Schwarz, B.T. Ulrich, Antenna-coupled infrared detectors. J. Appl. Phys. **48**(5), 1870–1873 (1977). https://doi.org/10.1063/1.323940
8. T.L. Hwang, S.E. Schwarz, D.B. Rutledge, Microbolometers for infrared detection. Appl. Phys. Lett. **34**(11), 773–776 (1979). https://doi.org/10.1063/1.90669
9. N. Oda et al., Detection of terahertz radiation from quantum cascade laser using vanadium oxide microbolometer focal plane arrays, in *Infrared Technology and Applications XXXIV*, vol. 6940, (International Society for Optics and Photonics, Bellingham, WA, 2008), p. 69402Y
10. A. Luukanen, A.J. Miller, E.N. Grossman, *Passive Hyperspectral Terahertz Imagery for Security Screening Using a Cryogenic Microbolometer (Defense and Security)* (SPIE, Bellingham, WA, 2005)
11. P. Khosropanah et al., Ultra-low noise TES bolometer arrays for SAFARI instrument on SPICA, in *Millimeter, Submillimeter, and Far-Infrared Detectors and Instrumentation for Astronomy VIII*, vol. 9914, (International Society for Optics and Photonics, Bellingham, WA, 2016), p. 99140B
12. B. Guillet, S. Liu, A. Aryan, C. Adamo, D.G. Schlom, L. Mechin, Uncooled phonon noise limited $La_{0.7}Sr_{0.3}MnO_3$ suspended bolometers, in *41st International Conference on Infrared, Millimeter, and Terahertz waves (IRMMW-THz)* (2016), pp. 1–2. https://doi.org/10.1109/IRMMW-THz.2016.7758405

13. A. Rogalski, F. Sizov, Terahertz detectors and focal plane arrays. Opto-Electron. Rev. **19**(3), 346–404 (2011)
14. D. Rutledge, S. Schwarz, Planar multimode detector arrays for infrared and millimeter-wave applications. IEEE J. Quantum Electron. **17**(3), 407–414 (1981). https://doi.org/10.1109/JQE.1981.1071097
15. A. Chynoweth, Dynamic method for measuring the pyroelectric effect with special reference to barium titanate. J. Appl. Phys. **27**(1), 78–84 (1956)
16. J. Cooper, A fast-response pyroelectric thermal detector. J. Sci. Instrum. **39**(9), 467 (1962)
17. S.T. Liu, D. Long, Pyroelectric detectors and materials. Proc. IEEE **66**(1), 14–26 (1978). https://doi.org/10.1109/PROC.1978.10835
18. A. Hossain, M.H. Rashid, Pyroelectric detectors and their applications. IEEE Trans. Ind. Appl. **27**(5), 824–829 (1991). https://doi.org/10.1109/28.90335
19. K.Y. Kim, A.J. Taylor, J.H. Glownia, G. Rodriguez, Coherent control of terahertz supercontinuum generation in ultrafast laser–gas interactions. Nat. Photon. **2**, 605 (2008). https://doi.org/10.1038/nphoton.2008.153
20. I. Kasalynas et al., Design and performance of a room-temperature terahertz detection array for real-time imaging. IEEE J. Sel. Top. Quant. Electron. **14**(2), 363–369 (2008). https://doi.org/10.1109/JSTQE.2007.912629
21. S.O. Kasap, *Principles of Electronic Materials and Devices* (McGraw-Hill, New York, 2006)
22. P. Sarro, H. Yashiro, A. Herwaarden, S. Middelhoek, An integrated thermal infrared sensing array. Sensors Actuators **14**(2), 191–201 (1988)
23. M.J. Golay, A pneumatic infra-red detector. Rev. Sci. Instrum. **18**(5), 357–362 (1947)
24. E. Bründermann, H.-W. Hübers, M.F. Kimmitt, *Terahertz Techniques* (Springer, Berlin, 2012)
25. T. Kenny et al., Micromachined infrared sensors using tunneling displacement transducers. Rev. Sci. Instrum. **67**(1), 112–128 (1996)
26. E. Ledwosinska, T. Szkopek, A. Guermoune, M. Siaj, *Application of Graphene Membrane in Micro-Golay Cell Array (SPIE OPTO)* (SPIE, Bellingham, WA, 2012)
27. D. Yoon, J. Kim, J. Yun, M. Kaynak, B. Tillack, J. Rieh, 300-GHz direct and heterodyne active imagers based on 0.13-μm SiGe HBT technology. IEEE Trans. Terahertz Sci. Technol. **7**(5), 536–545 (2017). https://doi.org/10.1109/TTHZ.2017.2715419
28. H.T. Friis, Noise figures of radio receivers. Proc. IRE **32**(7), 419–422 (1944). https://doi.org/10.1109/JRPROC.1944.232049
29. S. Gambini, N. Pletcher, J.M. Rabaey, Sensitivity analysis for AM detectors. Tech. Rep. UCB/EECS-31, EECS Department, University of California, Berkeley, 2008
30. E. Ojefors, B. Heinemann, U.R. Pfeiffer, Subharmonic 220- and 320-GHz SiGe HBT receiver front-ends. IEEE Trans. Microw. Theory Tech. **60**(5), 1397–1404 (2012). https://doi.org/10.1109/tmtt.2012.2190092
31. E. Brown, Fundamentals of terrestrial millimeter-wave and THz remote sensing. Int. J. High Speed Electron. Syst. **13**(04), 995–1097 (2003)
32. C.K. Walker, *Terahertz Astronomy* (CRC Press, Boca Raton, FL, 2015)
33. F. Braun, On the current conduction through metal sulphides. J Ann. Phys. Chem **153**, 556 (1874)
34. K.S. Champlin, G. Eisenstein, Cutoff frequency of submillimeter Schottky-barrier diodes. IEEE Trans. Microw. Theory Tech. **26**(1), 31–34 (1978). https://doi.org/10.1109/TMTT.1978.1129302
35. W.C.B. Peatman, T.W. Crowe, Design and fabrication of 0.5 micron GaAs Schottky barrier diodes for low-noise terahertz receiver applications. Int. J. Infrared Millimeter Waves **11**(3), 355–365 (1990). https://doi.org/10.1007/bf01010434
36. S.-M. Chen, Y.-K. Fang, F.R. Juang, W.-K. Yeh, C.-P. Chao, H.-C. Tseng, Terahertz Schottky barrier diodes with various isolation designs for advanced radio frequency applications. Thin Solid Films **519**, 471–474 (2010)
37. B. Sharma, *Metal-Semiconductor Schottky Barrier Junctions and their Applications* (Springer Science & Business Media, Berlin, 2013)

38. S. Sankaran, K.O. Kenneth, Schottky barrier diodes for millimeter wave detection in a foundry CMOS process. IEEE Electron Device Lett. **26**(7), 492–494 (2005). https://doi.org/10.1109/LED.2005.851127
39. U.R. Pfeiffer, C. Mishra, R.M. Rassel, S. Pinkett, S.K. Reynolds, Schottky barrier diode circuits in silicon for future millimeter-wave and terahertz applications. IEEE Trans. Microw. Theory Tech. **56**(2), 364–371 (2008)
40. R. Han et al., Active terahertz imaging using Schottky diodes in CMOS: Array and 860-GHz pixel. IEEE J. Solid State Circuits **48**(10), 2296–2308 (2013). https://doi.org/10.1109/JSSC.2013.2269856
41. A. Betz, R. Boreiko, A practical Schottky mixer for 5 THz (Part II), in *Proc. of the 7th Int. Symp. on Space THz Technology*, vol. 503 (1996)
42. A.H. Dayem, R.J. Martin, Quantum interaction of microwave radiation with tunneling between superconductors. Phys. Rev. Lett. **8**(6), 246–248 (1962). https://doi.org/10.1103/PhysRevLett.8.246
43. P.L. Richards, T.M. Shen, R.E. Harris, F.L. Lloyd, Quasiparticle heterodyne mixing in SIS tunnel junctions. Appl. Phys. Lett. **34**(5), 345–347 (1979). https://doi.org/10.1063/1.90782
44. G.J. Dolan, T.G. Phillips, D.P. Woody, Low-noise 115-GHz mixing in superconducting oxide-barrier tunnel junctions. Appl. Phys. Lett. **34**(5), 347–349 (1979). https://doi.org/10.1063/1.90783
45. B.-K. Tan, *Development of Coherent Detector Technologies for Sub-Millimetre Wave Astronomy Observations* (Springer International Publishing, Cham, 2016)
46. M.J. Wengler, Submillimeter-wave detection with superconducting tunnel diodes. Proc. IEEE **80**(11), 1810–1826 (1992)
47. S. Asayama, T. Noguchi, H. Ogawa, A fixed-tuned W-band waveguide SIS mixer with 4.0–7.5 GHz IF. Int. J. Infrared Millimeter Waves **24**(7), 1091–1099 (2003)
48. A. Karpov, D. Miller, F. Rice, J. Stem, B. Bumble, Low noise NbTiN 1.25 THz SIS mixer for Herschel Space Observatory, in *Proc. 16th Int. Symp. Space Terahertz Technol.* (2005), p. 450
49. D.H. Andrews, W.F. Brucksch Jr., W.T. Ziegler, E.R. Blanchard, Attenuated superconductors I. For measuring infra-red radiation. Rev. Sci. Instrum. **13**(7), 281–292 (1942). https://doi.org/10.1063/1.1770037
50. D.W. Floet, J. Baselmans, T. Klapwijk, J. Gao, Resistive transition of niobium superconducting hot-electron bolometer mixers. Appl. Phys. Lett. **73**(19), 2826–2828 (1998)
51. K.D. Irwin, G.C. Hilton, Transition-edge sensors, in *Cryogenic Particle Detection*, (Springer, Berlin, 2005), pp. 63–150
52. D.W. Floet, E. Miedema, T. Klapwijk, J. Gao, Hotspot mixing: A framework for heterodyne mixing in superconducting hot-electron bolometers. Appl. Phys. Lett. **74**(3), 433–435 (1999)
53. W. Zhang et al., Noise temperature and beam pattern of an NbN hot electron bolometer mixer at 5.25 THz. J. Appl. Phys. **108**(9), 093102 (2010). https://doi.org/10.1063/1.3503279
54. S.A. Maas, A GaAs MESFET mixer with very low intermodulation. IEEE Trans. Microw. Theory Tech. **35**(4), 425–429 (1987)
55. E. Ojefors, U.R. Pfeiffer, A. Lisauskas, H.G. Roskos, A 0.65 THz focal-plane array in a quarter-micron CMOS process technology. IEEE J. Solid State Circuits **44**(7), 1968–1976 (2009)
56. M.I. Dyakonov, M.S. Shur, Plasma wave electronics: Novel terahertz devices using two dimensional electron fluid. IEEE Trans. Electron Devices **43**(10), 1640–1645 (1996). https://doi.org/10.1109/16.536809
57. W. Knap et al., Plasma wave detection of sub-terahertz and terahertz radiation by silicon field-effect transistors. Appl. Phys. Lett. **85**(4), 675–677 (2004). https://doi.org/10.1063/1.1775034
58. W. Knap et al., Resonant detection of subterahertz radiation by plasma waves in a submicron field-effect transistor. Appl. Phys. Lett. **80**(18), 3433–3435 (2002). https://doi.org/10.1063/1.1473685
59. A. El Fatimy et al., Resonant and voltage-tunable terahertz detection in In GaAs/InP nanometer transistors. Appl. Phys. Lett. **89**(13), 131926 (2006)

60. F. Caster II, L. Gilreath, S. Pan, Z. Wang, F. Capolino, P. Heydari, Design and analysis of a W-band 9-element imaging array receiver using spatial-overlapping super-pixels in silicon. IEEE J. Solid State Circuits **49**(6), 1317–1332 (2014)
61. S. Emami, C.H. Doan, A.M. Niknejad, R.W. Brodersen, A 60-GHz down-converting CMOS single-gate mixer, in *IEEE Radio Frequency Integrated Circuits (RFIC) Symposium. Digest of Papers*, (IEEE, New York, 2005), pp. 163–166
62. M.T. Terrovitis, R.G. Meyer, Noise in current-commutating CMOS mixers. IEEE J. Solid State Circuits **34**(6), 772–783 (1999). https://doi.org/10.1109/4.766811
63. M.T. Terrovitis, R.G. Meyer, Intermodulation distortion in current-commutating CMOS mixers. IEEE J. Solid State Circuits **35**(10), 1461–1473 (2000). https://doi.org/10.1109/4.871323
64. K. Schmalz, W. Winkler, J. Borngräber, W. Debski, B. Heinemann, J.C. Scheytt, A subharmonic receiver in SiGe technology for 122 GHz sensor applications. IEEE J. Solid State Circuits **45**(9), 1644–1656 (2010). https://doi.org/10.1109/jssc.2010.2051474
65. A. Muller, M. Thiel, H. Irion, H. Ruob, A 122 GHz SiGe active subharmonic mixer, in *European Gallium Arsenide and Other Semiconductor Application Symposium* (2005), pp. 57–60
66. E. Ojefors, U. Pfeiffer, A 94-GHz monolithic front-end for imaging arrays in SiGe:C technology, in *2008 38th European Microwave Conference* (2008), pp. 1449–1452. https://doi.org/10.1109/EUMC.2008.4751739
67. R.M. Kodkani, L.E. Larson, An integrated 50-GHz SiGe sub-harmonic mixer/downconverter with a quadrature ring VCO, in *2007 Topical Meeting on Silicon Monolithic Integrated Circuits in RF Systems* (2007), pp. 223–226. https://doi.org/10.1109/SMIC.2007.322825
68. K. Nimmagadda, G.M. Rebeiz, A 1.9 GHz double-balanced subharmonic mixer for direct conversion receivers, in *2001 IEEE Radio Frequency Integrated Circuits (RFIC) Symposium*, (IEEE, New York, 2001), pp. 253–256
69. B.G. Perumana et al., A SiGe sub-harmonic mixer for millimeter-wave applications, in *2007 European Microwave Integrated Circuit Conference* (2007), pp. 80–83. https://doi.org/10.1109/EMICC.2007.4412652
70. D. Yoon, J. Kim, J.-S. Rieh, A 290-GHz CMOS heterodyne integrated imager, in *2016 IEEE International Symposium on Radio-Frequency Integration Technology (RFIT)*, (IEEE, New York, 2016), pp. 1–3
71. D.-H. Kim, J.-S. Rieh, A SiGe 140-GHz low power G_m-boosted down-conversion mixer, in *Asia-Pacific Microwave Conference Proceedings* (2011), pp. 1126–1129
72. D.H. Kim, J.-S. Rieh, A 135 GHz differential active star mixer in SiGe BiCMOS technology. IEEE Microw. Wirel. Compon. Lett. **22**(8), 409–411 (2012). https://doi.org/10.1109/lmwc.2012.2205909
73. N. Kim, V. Aparin, L.E. Larson, A resistively degenerated wide-band passive mixer with low noise figure and +60dBm IIP2 in 0.18μm CMOS, in *2008 IEEE Radio Frequency Integrated Circuits Symposium* (2008), pp. 185–188. https://doi.org/10.1109/RFIC.2008.4561414
74. X. Li, S. Shekhar, D.J. Allstot, Gm-boosted common-gate LNA and differential colpitts VCO/QVCO in 0.18-μm CMOS. IEEE J. Solid State Circuits **40**(12), 2609–2619 (2005). https://doi.org/10.1109/JSSC.2005.857426
75. I. Kallfass et al., 64 Gbit/s transmission over 850 m fixed wireless link at 240 GHz carrier frequency. J. Infrared Millimeter Terahertz Waves **36**(2), 221–233 (2015)
76. S. Bouvot et al., A D-band passive receiver with 10 dB noise figure for in-situ noise characterization in BiCMOS 55 nm, in *2017 IEEE 17th Topical Meeting on Silicon Monolithic Integrated Circuits in RF Systems (SiRF)*, (IEEE, New York, 2017), pp. 107–110
77. E. Ojefors, J. Grzyb, Y. Zhao, B. Heinemann, B. Tillack, U.R. Pfeiffer, A 820GHz SiGe chipset for terahertz active imaging applications, in *IEEE International Solid-State Circuits Conference* (2011), pp. 224–226
78. Y. Mao, K. Schmalz, J. Borngräber, J.C. Scheytt, A 245 GHz CB LNA in SiGe, in *2011 6th European Microwave Integrated Circuit Conference*, 10–11 Oct 2011 (2011), pp. 224–227

79. S. Kim et al., 300 GHz integrated heterodyne receiver and transmitter with on-chip fundamental local oscillator and mixers. IEEE Trans. Terahertz Sci. Technol. **5**(1), 92–101 (2015). https://doi.org/10.1109/TTHZ.2014.2364454
80. D. Kim, S. Kim, K. Song, J. Kim, J. Yoo, J.-S. Rieh, Characterization of a CMOS 135-GHz low noise amplifier with two different noise measurement methods. J. Semicond. Technol. Sci. **18**(4), 536–540 (2018)
81. Y. Yan et al., Monolithically integrated 200-GHz double-slot antenna and resistive mixers in a GaAs-mHEMT MMIC process. IEEE Trans. Microw. Theory Tech. **59**(10), 2494–2503 (2011). https://doi.org/10.1109/TMTT.2011.2161326
82. D.H. Kim, D. Kim, J.-S. Rieh, A D-band CMOS amplifier with a new dual-frequency interstage matching technique. IEEE Trans. Microw. Theory Tech. **65**(5), 1580–1588 (2017). https://doi.org/10.1109/TMTT.2017.2655508
83. D. Yoon, K. Song, J. Kim, M. Kaynak, B. Tillack, J.-S. Rieh, A D-band active imager in a SiGe HBT technology. J. Infrared Millimeter Terahertz Waves **36**(4), 335–349 (2015)
84. R. Smith, F. Jones, R. Chasmar, *The Detection and Measurement of Infra Red Radiation* (Oxford University Press, Oxford, UK, 1968)
85. K.M.K.H. Leong et al., A 0.85 THz low noise amplifier using InP HEMT transistors. IEEE Microw. Wirel. Compon. Lett. **25**(6), 397–399 (2015). https://doi.org/10.1109/LMWC.2015.2421336
86. P. Kangaslahti et al., Low noise amplifier receivers for millimeter wave atmospheric remote sensing, in *2012 IEEE/MTT-S International Microwave Symposium Digest*, 17–22 June 2012 (2012), pp. 1–3. https://doi.org/10.1109/MWSYM.2012.6259468
87. G. Moschetti et al., A 183 GHz metamorphic HEMT low-noise amplifier with 3.5 dB noise figure. IEEE Microw. Wirel. Compon. Lett. **25**(9), 618–620 (2015). https://doi.org/10.1109/LMWC.2015.2451355
88. Z. Wang, P. Chiang, P. Nazari, C. Wang, Z. Chen, P. Heydari, A CMOS 210-GHz fundamental transceiver with OOK modulation. IEEE J. Solid State Circuits **49**(3), 564–580 (2014). https://doi.org/10.1109/JSSC.2013.2297415
89. K. Schmalz, J. Borngraber, Y. Mao, H. Rucker, R. Weber, A 245 GHz LNA in SiGe technology. IEEE Microw. Wirel. Compon. Lett. **22**(10), 533–535 (2012). https://doi.org/10.1109/LMWC.2012.2218097
90. B. Thomas, A. Maestrini, G. Beaudin, A low-noise fixed-tuned 300–360-GHz sub-harmonic mixer using planar Schottky diodes. IEEE Microw. Wirel. Compon. Lett. **15**(12), 865–867 (2005). https://doi.org/10.1109/LMWC.2005.859992
91. P.J. Sobis, N. Wadefalk, A. Emrich, J. Stake, A broadband, low noise, integrated 340 GHz Schottky diode receiver. IEEE Microw. Wirel. Compon. Lett. **22**(7), 366–368 (2012). https://doi.org/10.1109/LMWC.2012.2202280
92. J. Treuttel et al., A 520–620-GHz Schottky receiver front-end for planetary science and remote sensing with 1070 K–1500 K DSB noise temperature at room temperature. IEEE Trans. Terahertz Sci. Technol. **6**(1), 148–155 (2016). https://doi.org/10.1109/TTHZ.2015.2496421
93. J. Winters, On the capacity of radio communication systems with diversity in a Rayleigh fading environment. IEEE J. Sel. Areas Commun. **5**(5), 871–878 (1987)
94. K. Song, J. Kim, D. Kim, J. Yoo, J.-S. Rieh, A 300-GHz CMOS 7-by-7 detector array for optics-less THz imaging with scan-less target object. J. Infrared Millimeter Terahertz Waves **41**, 1–13 (2020)
95. K. Song, Development of THz Imaging Arrays Based on Si CMOS Technology, Ph.D. Dissertation, Korea University, 2020
96. N. Oda, Uncooled bolometer-type terahertz focal plane array and camera for real-time imaging. C. R. Phys. **11**(7–8), 496–509 (2010)
97. D.L. Donoho, Compressed sensing. IEEE Trans. Inf. Theory **52**(4), 1289–1306 (2006). https://doi.org/10.1109/TIT.2006.871582

98. W.L. Chan, K. Charan, D. Takhar, K.F. Kelly, R.G. Baraniuk, D.M. Mittleman, A single-pixel terahertz imaging system based on compressed sensing. Appl. Phys. Lett. **93**(12), 121105 (2008)
99. H. Sherry et al., A 1kpixel CMOS camera chip for 25fps real-time terahertz imaging applications, in *2012 IEEE International Solid-State Circuits Conference*, (IEEE, New York, 2012), pp. 252–254
100. R. Al Hadi et al., A 1 k-pixel video camera for 0.7–1.1 terahertz imaging applications in 65-nm CMOS. IEEE J. Solid State Circuits **47**(12), 2999–3012 (2012)
101. N. Oda, A.W. Lee, T. Ishi, I. Hosako, Q. Hu, Proposal for real-time terahertz imaging system with palm-size terahertz camera and compact quantum cascade laser, in *Terahertz Physics, Devices, and Systems VI: Advanced Applications in Industry and Defense*, vol. 8363, (International Society for Optics and Photonics, Bellingham, WA, 2012), p. 83630A
102. A. Wootten, A.R. Thompson, The Atacama large millimeter/submillimeter array. Proc. IEEE **97**(8), 1463–1471 (2009)
103. K. Akiyama et al., First M87 event horizon telescope results. I. The shadow of the supermassive black hole. Astrophys. J. Lett. **875**, 1–17 (2019)

Chapter 4
THz Propagation and Related Topics

THz waves generated by signal sources (topic in Chap. 2) will be detected by detectors (topic in Chap. 3) after propagating through a medium. In this chapter, the propagation of THz waves and related topics will be discussed. A popular propagation method in the lower frequency ranges, such as RF and microwave, is through coaxial cables. However, the loss will grow with frequency, and it will reach a point where it become prohibitively large for practical uses. Beyond such a point, various types of waveguides, most notably the metallic rectangular waveguides, will be better suited for the propagation of electromagnetic waves. Nevertheless, they also have their own loss-dominated frequency upper limit. The favorable solution beyond this limit would be the propagation through free space, guided by optical components such as lenses and mirrors. The THz band is situated across the region that prefers the propagation through the waveguides or free space. The two propagation schemes will be the main topics in this chapter and described as individual sections. In the case of propagation through free space, usually in the form of Gaussian beam, a device that converts guided waves (or AC signal) to radiated electromagnetic waves (and vice versa) will be needed. The device, or antenna, will be also treated in this chapter as a separate section with focus on its THz application.

4.1 Gaussian Beam

Probably the simplest optical principles can be provided by the geometrical optics, or ray optics, in which the propagation of electromagnetic waves, or light as in most optics domain, follows a straight-line path in the homogenous medium as well as across the interface between different media that involves reflection and refraction. Many basic optical behaviors of electromagnetic waves can be successfully explained with geometrical optics. While it nicely applies to the cases where the wavelength is much smaller than the dimension of interacting optical components, it cannot explain some phenomena such as diffraction. When the size of the

© Springer Nature Switzerland AG 2021
J.-S. Rieh, *Introduction to Terahertz Electronics*,
https://doi.org/10.1007/978-3-030-51842-4_4

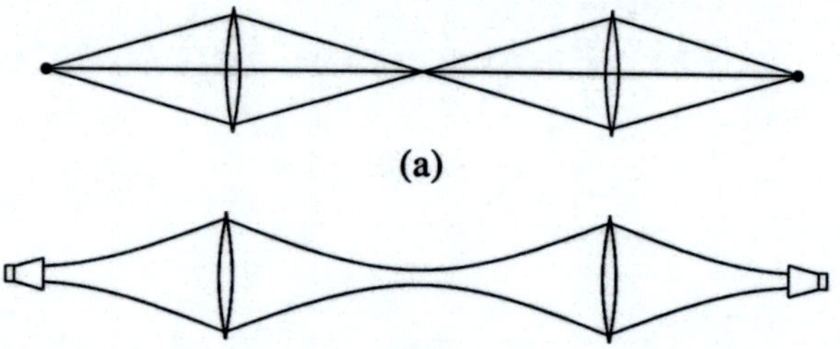

Fig. 4.1 Conceptual comparison of the beam propagation through a couple of lenses: (**a**) Geometrical optics view (from point to point), (**b**) Gaussian beam view (from antenna to antenna)

components becomes comparable to the wavelength, which belongs to the regime of quasi-optics, the geometrical optics comes short of correctly describing the propagation of electromagnetic waves. In those cases, the concept of Gaussian beams has proven to be very useful in describing various propagation cases, including the THz wave propagation. A conceptual comparison can be made with Fig. 4.1. In geometrical optics, a light ray emitted from a point source follows a straight path and can be focused into a point after passing through a lens. On the other hand, in the Gaussian beam picture, a beam radiated from a directional source with a finite dimension gradually spreads out, mainly due to diffraction, and narrows down after passing through a lens. The beam after the lens, however, never converges into a point, but rather forms a minimum beam with a finite radius spread around the point of focus, after which it resumes spreading out. This minimum beam spread is called the beam waist, the size of which depends on various propagation parameters as will be discussed later. A beam waist occurs near the radiation plane of the source as well, if the radiation is effectively coupled into a Gaussian beam. In this section, the formalism of Gaussian beam will be first introduced, and then its properties in free space propagation and the transformation by lens components will be discussed.

4.1.1 Gaussian Beam Basics

For many practical applications, the final expression of Gaussian beam given as a function of various propagation parameters will be certainly useful, while its derivation may not for immediate engineering needs. Apart from the practical causes, however, understanding of Gaussian beam should be greatly enhanced by reviewing its derivation steps, which is presented below.

A formal derivation should start from the time-dependent Maxwell's equations, but a simpler approach may start with the Helmholtz equation, in which the time-dependent term is eliminated assuming a sinusoidal waveform over the time. That is:

$$(\nabla^2 + k^2)E = 0, \tag{4.1}$$

where k is the wave number ($=2\pi/\lambda$), and E is the electric field with the time dependent factor dropped out. Note that the expression will be also valid when the electric field E is replaced by magnetic field H. However, we will stay with the electric field E throughout the analysis and express the Gaussian beam as a function of E. Equation (4.1) is general and valid for any type of sinusoidal waves, including both plane wave and Gaussian beam. For *plane waves*, as is treated in college level physics courses, E does not have x- or y-dependence assuming the wave propagates along z-axis, and the equation is reduced to:

$$\frac{\partial^2 E}{\partial z^2} + k^2 E = 0, \tag{4.2}$$

and E is given as:

$$E = Ae^{-jkz} + Be^{jkz}, \tag{4.3}$$

which is basically a wave with a well-defined phase without variation over the transverse direction. How about Gaussian beams? One of the major properties of Gaussian beam that distinguishes itself from the plane wave is that E has x- and y-dependence, or radial dependence, as well as z-dependence in addition to the normal propagation factor, $\exp(-jkz)$. Hence, E for can be expressed as:

$$E(x,y,z) = u(x,y,z)\exp(-jkz). \tag{4.4}$$

With Eq. (4.4) substituted into the Helmholtz equation of Eq. (4.1), the following equation is obtained for *Gaussian beams* where $\exp(-jkz)$ dependence is factored out:

$$\frac{\partial^2 u}{\partial x^2} + \frac{\partial^2 u}{\partial y^2} + \frac{\partial^2 u}{\partial z^2} - 2jk\frac{\partial u}{\partial z} = 0. \tag{4.5}$$

This equation is called the reduced wave equation, where both t-dependent and z-dependence sinusoidal factors, $e^{j\omega t}$ and e^{-jkz}, are taken out. If we assume the z-dependence, which is mostly due to the diffraction, is small, the third term in Eq. (4.5) can be ignored. This is called the paraxial approximation, which leads to:

$$\frac{\partial^2 u}{\partial x^2} + \frac{\partial^2 u}{\partial y^2} - 2jk\frac{\partial u}{\partial z} = 0. \tag{4.6}$$

Equation (4.6) is called the paraxial wave equation and the derivation of Gaussian beam usually starts from here. The paraxial approximation may not appear obvious to many readers, but there are many different ways to prove its validity in practical cases. Here, we will make yet another assumption that u has only radial dependence, invariant over the angle change. In fact, this is not the general case, and the angular dependence should be considered for higher order modes. However, we will confine the current analysis to this axially symmetric case only, which will correspond to lower order modes including the fundamental mode. In this case, u will be governed by the axially symmetric paraxial wave equation, which is better presented in cylindrical coordinate such that:

$$\frac{\partial^2 u}{\partial r^2} + \frac{1}{r}\frac{\partial u}{\partial r} - 2jk\frac{\partial u}{\partial z} = 0, \tag{4.7}$$

where r is the radial distance from the axis of propagation (z-axis). Note that there is no angular dependence in Eq. (4.7). A nice trial solution to the axially symmetric paraxial equation can be written as follows:

$$u(r, z) = A(z)\exp\left[-jk\frac{r^2}{(2qz)}\right]. \tag{4.8}$$

A and q are parameters with complex values that have only z-dependence and their actual expressions are to be obtained shortly after. The exponent in Eq. (4.8) should contain the information on the radius of curvature of the beam and the decay of the field magnitude in the radial direction. Hence, Eq. (4.8) can be expressed in terms of parameters with more direct physical meaning such that:

$$u(r, z) = A(z)\exp\left(-jk\frac{r^2}{2R(z)} - \frac{r^2}{\omega^2(z)}\right) = A(z)\exp\left(-j\frac{\pi r^2}{\lambda R(z)} - \frac{r^2}{\omega^2(z)}\right), \tag{4.9}$$

which is obtained by assuming the following relation for $q(z)$:

$$\frac{1}{q(z)} = \frac{1}{R(z)} - j\frac{\lambda}{\pi\omega^2(z)}, \tag{4.10}$$

where $R(z)$ is the radius of curvature (for the equiphase surface of the beam) and $\omega(z)$ is the beam radius, which is defined as the radial distance where the electric field magnitude drops to $1/e$ of the value on the propagation axis. The factor $\pi r^2/\lambda R$ is the off-axis phase shift observed at a radial distance of r, which is a result of the finite radius of curvature of Gaussian beam. The detailed procedure to arrive at Eqs. (4.9) and (4.10) will not be given here, but can be found elsewhere (for example in [1]). The physical meaning of q may not appear as straightforward as those of R or ω to the readers, but it does contain lots of related key information in it. It is called in various ways such as the complex beam parameter, Gaussian beam parameter, or simply beam parameter. Sometimes it is also called the complex radius of curvature, indicating it is a complex generalization of R because it is reduced to R if purely real, which corresponds to the spherical wave case. By the same token, it can be stated that the Gaussian beam is a spherical wave with a complex radius of curvature of q [2].

Equation (4.9) already shows the r-dependence of $u(r, z)$ as the Gaussian function (i.e., including a factor of $\exp(-(x/x_0)^2)$, but still is not the final form and will be developed further below. Before moving forward, however, a few practical discussions will be useful. With the help from the boundary condition of q, R and ω can be shown as an explicit function of z, such that:

$$R(z) = z + \frac{1}{z}\left(\frac{\pi\omega_0^2}{\lambda}\right)^2 \tag{4.11}$$

$$\omega(z) = \omega_0\sqrt{1 + \left(\frac{\lambda z}{\pi\omega_0^2}\right)^2}, \tag{4.12}$$

where ω_0 is the beam radius at $z = 0$, which is also called the beam waist since it is the minimum beam radius available as mentioned earlier. One may note that both expressions include a common factor of $\pi\omega_0^2/\lambda$. This factor is called the confocal distance z_c with the following definition:

$$z_c = \frac{\pi\omega_0^2}{\lambda}. \tag{4.13}$$

It can be interpreted as a distance for a beam to travel from a beam waist until the beam radius grows by a factor of $\sqrt{2}$ or the beam area is doubled, according to Eq. (4.12). Or, probably with a greater physical significance, it can be regarded an approximate boundary between the near-field and far-field ranges. For this reason, it is sometimes called Rayleigh range or Rayleigh length, especially in the laser community. With z_c defined, more compact expressions can be obtained for R and ω:

$$R(z) = z + \frac{z_c^2}{z} \tag{4.14}$$

$$\omega(z) = \omega_0\sqrt{1 + \left(\frac{z}{z_c}\right)^2}. \tag{4.15}$$

Now let us get back to the expression for the Gaussian beam given as Eq. (4.9). The only remaining undetermined is A, a z-dependent complex coefficient of u. From the relation between A and q, which can be obtained by substituting the trial solution (Eq. 4.8) into the paraxial equation (Eq. 4.6), one can show that the following relation can be reached:

$$A(z) = A(0)\frac{\omega_0}{\omega}\exp\left(\, j\phi_0\right), \tag{4.16}$$

where ϕ_0 is the Gaussian beam phase shift, or the Gaussian beam phase slippage, which is given as:

$$\phi_0 = \tan^{-1}\left(\frac{\lambda z}{\pi \omega_0^2}\right) = \tan^{-1}\left(\frac{z}{z_c}\right). \tag{4.17}$$

The positive value of ϕ_0 implies that the phase of Gaussian beam on the propagation axis slightly lags the plane waves. This on-axis phase shift should not be confused with the off-axis phase shift, $\pi r^2/\lambda R$, as described above. With the expression for A as obtained in Eq. (4.16), the final form of u is given as:

$$u(r,z) = A(0)\frac{\omega_0}{\omega}\exp\left(\frac{-r^2}{\omega^2} + j\phi_0 - \frac{j\pi r^2}{\lambda R}\right). \tag{4.18}$$

Then, the electric field is given as:

$$E(r,z) = u(r,z)\exp(-jkz) = A(0)\frac{\omega_0}{\omega}\exp\left(\frac{-r^2}{\omega^2} - jkz + j\phi_0 - \frac{j\pi r^2}{\lambda R}\right). \tag{4.19}$$

Equation (4.19) could be regarded as the final form of the Gaussian beam, but one may wonder about $A(0)$, the value of A at the origin, which still remains undetermined. As the last step, a normalization can be made to finalize the form. Here, the normalization condition suggested in [1] will be taken, which assumes that the integration of the square of the electric field over the beam area equals unity, or $\int 2\pi r|E|^2 dr = 1$. Although the unit is not exactly physical, the coefficient for the electric field can be determined with this normalization, leading to:

$$E(r,z) = \frac{\sqrt{2/\pi}}{\omega}\exp\left(\frac{-r^2}{\omega^2} - jkz + j\phi_0 - \frac{j\pi r^2}{\lambda R}\right). \tag{4.20}$$

This is the final expression for the Gaussian beam in the fundamental mode. The interpretation of this form will be discussed in the following section. Higher order solutions to Gaussian beams can be found in other literatures that focus on this topic.

4.1.2 Gaussian Beam in Free Space

Understanding the behavior of Gaussian beams in free space (or, more precisely, in homogeneous media) can start with inspecting the expression for Gaussian beam as is given in Eq. (4.20). There are four terms in the exponent, and the beam behavior is dictated by these four exponential factors. The first term relates the magnitude, and the last three terms concern the phase of the Gaussian beam, each of which bears its own physical meaning as follows:

- $-\frac{r^2}{\omega^2}$: A term indicating that the magnitude of the field decays with the Gaussian function along the radial direction, from which the name of the beam is derived.

- $-jkz$: A term corresponding to the on-axis phase progress as of normal plane waves. If not for the other two phase terms, the phase of Gaussian beam would have been the same as that of plane waves with this term.
- $j\phi_0$: A term representing the on-axis phase shift against the plane waves, with the definition of ϕ_0 given in Eq. (4.17).
- $-\frac{j\pi r^2}{\lambda R}$: A term denoting the off-axis phase shift against the on-axis wave front, which results from the finite radius of curvature of the equiphase surface of Gaussian beam.

In terms of these properties as indicated by the expression, together with the implicit z-dependence of the beam radius ω and the radius of curvature R, the properties of Gaussian beam propagating in free space will be now discussed. The discussion will deal with its radial dependence (r-dependence) first and then its axial dependence (z-dependence).

The most prominent radial dependence of Gaussian beam is the exponential decay of the field magnitude driven by the Gaussian function. To better show the dependence, the relative magnitude of the field against the on-axis value can be obtained by normalization:

$$\frac{|E(r)|}{|E(0)|} = \exp\left(-\left(\frac{r}{\omega}\right)^2\right), \tag{4.21}$$

where its z-dependence is not explicitly shown because the profile is governed by the ratio r/ω regardless the position on the z-axis (as well as the actual value of ω). While the electrical field is the main parameter we are interested in, the radial spread of the beam may be better understood in terms of the power from a practical viewpoint. For this purpose, the relative power normalized over the on-axis value is given in Eq. (4.22). In fact, this ratio is called the edge taper T_e, serving as a useful parameter to present the radial spread profile of Gaussian beams.

$$T_e = \frac{P(r)}{P(0)} = \frac{|E(r)|^2}{|E(0)|^2} = \exp\left(-2\left(\frac{r}{\omega}\right)^2\right). \tag{4.22}$$

One can easily notice that only difference from the normalized field is the factor of 2 introduced in the exponent, indicating a faster decay of the normalized power. The edge taper is plotted in Fig. 4.2 as a function of the radial distance normalized to the beam waist, or r/ω. An inspection of the plot shows that T_e is reduced by half at $r/\omega = 0.589$, a value that corresponds to HWHM (half-width-half-maximum) of Gaussian beams. T_e is further reduced down to 0.135 at $r/\omega = 1$, which implies that a substantial power level is still maintained at the beam radius. But it will quickly fade out and become negligibly small as r grows, as indicated by $T_e = 0.0111$ and 0.0003 at $r/\omega = 1.5$ and 2, respectively (outside the range of Fig. 4.2). The phase of the field also varies over the radial distance, a clear difference from the plane wave, in which case the same phase is retained over the any direction perpendicular to the

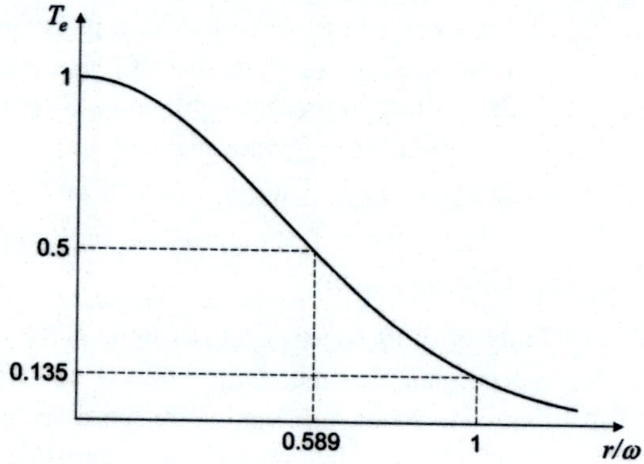

Fig. 4.2 Edge taper T_e shown as a function of the radial distance r normalized to the beam waist ω

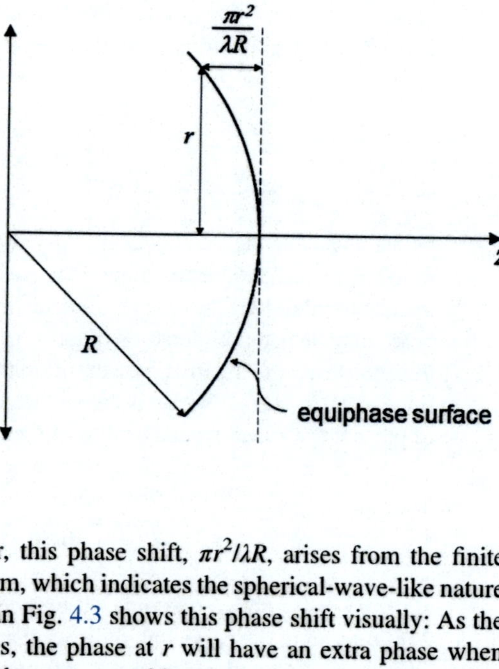

Fig. 4.3 Gaussian beam phase shift (off-axis) due to the finite radius of curvature of the equiphase surface

propagation. As mentioned earlier, this phase shift, $\pi r^2/\lambda R$, arises from the finite radius of curvature of Gaussian beam, which indicates the spherical-wave-like nature that exists in the beam. A cartoon in Fig. 4.3 shows this phase shift visually: As the wave front sets back as r increases, the phase at r will have an extra phase when compared to the on-axis case with the same z position.

Now let us consider the axial dependence of the Gaussian beam along the z-axis. The dominating feature in this consideration is the z-dependence of the radius of curvature R and beam radius ω. Based on Eqs. (4.14) and (4.15), Fig. 4.4 shows how R and ω develops as the beam propagates along z-direction. The radius of curvature, which is infinite at $z = 0$, exhibits an initial reduction down to a minimum value of $2z_c$ at $z = z_c$, then snaps back upward and keeps increasing afterwards. Eventually, the curve converges to a straight line of $R = z$, as the second term in Eq. (4.14) becomes negligible. This behavior reveals an important aspect of the beam: After a long-distance propagation, a Gaussian beam can be approximated as a spherical

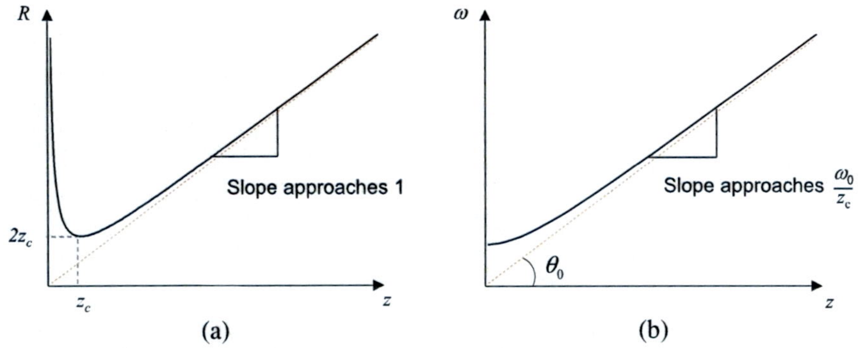

Fig. 4.4 Dependence of beam parameters on the propagation distance z: (**a**) Radius of curvature R, (**b**) Beam waist ω

wave centered at the origin, as indicated by the radius of curvature that equals z. This basically means that the source can be regarded as a point source for a very far-field location along the axis, which makes good sense. The development of the beam radius ω over the propagation would be of more practical interest to those readers who work on optical setups to manipulate Gaussian beams. As shown in Fig. 4.4b, the beam waist gradually spreads out with a monotonic increase, which is mainly a diffraction effect caused by the source aperture size being comparable to the wavelength. After a long propagation, ω eventually converges to a linear line with a slope of ω_0/z_c, as the term z/z_c in the square root in Eq. (4.15) becomes much larger than 1. This indicates that, when far away from the source, the beam spreads in a linear fashion, from which we can define a beam spread angle, or the far-field divergence angle as is officially termed:

$$\theta_0 \approx \tan^{-1}\left(\frac{\omega_0}{z_c}\right) = \tan^{-1}\left(\frac{\lambda}{\pi\omega_0}\right). \tag{4.23}$$

One useful observation is that the angle will be larger if the aperture size of the source is smaller in terms of the wavelength, assuming the minimum beam waist ω_0 is about the same size as the aperture. This also aligns with our common sense because a stronger diffraction effect would be expected for a smaller aperture, leading to a faster beam spread. The Gaussian beam phase shift, $\phi_0 = \tan^{-1}(z/z_c)$, is also given as a function of z. Being the on-axis phase delay experienced by the Gaussian beam against the plane wave, it grows with the beam propagation and eventually converges to $\pi/2$ at very far-field locations. Hence, there would be a maximum phase shift for Gaussian beams, which is a quarter-wavelength. A graphical representation of the properties of Gaussian beam propagation based on the discussion above is shown in Fig. 4.5.

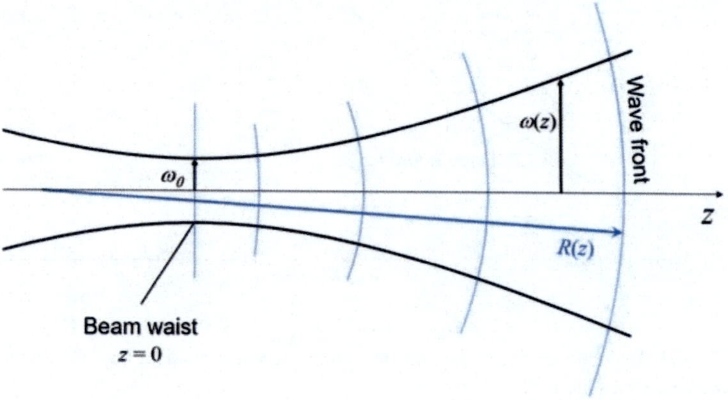

Fig. 4.5 Radius of curvature R and beam width ω of Gaussian beam along the propagation axis

Fig. 4.6 Definition of r and r' for the ray transfer matrices

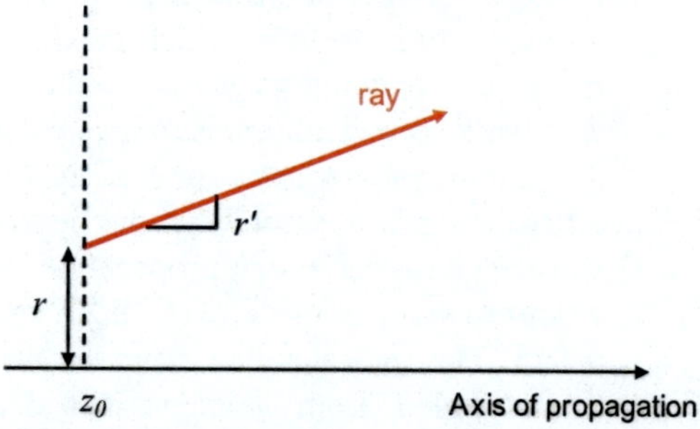

4.1.3 *Gaussian Beam Transformation by Lenses*

With the understanding of Gaussian beam in the free space as reviewed in the previous section, it will be useful to observe the way Gaussian beams are coupled with optical or electromagnetic components. In particular, the coupling between Gaussian beams and lenses is of high importance as it is widely practiced in practical applications, and thus will be discussed in this section. The transformation of Gaussian beams with lenses can be nicely described by introducing the ray transfer matrices that originate from the geometrical optics. With this fact given, in this subsection, the ray transfer matrix will be first introduced with various example cases, and then its application to Gaussian beam transformation by lenses will be discussed.

4.1.3.1 Ray Transfer Matrix (RTM)

In the frame of geometrical optics, a ray can be described by two parameters, the position from the propagation axis r and the angle against the axis r' as depicted in Fig. 4.6. Hence, naturally, the transformation of a ray by an optical component can be described by a 2 × 2 matrix, which is called the ray transfer matrix (RTM) or *ABCD* matrix:

$$\begin{bmatrix} r_{\text{out}} \\ r'_{\text{out}} \end{bmatrix} = \begin{bmatrix} A & B \\ C & D \end{bmatrix} \begin{bmatrix} r_{\text{in}} \\ r'_{\text{in}} \end{bmatrix}, \tag{4.24}$$

where the subscripts "in" and "out" represent input and output, respectively. This *ABCD* matrix configuration allows a convenient description of the commonly encountered cases of the cascaded multiple optical components, which can be simply represented by a progressive multiple product of the corresponding matrices. It is noted that the determinant of the ray transfer matrix is unity, or $\|ABCD\| = 1$, if the starting and ending points are with the same refractive index, a property that will be used later in the beam analysis. A few example cases relevant to practical cases with lens transformation are introduced below.

The simplest case would be propagation through a homogeneous medium with a length of *L*, as shown in Fig. 4.7a. In this case, one can see that very simple relations will be held between input and output rays such that:

$$r_{\text{out}} = r_{\text{in}} + r'_{\text{in}} L, \tag{4.25}$$

$$r'_{\text{out}} = r'_{\text{in}}. \tag{4.26}$$

This pair of relations can be squeezed into one matrix expression:

$$\begin{bmatrix} r_{\text{out}} \\ r'_{\text{out}} \end{bmatrix} = \begin{bmatrix} 1 & L \\ 0 & 1 \end{bmatrix} \begin{bmatrix} r_{\text{in}} \\ r'_{\text{in}} \end{bmatrix}. \tag{4.27}$$

So we can define a matrix for a free space propagation as:

$$\mathbf{M}_{\text{free space}} = \begin{bmatrix} 1 & L \\ 0 & 1 \end{bmatrix}. \tag{4.28}$$

Note that this matrix is valid regardless the refractive index of the propagation medium.

4.1.3.2 RTM for Interface

Another important case is the curved interface between media with different refractive indices, say, n_1 and n_2, as is described in Fig. 4.7b. The corresponding ray transfer matrix $\mathbf{M}_{\text{interface}}$ can be obtained by considering the angles shown in the figure. If we assume the incident angle is small, Snell's law at the interface is obtained as:

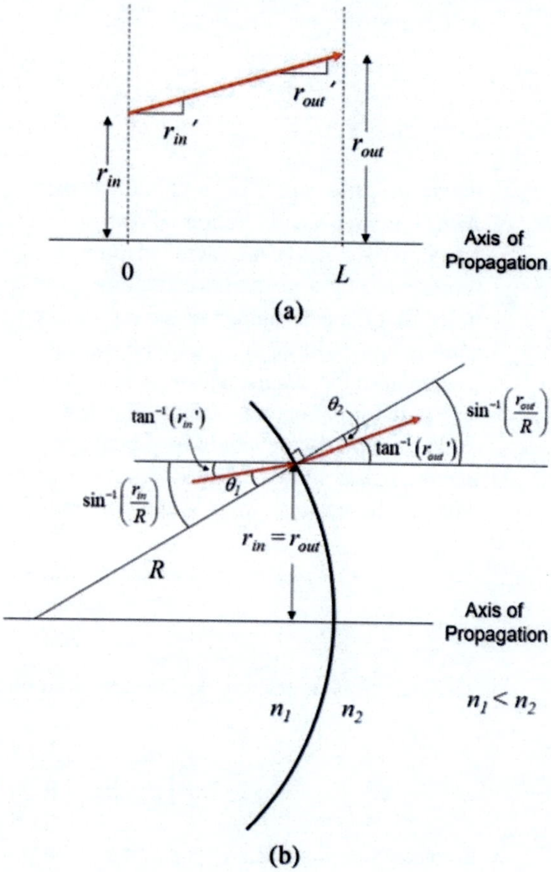

Fig. 4.7 Ray propagation: (**a**) In a homogeneous medium, (**b**) Across a curved interface between media with different refractive indices n_1 and n_2

$$\frac{n_2}{n_1} = \frac{\sin\theta_1}{\sin\theta_2} \approx \frac{\theta_1}{\theta_2} \quad \text{or} \quad n_1\theta_1 \approx n_2\theta_2. \tag{4.29}$$

With the angles shown in Fig. 4.7b, Eq. (4.29) becomes:

$$n_1\left(\sin^{-1}\left(\frac{r_{\text{in}}}{R}\right) - \tan^{-1}(r'_{\text{in}})\right) = n_2\left(\sin^{-1}\left(\frac{r_{\text{out}}}{R}\right) - \tan^{-1}(r'_{\text{out}})\right). \tag{4.30}$$

Based on the equality $r_{\text{in}} = r_{\text{out}}$ at the interface and the assumption that the angle components included in the expression are small, Eq. (4.30) is reduced to:

$$n_1\left(\frac{r_{\text{in}}}{R} - r'_{\text{in}}\right) = n_2\left(\frac{r_{\text{in}}}{R} - r'_{\text{out}}\right) \quad \text{or} \quad r'_{\text{out}} = \frac{n_2 - n_1}{n_2}\frac{1}{R}r_{\text{in}} + \frac{n_1}{n_2}r'_{\text{in}}, \tag{4.31}$$

where the following approximations were used: $\sin^{-1}x \approx \tan^{-1}x \approx x$ for small x. Based on these relations, the transfer matrix becomes:

$$\mathbf{M}_{\text{interface}} = \begin{bmatrix} 1 & 0 \\ \dfrac{n_2 - n_1}{n_2}\dfrac{1}{R} & \dfrac{n_1}{n_2} \end{bmatrix}, \tag{4.32}$$

where near-normal and near-parallel incident beam was assumed. For an arbitrary incident angle, one can find information from other references such as [3]. Note that the sign of the radius of curvature is defined so that it is positive for the case shown in Fig. 4.7b (i.e., $R > 0$ if concave seen from left). Of course, this matrix is valid for a negative radius of curvature as well ($R < 0$, convex seen from left), and is also true for a flat interface where the radius of curvature is infinity ($R = \infty$).

4.1.3.3 RTM for Lenses

With these two basic ray transfer matrices, we are now ready to work on matrices for lenses. As a simpler case, we can first consider a thin lens where the propagation interior to the lens can be neglected. With this simplification, a thin lens can be considered as two interfaces with different radius of curvatures located in immediate proximity. Accordingly, the ray transfer matrix for a thin lens can be obtained by applying Eq. (4.32) twice with different radius of curvatures, say R_1 and R_2, in sequence. Then the matrix becomes:

$$\begin{aligned} \mathbf{M}_{\text{thin lens}} &= \begin{bmatrix} 1 & 0 \\ \dfrac{n_1 - n_2}{n_1}\dfrac{1}{R_2} & \dfrac{n_2}{n_1} \end{bmatrix} \begin{bmatrix} 1 & 0 \\ \dfrac{n_2 - n_1}{n_2}\dfrac{1}{R_1} & \dfrac{n_1}{n_2} \end{bmatrix} \\ &= \begin{bmatrix} 1 & 0 \\ \dfrac{n_2 - n_1}{n_1}\left(\dfrac{1}{R_1} - \dfrac{1}{R_2}\right) & 1 \end{bmatrix}. \end{aligned} \tag{4.33}$$

Although a convex lens was assumed in Fig. 4.8a, Eq. (4.33) is general and can be applied to other cases with arbitrary two radius of curvatures. Also note that when multiple optical components are cascaded, the matrix for the first interacting component will appear on the rightmost side of the matrix product since the matrix application starts from the *right*. As we follow the convention of beam propagation from the *left*, the readers need to be careful about the order of matrices in the matrix product. It is also interesting to note that we can obtain a relation between the focal length f and its radius of curvature and refractive index for thin lenses with the help of Eq. (4.33), because it is known that the ray transfer matrix of a thin lens can also be expressed as a function of f as follows:

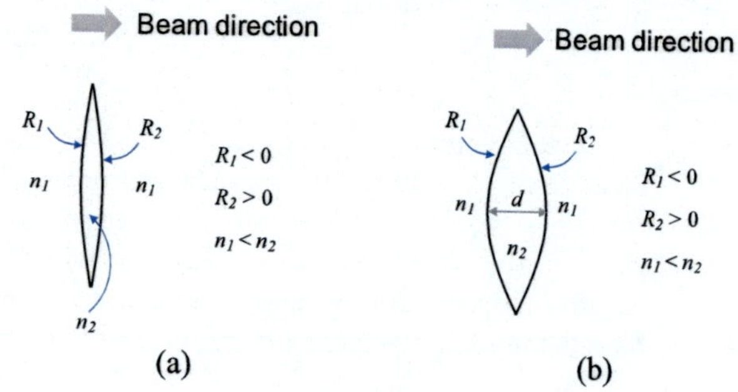

Fig. 4.8 (**a**) Thin convex lens assumed for Eq. (4.33). (**b**) Thick convex lens assumed for Eq. (4.36)

$$\mathbf{M}_{\text{thin lens}} = \begin{bmatrix} 1 & 0 \\ -\frac{1}{f} & 1 \end{bmatrix}. \tag{4.34}$$

The procedure to arrive at Eq. (4.34) would be a good chance for the readers to be reminded of the basic optical principles related to the focal length of a lens. The four matrix elements for a thin lens can be decided if we consider two special cases shown in Fig. 4.9, where the incident beam is parallel to the propagation axis (Fig. 4.9a) and the exiting beam is parallel to the axis (Fig. 4.9b). In either case, the equality $r_{\text{in}} = r_{\text{out}}$ is valid, leading to the first-row elements of 1 and 0. To determine the second row, recall a basic principle that a parallel beam incident on a lens will converge at the focal length f on the other side of the lens (if we assume a convex lens without loss of generality). Then, as is clear from Fig. 4.9a, we obtain $r'_{\text{in}} = 0$ and $r'_{\text{out}} = -r_{\text{in}}/f$. A reciprocal statement is that a parallel exiting beam will be obtained if a source is located at the focal length on the opposite side. With the help of Fig. 4.9b, one can infer $r'_{\text{out}} = 0$ and $r'_{\text{in}} = r_{\text{in}}/f$. From these relations, one can obtain the second-row elements of $-1/f$ and 1, completing the matrix shown in Eq. (4.34). It is noted that Eq. (4.34) is valid for an arbitrary incident angle although the two special cases shown in Fig. 4.9 were used for the derivation above. Now, comparing Eqs. (4.33) and (4.34), one can obtain the focal length f of a thin lens as a function of geometrical parameters (R_1 and R_2) and material parameters (n_1 and n_2) as:

$$\frac{1}{f} = \frac{n_2 - n_1}{n_1}\left(\frac{1}{R_2} - \frac{1}{R_1}\right). \tag{4.35}$$

For the analysis of thick lenses as is shown Fig. 4.8b, we need to consider the finite thickness of the lens and propagation across it. It can be achieved by inserting the length matrix of Eq. (4.28) between the two interface matrices that appear in Eq. (4.33). Assuming the lens thickness as d, the matrix becomes:

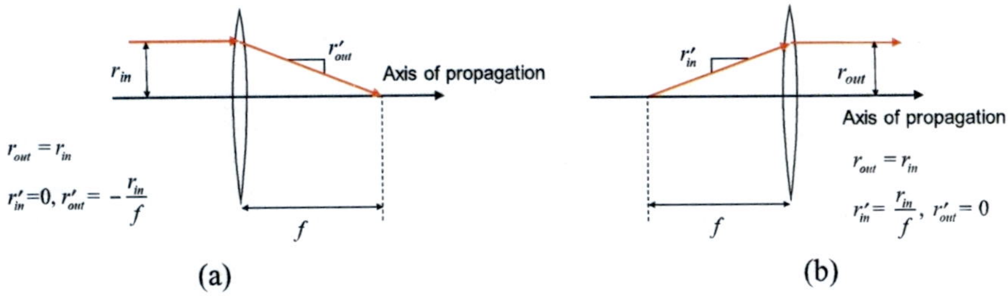

Fig. 4.9 Two cases of a ray passing through a focal point of a thin lens: (**a**) *Incident* beam is parallel to the propagation axis, (**b**) *Exiting* beam is parallel to the propagation axis

$$\mathbf{M}_{\text{thick lens}} = \begin{bmatrix} 1 & 0 \\ \frac{n_1 - n_2}{n_1}\frac{1}{R_2} & \frac{n_2}{n_1} \end{bmatrix} \begin{bmatrix} 1 & d \\ 0 & 1 \end{bmatrix} \begin{bmatrix} 1 & 0 \\ \frac{n_2 - n_1}{n_2}\frac{1}{R_1} & \frac{n_1}{n_2} \end{bmatrix} = \begin{bmatrix} 1 + \frac{n_2 - n_1}{n_2}\frac{d}{R_1} & \frac{n_1}{n_2}d \\ -\frac{1}{f} - \frac{(n_2 - n_1)^2}{n_1 n_2}\frac{d}{R_1 R_2} & 1 + \frac{n_1 - n_2}{n_2}\frac{d}{R_2} \end{bmatrix}. \tag{4.36}$$

One can see that Eq. (4.36) approaches Eq. (4.34) for small d. In view of the complexity in Eq. (4.36), we will maintain the assumption of the thin lens for remaining part of this discussion, which will provide reasonably accurate results.

4.1.3.4 Application of RTM to Gaussian Beam

Up to this point, we have walked through the ray transfer matrices for various practical cases. Now, we will proceed to combine this matrix representation with the Gaussian beam scheme, so that the transformation of Gaussian beam by an optical component can be described by the corresponding ray transfer matrix. The first task will be relating the input and output complex beam parameters of an optical component, q_{in} and q_{out}, in terms of the matrix elements of the optical component, A, B, C, and D. From the definition of the ray transfer matrix shown in Eq. (4.24), we have $r_{\text{out}} = Ar_{\text{in}} + Br'_{\text{in}}$ and $r'_{\text{out}} = Cr_{\text{in}} + Dr'_{\text{in}}$. If we take the ratio of these two expressions,

$$\frac{r_{\text{out}}}{r'_{\text{out}}} = \frac{Ar_{\text{in}} + Br'_{\text{in}}}{Cr_{\text{in}} + Dr'_{\text{in}}}. \tag{4.37}$$

Now, from Fig. 4.10, we can see that the radius of curvature of Gaussian beam can be simply given as r/r' in the far field. Hence, Eq. (4.37) can be converted to an equation relating R_{in} and R_{out} as follows:

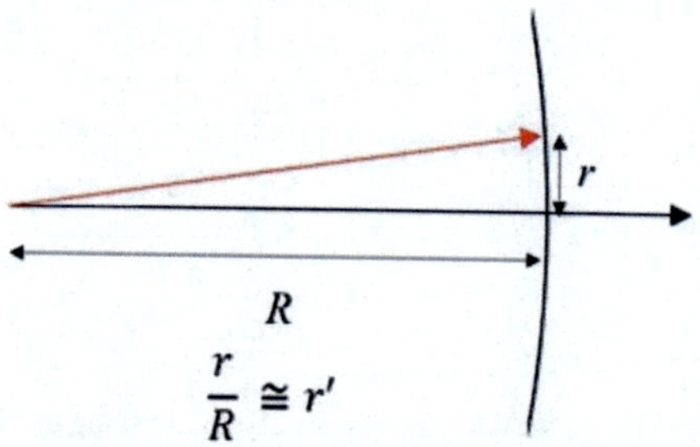

Fig. 4.10 Radius of curvature R of Gaussian beam and its relation with ray parameters r and r'

$$R_{\mathrm{out}} = \frac{AR_{\mathrm{in}} + B}{CR_{\mathrm{in}} + D}. \tag{4.38}$$

Generalization of Eq. (4.38) can be obtained by replacing the radius of curvatures by the complex radius of curvatures, or the complex beam parameters [4]:

$$q_{\mathrm{out}} = \frac{Aq_{\mathrm{in}} + B}{Cq_{\mathrm{in}} + D}. \tag{4.39}$$

Given Eq. (4.39), we are now ready to take on the transformation of Gaussian beam by a thin lens. As is depicted in Fig. 4.11, let us assume the input beam waist, $\omega_{0\mathrm{in}}$, is located d_{in} away from the lens and the output beam waist, $\omega_{0\mathrm{out}}$, is located at d_{out}. From a practical point of view, it is important to be able to express $\omega_{0\mathrm{out}}$ and d_{out} as a function of $\omega_{0\mathrm{in}}$ and d_{in} as well as the ray transfer matrix elements of the lens. This is the short-term object of the following analysis. The ray transfer matrix of the optical system described in Fig. 4.11 is:

$$\begin{aligned} \mathbf{M} &= \begin{bmatrix} 1 & d_{\mathrm{out}} \\ 0 & 1 \end{bmatrix} \begin{bmatrix} 1 & 0 \\ -\frac{1}{f} & 1 \end{bmatrix} \begin{bmatrix} 1 & d_{\mathrm{in}} \\ 0 & 1 \end{bmatrix} \\ &= \begin{bmatrix} 1 - \frac{d_{\mathrm{out}}}{f} & d_{\mathrm{in}} + d_{\mathrm{out}}\left(1 - \frac{d_{\mathrm{in}}}{f}\right) \\ -\frac{1}{f} & 1 - \frac{d_{\mathrm{in}}}{f} \end{bmatrix}. \end{aligned} \tag{4.40}$$

If the matrix elements of Eq. (4.40) is substituted into Eq. (4.39), we obtain:

$$q_{out} = \frac{(1 - d_{out}/f)jz_c + d_{in} + d_{out}(1 - d_{in}/f)}{(1 - d_{\mathrm{in}}/f) - jz_c/f}. \tag{4.41}$$

where the relation $q_{\mathrm{in}} = jz_c$ is used, which is valid because R is infinite at the input beam waist location (recall Eq. (4.10)). It would be worthwhile to note that Eq. (4.41) is in fact valid for any point located right to the lens in Fig. 4.11 by

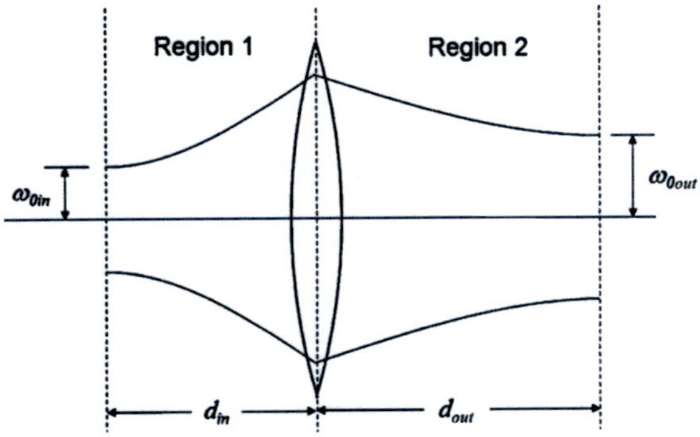

Fig. 4.11 Optical system considered to describe the Gaussian beam transformation by a thin lens

substituting the distance from the lens for d_{out} in the expression, although we will confine our discussion to d_{out} where the output beam waist is located.

By examining the real and imaginary parts of $1/q_{\text{out}}$, the inverse of Eq. (4.41), we can retrieve two constraints that can be used to obtain d_{out} and $\omega_{0\text{out}}$. First, at the output beam waist, the radius of curvature R is again infinity, leaving the real part of $1/q_{\text{out}}$ zero. Based on Eq. (4.41), one can derive:

$$\frac{d_{\text{out}}}{f} = 1 + \frac{d_{\text{in}}/f - 1}{(d_{\text{in}}/f - 1)^2 + (z_c/f)^2}. \tag{4.42}$$

Hence, d_{out} can be expressed as a function of d_{in} and z_c as well as f, which will be more intuitively understood if d_{out}, d_{in}, and z_c are shown normalized to f as is the case in Eq. (4.42). Second, the imaginary part of $1/q_{\text{out}}$ becomes $-\lambda/\pi\omega_{0\text{out}}^2$ at the output beam waist (see Eq. 4.10). This relation can be equally expressed as $\text{Im}(1/q_{\text{out}}) = -(\omega_{0\text{in}}/\omega_{0\text{out}})^2/z_c$ since z_c was defined with respect to the input beam waist $\omega_{0\text{in}}$ such that $z_c = \pi\omega_{0\text{in}}^2/\lambda$ (see Eq. 4.13). With this relation, along with $\|ABCD\| = 1$, which is valid as the refractive indices are the same in the regions d_{in} and d_{out} belong to, one can obtain the following:

$$\omega_{0\text{out}} = \frac{\omega_{0\text{in}}}{\sqrt{(d_{\text{in}}/f - 1)^2 + (z_c/f)^2}}. \tag{4.43}$$

Therefore, we can express the output beam waist in terms of d_{in} and z_c as well as f. Occasionally, one will find a parameter called the "system magnification" useful, which is defined as the ratio between output and input beam waist:

$$\mathfrak{M} \equiv \frac{\omega_{0\text{out}}}{\omega_{0\text{in}}} = \frac{1}{\sqrt{(d_{\text{in}}/f - 1)^2 + (z_c/f)^2}}. \tag{4.44}$$

One should not confuse this *system magnification* with the other more popular definition of the *magnification* in optics, which is the ratio between the object and image sizes.

Fig. 4.12 Relations between key parameters with various z_c values for a transformation by a thin lens with a focal length f: **(a)** Relation between d_{in} and d_{out}, **(b)** Relation between magnification (the ratio of input and output beam waists) and d_{in}

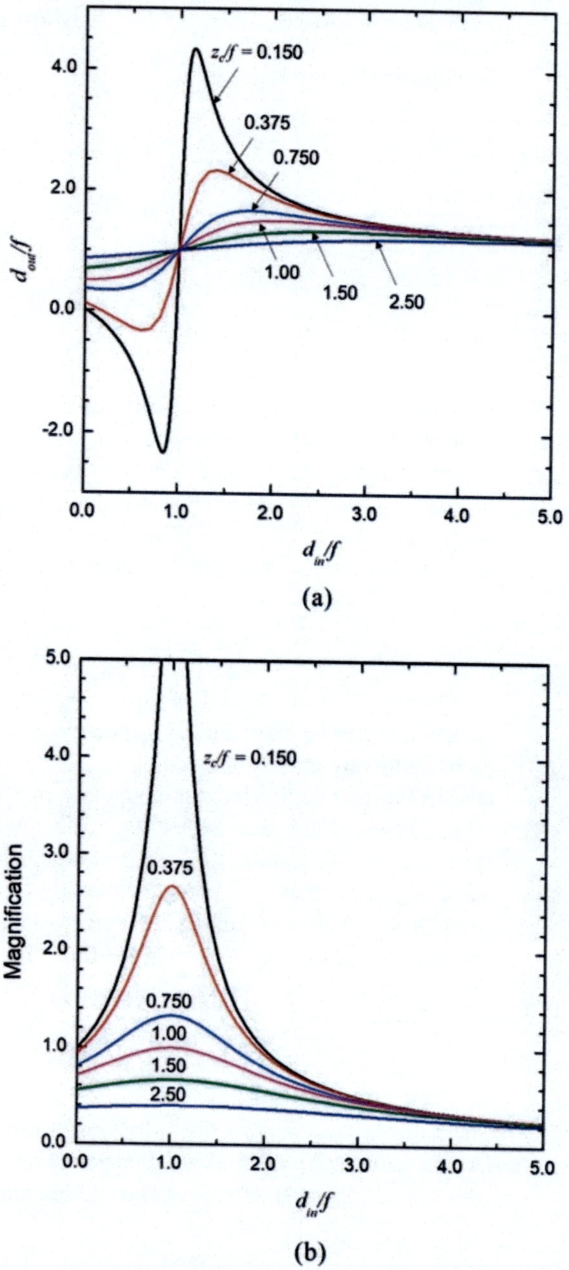

Now let us make some observations on the obtained output beam waist and its location, starting with its location, d_{out}. The relation between d_{out} and d_{in}, as is presented by Eq. (4.42), is depicted in Fig. 4.12a with z_c as a parameter. This plot is extremely useful for those dealing with Gaussian beams, which they may want to keep near at hand. The key observation from the plot is that $d_{out} = f$ if $d_{in} = f$, or the output beam waist is located at the focal length of the lens if the input beam waist (near the source position) is also located at the focal length on the other side, which is also obvious from Eq. (4.42). If the input beam waist is pushed farther away from the lens than f (i.e., $d_{in} > f$), d_{out} will also move farther away from the lens initially (i.e., $d_{out} > f$), then snap back and eventually approach the focal length for very large d_{in}/f. If the input beam waist is brought closer to the lens than the focal length (i.e., $d_{in} < f$), the output beam waist will be also pulled closer to the lens (i.e., $d_{out} < f$). The response of d_{out} to the change in d_{in} will be more sensitive with smaller z_c, or larger wavelength and smaller input beam waist. One comment on Fig. 4.12a can be made regarding the negative values of d_{out} obtained for some small z_c cases when $d_{in} < f$. This implies that the beam will diverge immediately after passing through the lens without forming a beam waist in those special cases. Next, let us make observation on the output beam waist size. Figure 4.12b shows the ratio of the input and output beam waist, or the system magnification, as a function of d_{in} normalized to f, following Eq. (4.43). It is clear the output beam waist will be largest when $d_{in} = f$, or when the input beam waist is located at the focal length. As d_{in} is pushed away from f, toward either direction from f, the output beam waist will be reduced. When d_{in} becomes very large, the magnification converges zero, or the output beam waist will become infinitesimally small, similar to what the geometrical optics would predict when the point source is at infinity. The reaction of ω_{0out} to the change in d_{in} is again more sensitive when z_c is smaller. A system magnification larger than unity will be obtained only for z_c smaller than f, because the maximum value of the system magnification that is achieved when $d_{in} = f$ is f/z_c, as indicated by Eq. (4.44).

4.2 Antennas

The antenna can be defined as a device that converts guided waves into electromagnetic waves in open space or vice versa. For the propagation in the open media discussed in the previous section, it was implied that antennas are employed for radiation and detection. As many of the THz applications involve propagation in the open space, it became clear that THz researchers are required to have some basic knowledge of antennas. The topic of antenna is extensive, and an enormous amount of related research results are being published in the journals and conferences dedicated to the topic. A plenty of textbooks are also available that solely focus on antennas, providing detailed description of the topic. However, for THz researchers trying to get acquainted with the basics of antennas, those dedicated sources may appear overwhelming. This section is intended to provide a very quick introduction to the antenna basics, together with a few types of antennas that are available for integration with planar THz circuits. Those readers who need a deeper look into antennas are referred to more comprehensive coverages of this broad topic such as [5, 6]. For the focused treatment of mm-wave and THz antennas, there is an excellent review article as well [7].

4.2.1 Antenna Basics

One of the basic consequences of the electromagnetic theory is that accelerated charges radiate electromagnetic waves. The acceleration may be on the speed (magnitude of the velocity) or the direction, or the two combined. A situation relevant to antennas is charges confined within a conductor, which can be a waveguide or circuit, moving back and forth periodically, in the form of a sinusoidal AC current in most cases. As those charges are under acceleration due to the periodic change of the direction and the speed, any conductor with an AC current component inside would radiate and can be loosely considered as an antenna. When such a radiation is intended by design and the structure is optimized for a great efficiency, we call the device an antenna in a conventional sense. When it is not intended, it falls on the category of lossy waveguide or circuit. By reciprocity, a radiating antenna will be also able to detect electromagnetic waves, converting it to a sinusoidal AC current. In principle, any disturbance in electromagnetic field can propagate and thus can be considered as radiation. In most practical situations, however, the electromagnetic field has a sinusoidal time variation (and thus spatial variation) as it propagates, which we will assume throughout the discussion below. In this subsection, some basic concepts and expressions related to antennas will be reviewed, which include radiation pattern, antenna efficiency, directivity and gain, effective aperture, Friis transmission equation, EIRP, polarization, and bandwidth.

4.2.1.1 Radiation Pattern

The radiation pattern, also called the antenna pattern, is a mathematical function or a graphical representation of the radiation properties of an antenna, usually taken at the far-field regime of the radiation [8]. While radiation patterns can be obtained for various antenna properties, in most cases the patterns are presented for the magnitude of the field or power, in either linear or dB scale. An example would be useful to describe the radiation pattern in more detail. Suppose an infinitesimal dipole located at the origin of a coordinate system, which is basically a charge moving up and down periodically along a small metallic wire aligned along the z-axis (Fig. 4.13). As many of antennas can be modeled with a line source, this structure should be a nice example to consider. With a simple theory but a rather lengthy derivation, one can arrive at the expressions of the electric and magnetic fields at the far-field region as:

$$\mathbf{E} = j\omega\mu\frac{I\Delta z e^{-j\beta r}}{4\pi r}\sin\theta\mathbf{a}_\theta, \tag{4.45}$$

$$\mathbf{H} = j\beta\frac{I\Delta z e^{-j\beta r}}{4\pi r}\sin\theta\mathbf{a}_\phi, \tag{4.46}$$

where ω is the angular frequency of the dipole oscillation and thus the radiated field, μ is the permeability of the propagation medium, I and Δz are the current and the dipole size, β is the propagation constant, r is the distance from the origin, θ is the angle from the z-axis in the spherical coordinate, $\mathbf{a}_\theta$ and $\mathbf{a}_\phi$ are respectively the unit vector for the θ and ϕ direction. It was assumed that the medium is lossless, in which case the phases of E-field and H-field are the same, leading to the ratio between the field magnitudes being a real number. This ratio, given as $\eta = \omega\mu/\beta = \sqrt{\mu/\varepsilon}$ with ε as the permittivity of the medium, is known as the intrinsic impedance with a unit of Ω (376.7 Ω in free space). Because the magnitudes of the E-field and H-field are the same except for a constant factor (i.e., η), the field radiation patterns, typically presented as normalized over the maximum value, are identical for the two fields. Hence, the field radiation pattern can be created based on either E-field or H-field.

The 3D radiation pattern of the infinitesimal dipole is shown in Fig. 4.13b. It takes on a doughnut shape and does not have ϕ dependence, thus is often referred to as omnidirectional (different from isotropic, in which case the pattern is a perfect sphere). While 3D patterns best depict the radiation pattern, 2D patterns that are cut along reference planes are often preferred for practical purposes. In this case, it is conventional to have 2D patterns obtained on the E-plane and H-plane, which are

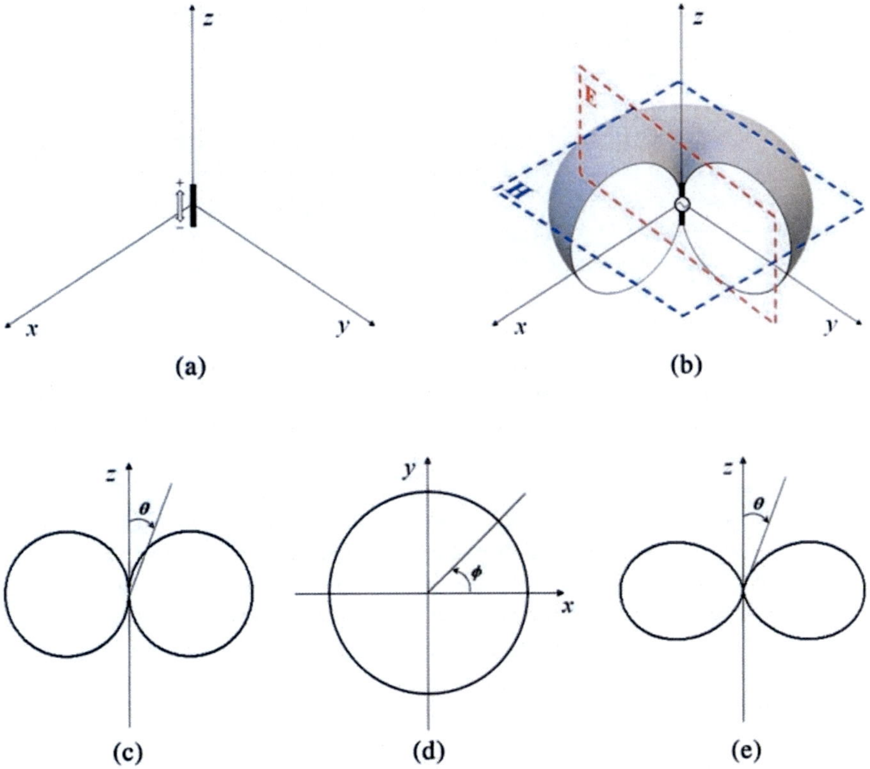

Fig. 4.13 (**a**) A small metallic wire as an infinitesimal dipole. (**b–e**) Radiation patterns for the infinitesimal dipole: (**b**) 3D radiation pattern for field, (**c**) E-plane field radiation pattern, (**d**) H-plane field radiation pattern, (**e**) E-plane power radiation pattern

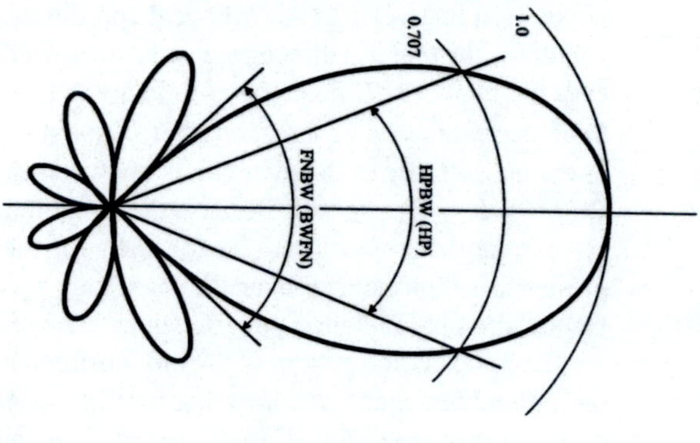

Fig. 4.14 Field radiation pattern example and the definition of half-power beamwidth (HPBW or HP) and first null beamwidth (FNBW or BWFN). (Adapted with permission from [6], © 2012 Wiley)

called the *E*-plane radiation pattern and the *H*-plane radiation pattern, respectively. These two radiation patterns are collectively called the principal plane patterns. Here, the *E*-plane refers to the plane that contains the electric field vector of the radiation. For instance, with the infinitesimal dipole depicted in Fig. 4.13, any plane with a fixed value of ϕ (constant-ϕ plane) can be an *E*-plane as indicated in the figure, because the electric field in this example does not have any ϕ-dependence. Likewise, *H*-plane refers to the plane that contains the magnetic field vector of the radiation. In the given example, any constant-z plane can be regarded as a *H*-plane, but the patterns are typically taken for a plane that includes the maximum magnitude, which is $z = 0$ plane or the x–y plane in this example. 2D radiation patterns for field magnitudes obtained on the *E*-plane and *H*-plane are shown in Fig. 4.13c, d, respectively. A 2D radiation for *power* is also shown in Fig. 4.13e, on *E*-plane as an example. The readers are cautioned not to make a naïve mistake of relating *E*-plane and *H*-plane patterns to *E*-field and *H*-field patterns. Again, the (normalized) field pattern should be the same for the two fields, and *E*-plane and *H*-plane simply denote the *reference planes* on which the 2D radiation patterns are obtained.

Another example for a radiation pattern would be helpful to introduce a few more definitions related to radiation patterns. Figure 4.14 shows a typical radiation pattern from an antenna with a finite dimension, in which case the phase differences in different parts of the antenna result in multiple lobes as shown in the pattern. The pattern shows a main lobe (or major lobe) and multiple side lobes (or minor lobes). The side lobe in the opposite direction of the main lobe is particularly called the back lobe. (Sometimes a narrower definition of the "side lobe" may be used, in which a side lobe indicates a minor lobe that is not the back lobe, or even indicates the lobes immediately next to the main lobe. In many cases, however, side lobes and minor lobes can be used interchangeably.) Since normalized to the maximum point value, the main lobe tip corresponds to a value of 1 in terms of the magnitude. The angle between the two directions where the field magnitude equals to $\sqrt{2}$, or 0.707 (which corresponds to 0.5 in terms of power magnitude), is called the half-power beamwidth (HPBW or HP). Another beamwidth is defined as the angle between the first nulls, which is called the first null beamwidth (FNBW) (sometimes, but less often, called the beamwidth between first nulls (BWFN)). There may be other definitions for beamwidth such as 10-dB beamwidth or others. However, when referred to simply as the beamwidth, it usually indicates HPBW. This is similar to the case for the *bandwidth* of amplifiers or many general electric circuits and systems, where the 3-dB power-roll off points are taken as the boundary of the operation bandwidth.

4.2.1.2 Directivity, Radiation Efficiency, and Antenna Gain

Antennas usually have a preferred direction in which the radiation is stronger than other directions, unless the antenna is perfectly isotropic. Such a property is quantified with a parameter called the directivity D, which is defined as the ratio of the radiation intensity in the given direction to the radiation intensity averaged over all directions. Unless stated otherwise, it is implied that the directivity is made for the direction of *maximum* radiation intensity, with the following expression:

$$D = \frac{U_{\mathrm{m}}}{U_{\mathrm{avg}}}, \tag{4.47}$$

where U_{m} is the maximum radiation intensity, and U_{avg} is the radiation intensity averaged over all directions assuming the source is isotropic with the same total radiation power. The definition will be better visualized with the help of Fig. 4.15, which compares the actual radiation pattern and the isotropic equivalence with the

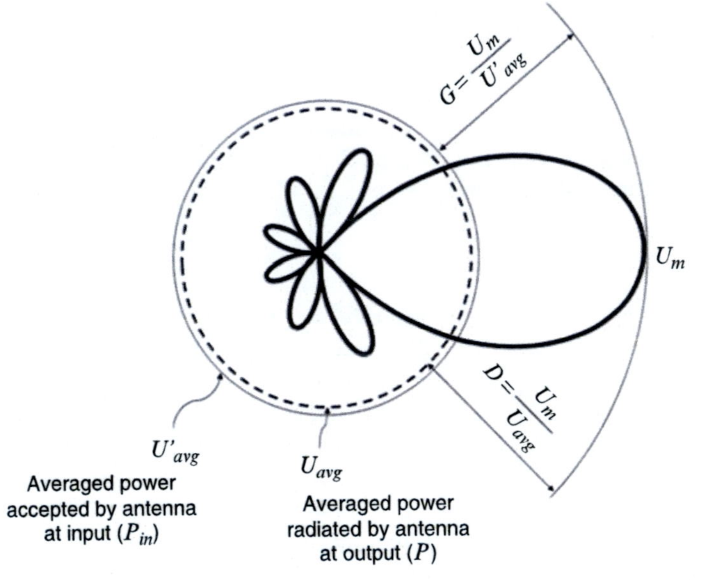

Fig. 4.15 Definition of directivity D and the gain G. (Adapted with permission from [6], © 2012 Wiley)

same total radiation power. It is noted that the directivity is a unitless parameter, often presented in dB.

When an antenna is used as a part of a system, not all of the power delivered from the system to the antenna is radiated by the antenna. In realistic situations, there is always a finite loss in the process of the power conversion due to the internal ohmic loss, both conduction and dielectric. The radiation efficiency e_r is introduced to account for this loss between the input and the output of antennas with the following definition:

$$e_r = \frac{P}{P_{in}}, \tag{4.48}$$

where P is the radiation power and P_{in} is the power delivered to the antenna at the input. It is noted that the power reflection due to the input mismatch is not considered as part of the loss in the definition of the radiation efficiency, since P_{in} denotes the power eventually *accepted* by the antenna. Another parameter, the total antenna efficiency, less popular, accounts for the loss due to the reflection as well, leading to a slightly smaller value than the radiation efficiency. Sometimes, it is useful to express the radiation efficiency in terms of the radiation resistance R_r, which is defined as $R_r = 2P_r/|I_A|^2$ where P_r is the averaged radiation power and I_A is the current supplied to the antenna. For the antenna impedance given by $Z_A = R_A + jX_A$, R_A is composed of two components such that $R_A = R_r + R_0$, where R_0 represents the resistance due to ohmic loss. In terms of these parameters, the radiation efficiency can be also expressed as:

$$e_r = \frac{R_r}{R_r + R_0}. \tag{4.49}$$

With the definition of the radiation efficiency given, we can now discuss the antenna gain. The antenna gain G is defined as the directivity multiplied by the radiation efficiency, such that:

$$G = e_r D. \tag{4.50}$$

In more formal terms, the antenna gain refers to the ratio of the radiation intensity in the given direction to the radiation intensity that would be obtained if the power *accepted* by the antenna were (entirely) radiated isotopically [8]. In a plain language, it is basically a directivity that also takes into account the antenna loss. A visual comparison of gain and directivity can be found in Fig. 4.15. As is the directivity, the antenna gain is a unitless parameter and in many cases shown in dB scale, particularly with a unit of dBi (i for isotropic). It should be remembered that there is no true power gain obtained even if the antenna gain is larger than 1. An antenna is not an amplifier, and, in fact, it is a passive device, which causes overall loss of power. The antenna "gain" is achieved over the ideal isotropic radiation, not over the input power.

4.2.1.3 Effective Aperture

Now let us shift out attention to an antenna at the receiver side and consider the relation between the power density incident on the antenna and the actual power received by the antenna. By common sense, the received power will be the product of the incident power density and the area. The issue here is how we should define the *area*. Is this the physical area of the antenna? The answer is apparently *no*, because there is not a clear definition of the physical area for many antenna types. Even for those with a fairly well-defined physical area (such as aperture antennas), it is not the entire power incident on the physical area that is accepted by the antenna. For this purpose, we define the *effective aperture*, or the effective area, which is defined with a relation:

$$P_{\mathrm{A}} = SA_{\mathrm{e}}, \tag{4.51}$$

where P_{A} is the available power for the receiving antenna, S is the incident power density, and A_{e} is the effective aperture. Hence, A_{e} can be regarded as the virtual area that relates the incident power density and the actual received power. The expression given in Eq. (4.51) is based on the assumption that the antenna is oriented to the maximum response, polarization-matched to the incoming wave, and impedance-matched to the load. If we further assume that the internal ohmic loss of the antenna is zero, we can introduce the maximum available received power P_{rm} and the maximum effective aperture A_{em} such that:

$$P_{\mathrm{rm}} = SA_{\mathrm{em}}. \tag{4.52}$$

Here, A_{em} is related to A_{e} with $A_{\mathrm{e}} = e_{\mathrm{r}}A_{\mathrm{em}}$ (the relation between A_{em} and A_{e} may appear differently for some literature [5], but we will follow that in [6]). Interestingly, there is a simple relation between the directivity and the maximum effective aperture as follows:

$$D = \frac{4\pi}{\lambda^2}A_{\mathrm{em}}. \tag{4.53}$$

This relation is general and holds for any type of antenna. (The proof can be found in general antenna textbooks.) If we recall the relation between the directivity and the gain given in Eq. (4.50), we also have:

$$G = \frac{4\pi}{\lambda^2}A_{\mathrm{e}}. \tag{4.54}$$

4.2.1.4 Friis Transmission Equation

A set of equations introduced above leads us to the famous Friis transmission equation, which establishes the relation between the transmitted power and the received power of a link between two antennas. Let us first consider the power density S incident on the receiving antenna located at a distance R from the transmitting antenna with an antenna gain of G_t and transmitting power P_t, which is given as:

$$S = \frac{G_t P_t}{4\pi R^2}. \tag{4.55}$$

This relation can be understood intuitively, because an isotropic radiation would lead to $S = P_t/4\pi R^2$ for a point at R, and a factor of G_t should be multiplied for a directional radiation. With the given incident power density, the received power P_r can be easily obtained with the help of Eq. (4.51):

$$P_r = \frac{G_t P_t}{4\pi R^2} A_{er}, \tag{4.56}$$

where A_{er} is the effective aperture of the receiving antenna and the assumptions for Eq. (4.51) still hold. Finally, by applying Eq. (4.54) to Eq. (4.56) with receiving antenna gain of G_r, we obtain the final form of the Friis transmission equation:

$$P_r = G_t G_r \left(\frac{\lambda}{4\pi R}\right)^2 P_t. \tag{4.57}$$

According to this equation, the received power reduces with R^2 but can be improved with antenna gain on either side. It is notable that the received power is proportional to λ^2, which stems from the fact that the effective aperture is proportional to λ^2 for a given antenna gain as can be seen from Eq. (4.54). Equivalently, it can be said that the antenna gain is proportional to $1/\lambda^2$ for a given effective aperture. This is a natural result from the diffraction viewpoint, because less diffraction will happen with a smaller wavelength for a given aperture, leading to a higher directivity. However, it will serve as one of the major challenges for THz communication systems, as the radiated power decays increasingly faster with higher carrier frequency.

4.2.1.5 Equivalent Isotropic Radiation Power

Equivalent isotropic radiation power (EIRP), which was introduced earlier in Chap. 2, also has a simple relation to the transmitting antenna gain and power. EIRP can be defined as the power emitted from an isotropic antenna to obtain the same power density level as measured at the direction of the maximum radiation. From this definition, the following relation is valid:

$$\text{EIRP} = P_t G_t, \tag{4.58}$$

where P_t is the transmission power and G_t is the transmitting antenna gain. This relation can be explained again with the help of Fig. 4.15. If the total radiation power P_t of a directive antenna is averaged over all directions to be isotropic, the power level is down from the peak by G_t. If we want to recover the peak power level with an isotropic antenna, the power needs to be increased by G_t, leading to Eq. (4.58). Here, to be precise, P_t is the *accepted* power by the transmitting antenna from the system attached at the input, which may be slightly larger than the *radiated* power considering the antenna loss.

Equation (4.58) can be substituted into Eq. (4.57), leading to:

$$P_r = G_r \left(\frac{\lambda}{4\pi R} \right)^2 \cdot \mathrm{EIRP}. \tag{4.59}$$

This relation implies that EIPR is a useful parameter to describe the transmitter when the detected power at the receiver is available while the gain of the transmitting antenna is unknown. For example, when characterizing a radiating on-chip THz signal source, they often report EIRP instead of the source output power, since EIRP can be directly obtained with received power from Eq. (4.59). Note that the receiving antenna gain is usually well defined (such as the case of a commercial horn antenna) unlike the gain of the transmitting antenna integrated with the source. In this case, the actual source output power can be obtained from EIRP only with an estimated gain of the on-chip transmitting antenna.

4.2.1.6 Polarization

The polarization is another main parameter that describes an antenna. Let us take a review on the concept of polarization first. In a general sense, the polarization can be understood as the time-varying behavior of the electric field vector of propagating electromagnetic waves at a given location. Assume you are measuring the electric field at a certain point inside the radiation field as a function of time. If the radiation is polarized, there can be three scenarios for the behavior of the electric field vector. First, the direction of the vector is fixed along a certain axis, but its magnitude periodically changes (Fig. 4.16a). This case is called the linear polarization. Second, the direction of the vector makes a circular rotation, while the magnitude remains constant (Fig. 4.16b). This is called the circular polarization. Finally, the direction of the vector makes a circular rotation simultaneously with a periodic change of the magnitude, resulting in an elliptical trace of the point vector (Fig. 4.16c). This is a general case positioned between the first and second extreme cases (i.e., linear and circular cases), called the elliptical polarization. The three cases are illustrated in Fig. 4.16 in terms of the trace of the electric field vector. The circular and elliptical polarization can be sub-divided into left-handed and right-handed polarization, which correspond to clockwise and counter clockwise rotation, respectively, when propagation is toward the observer.

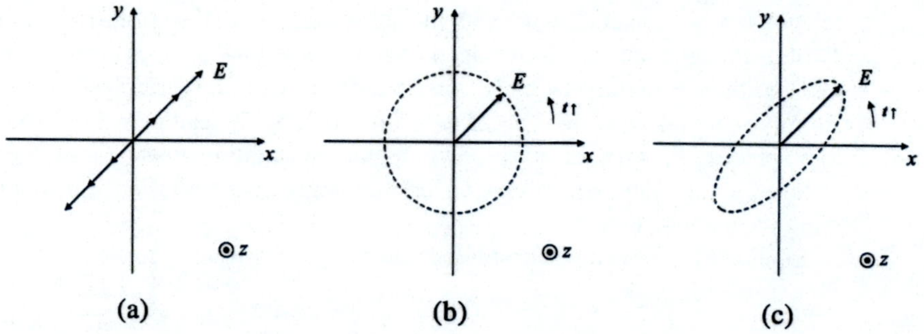

Fig. 4.16 Three polarization cases in terms of the electric field vector: (**a**) Linear polarization, (**b**) Circular polarization, (**c**) Elliptical polarization. Right-handed polarization is assumed for the circular and elliptical polarizations with propagation toward +z direction

For the readers who are not fully comfortable with the concepts of the three types of polarization with Fig. 4.16, it may help to compare the three cases in terms of the relative phase of two orthogonal components as presented in Fig. 4.17. For the linear polarization, let us first take the orientation of the two mutually perpendicular axes on the wave front, so that the magnitude of the field is equally divided into the two axes with 45° angle against both axes (x- and y-axis in the example shown in the figure, with wave propagation along the z-axis). In this case, the phases of the x and y components are perfectly aligned as depicted in Fig. 4.17a, which is shown along with the vector diagram for reference. Also, the peak intensity will be the same for the two components. If we relax the condition for the orientation selection of x- and y-axis, so that the magnitude is not necessarily divided equally into the two axes, which is in fact the general case, the waveform plot will now appear as shown in Fig. 4.17b. For the circular polarization, there will be a phase difference of $\pi/2$ as is described in Fig. 4.17c. It is noted that the waveform and phase difference will be independent of the choice of x- and y-axis, which is a unique property of the perfect circular polarization. As for the elliptical polarization, the phase difference will show an intermediate value between 0 and $\pi/2$ in general, as is presented by Fig. 4.17d. However, depending on the choice of polarization axis with respect to x- and y-axis, the phase difference and relative intensity may appear differently. In the special case when the polarization axis is aligned with either x- or y-axis, the phase difference will be reduced to $\pi/2$, with the difference in the peak intensities maximized, which is shown in Fig. 4.17e.

Although unpolarized electromagnetic waves do exist in the nature where the multiple polarization states co-exist in a mixed manner, the radiation from an antenna is always polarized with one of the polarization states. Accordingly, the polarization of an *antenna* is defined as the polarization of the wave radiated by the antenna. When the polarization varies with the radiation direction, which is often the case, the polarization of the main beam is taken. The transmitting antenna and receiving antenna should be matched with the same polarization states (in terms of types as well as the axis orientation) for maximum transmission efficiency. If not matched, it may result in a partial or even total loss of the transmission.

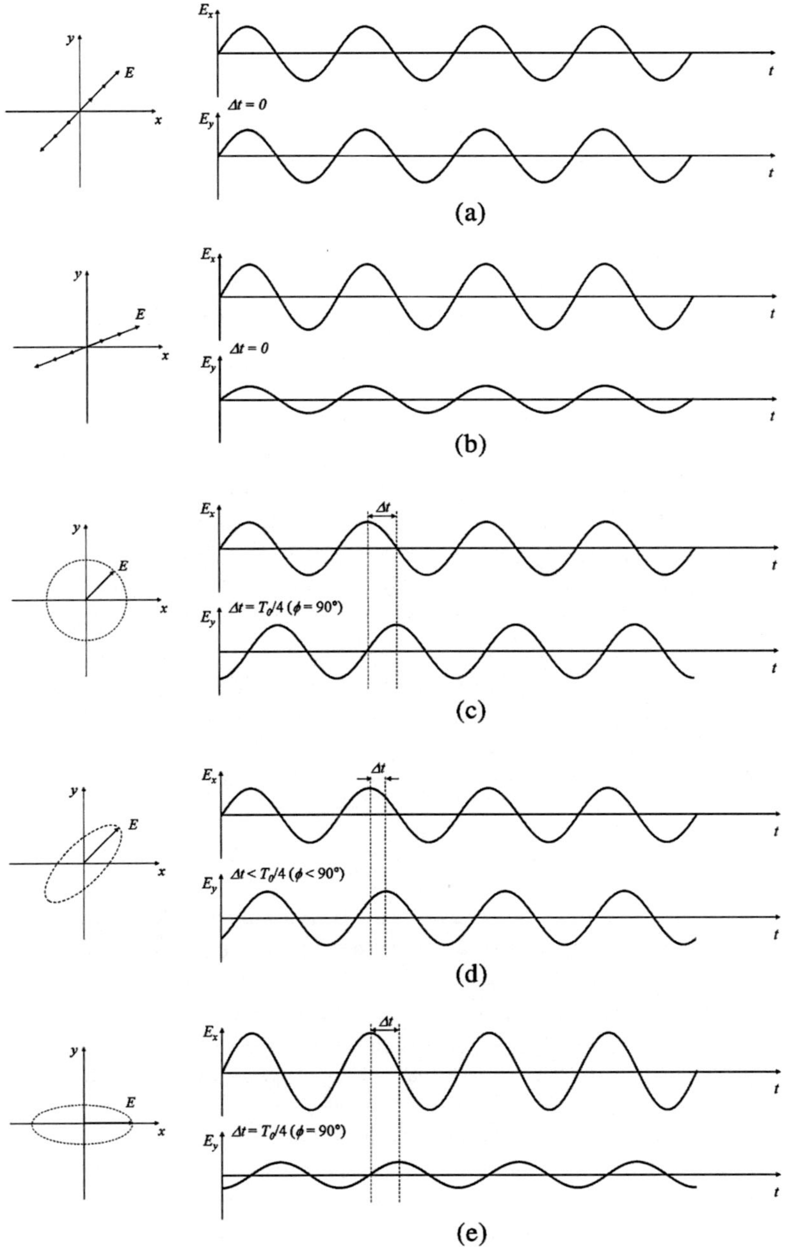

Fig. 4.17 (**a**) Linear polarization with the field magnitude equally divided into x- and y-axis. (**b**) Linear polarization with an arbitrary angle of the field with respect to x- and y-axis. (**c**) Circular polarization. (**d**) Elliptical polarization with the field magnitude equally divided into x- and y-axis. (**e**) Elliptical polarization with the polarization axis aligned with x-axis. Right-handed polarization is assumed for the circular and elliptical polarizations

4.2.1.7 Bandwidth

Bandwidth is also a key parameter of antennas. There are a few different ways to define the bandwidth of antennas. The *antenna bandwidth*, simply denoted as BW, is given as the difference between the upper boundary (f_U) and lower boundary (f_L) of operation, or BW $= f_U - f_L$. However, a more practical measure of the frequency range covered by an antenna is provided when the bandwidth is expressed as a ratio, rather than in terms of the absolute difference. For narrow band antennas, the *fractional bandwidth* is widely used, which is defined as $(f_U - f_L)/f_C$ where f_C is the center frequency. It is often presented in terms of percentage (%) as well. Here, f_C can be defined by either the arithmetic average $(f_U - f_L)/2$ or the geometric average $\sqrt{f_U/f_L}$. The two averages are not much different for narrow-band antennas. For broadband antennas, the *ratio bandwidth* is preferred, which is simply defined as f_U/f_L. It is also common to present the ratio bandwidth in the form of f_U/f_L:1 (e.g., 10:1 or 5:1). Note that there is no standard way of determining f_U and f_L. They can be generally regarded as boundaries for acceptable or satisfactory operation, for which many different characteristics of antennas can be considered depending on the focus of the performance. For example, the frequency boundaries can be determined based on the input impedance, gain, side lobe level, beam width, polarization, beam direction, and so forth [5].

4.2.2 *Types of Antennas*

If you look around, you will soon notice so many antennas are surrounding our daily lives. They can be found in all kinds of the wireless mobile devices as well as base stations, also with traditional devices receiving broadcast signals such as radios or TVs. If you happen to work on experiments based on air radiation of electromagnetic waves such as wireless links, lots of test antennas will be populating your laboratory. Those antennas take on various sizes and shapes, and you can easily come up with a long list of different types of antennas. Overviewing all those individual types of antenna will be overwhelming, and we can instead group them into certain categories based on common features and review each category. There can be many different ways for categorization, while we will follow here the one suggested in [6], which groups antennas into four different basic types: electrically small antennas, resonant antennas, broadband antennas, and aperture antennas.

Electrically small antennas refer to those with a dimension much smaller than the wavelength. Technically, they may include any antennas with a dimension that fit into a sphere of radius a smaller than $1/k$, where k is the wave number, or $2\pi/\lambda$ [9]. However, for most practical electrically small antennas, a is much smaller than $1/k$, by at least a factor of an order. As the antenna performance is generally degraded if the dimension shrinks far smaller than the wavelength, one may want to avoid too much scaling of the antenna sizes. Nevertheless, in many cases, there exist practical

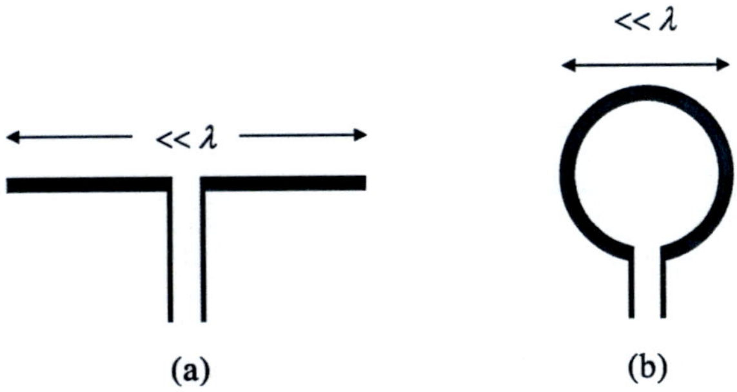

Fig. 4.18 Examples of electrically small antennas: (**a**) Short dipole antenna, (**b**) Small loop antenna

constraints that limit the antenna size, especially when the wavelength is larger than the dimension that can be properly handled in a practical way. For example, AM radio signal spans a wavelength range of a few hundred meters, but antennas with a half of these wavelengths will not be a practical solution, in which case electrically small antennas need to be used. They typically exhibit a low directivity, which is somewhat expected from Eq. (4.53) as the effective aperture of small antennas is much smaller than the wavelength. The low directivity is not necessarily a shortcoming and sometimes may be preferred. However, when it comes to the impedance characteristics, the small antennas pay the price for the compactness as explained below. Most of all, the radiation resistance is very small, which limits the radiation power level with a relation $P_r = R_r|I_A|^2/2$. For a *short dipole antenna*, which is one of the typical examples of electrically small antennas that also include *small loop antennas* (see Fig. 4.18), R_r is proportional to $(\Delta z/\lambda)^2$ where Δz is the length of the dipole. This leads to radiation power also proportional to $(\Delta z/\lambda)^2$ if I_A remains the same. It is clear from this relation that the radiation power will drop significantly for small antennas where $\Delta z \ll \lambda$ and so will the radiation efficiency. The radiation efficiency of electrically small antennas may easily fall below 10%. The small radiation resistance also results in a small real part of the antenna impedance (i.e., resistance), making the impedance match with the adjacent system, typically terminated around 50 Ω, quite difficult. The large reactance component of electrically small antennas also adds to the difficulty in the impedance matching. Another downside of electrically small antennas is the small bandwidth as limited by the large Q-factor, which rapidly increases with a smaller electrical dimension [10]. The improvement of the bandwidth with reduced Q-factor needs a larger dimension or enhanced structural optimization. As mentioned above, however, there are cases where the adoption of small antennas cannot be avoided, especially for applications with longer wavelengths.

Resonant antennas have dimensions that support the resonant modes of the corresponding wavelengths. They typically involve a dimension of a half wavelength or its integer multiples, which allows standing waves formed across the antenna structure. Being based on the resonant mode, they naturally show narrow bandwidth characteristics. On the other hand, resonant antennas in many cases exhibit impedance characteristics that are favored for impedance matching, with resistance not far from 50 Ω and small reactance values. The examples of the

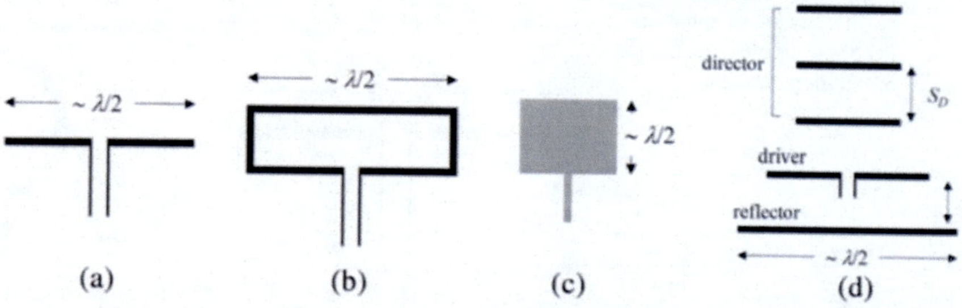

Fig. 4.19 Examples of resonant antennas: (**a**) Half-wave dipole antenna, (**b**) Half-wave folded dipole antenna, (**c**) Patch antenna, (**d**) Yagi-Uda antenna

resonant antennas are shown in Fig. 4.19, which include *half-wave dipole antennas*, *patch antennas*, and *Yagi-Uda antennas*. The first two antenna types will be described in more detail separately in Sects. 4.2.3 and 4.2.4. The Yagi-Uda antenna will be briefly described here. It is composed of a dipole-based driver and parasitic elements, director and reflector, which are added to improve the directivity (see Fig. 4.19d). The directors are typically composed of multiple conductor bars, placed in the direction of the radiation. They significantly improve the directivity, which increases with the number of the director elements, invoking a trade-off between size and performance. The reflector, usually a single element bar, is located on the other side of radiation, which reflects the power radiated to the opposite side of the main lobe back to the main direction. The spacing between the reflector and the dipole driver (S_R) as well as between the director elements (S_D) also affect the directivity, the optimum value ranging around 0.15–0.25λ for S_R and 0.2–0.35λ for S_D. The directors are usually slightly shorter than the dipole driver, while the reflector is longer, typically around 0.5λ. Although not typically practiced, there are reports on on-chip Yagi-Uda antenna for end-fire-type radiation as well [11, 12].

Broadband antennas literally indicate a group of antennas that show broadband characteristics. Examples of broadband antennas include traveling-wave antenna, helical antenna, biconical antenna, bow-tie antenna, log-periodic antenna, and spiral antenna, as sketched in Fig. 4.20. When an extremely large bandwidth of larger than 10:1 is achieved, those antennas are particularly called frequency-independent antennas. Spiral antennas and log-periodic antenna belong to this category. As spiral antennas and log-periodic antenna are widely employed for planar THz circuits, they will be described in more detail later in Sect. 4.2.5. Here a brief introduction to other broadband antennas will be made. The *traveling-wave antenna* (Fig. 4.20a) is an antenna that supports a traveling wave without reflection wave, which typically covers a bandwidth larger than 2:1. A long conducting wire, longer than a wavelength, terminated with a matched load may serve as a traveling-wave antenna. It shows a main lobe formed with a certain angle off the longitudinal direction of the wire with an azimuthal symmetry around it as depicted in the figure. There are variants such as V antenna and rhombic antenna, designed to provide a main lobe aligned toward the longitudinal direction of the wire. The *helical antenna*

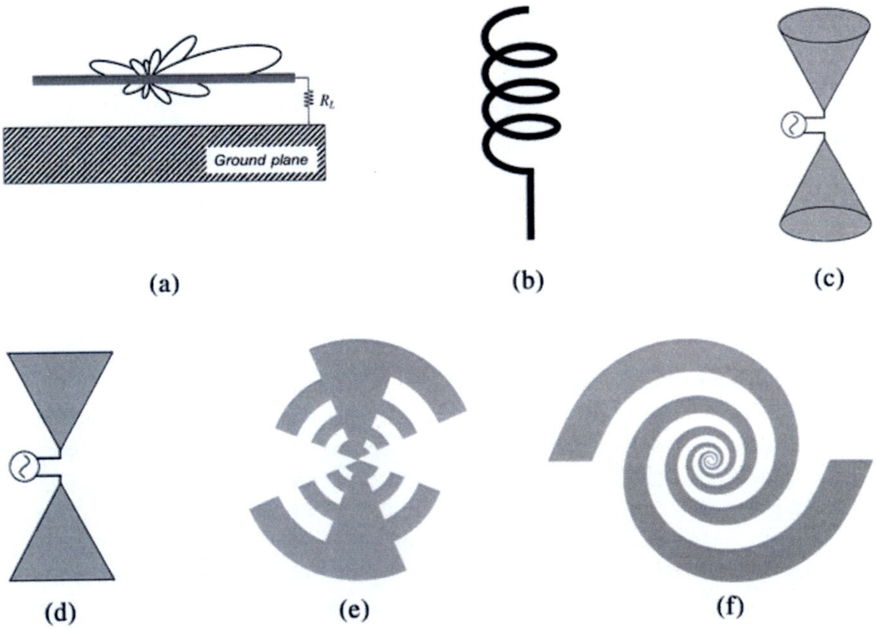

Fig. 4.20 Examples of broadband antennas: (**a**) Traveling-wave antenna (shown with the radiation pattern), (**b**) Helical antennal, (**c**) Biconic antenna, (**d**) Bow-tie antenna, (**e**) Log-periodic antenna, (**f**) Spiral antenna

(Fig. 4.20b) is basically a conductor wound to form a helix. When the radius of the helix is small compared to the wavelength, it operates in the normal (broadside) mode and the radiation is made toward the side of the main axis of the helix (normal to the helix axis). When the helix diameter is comparable to or larger than the wavelength, its operation is in the axial (end-fire) mode with the main lobe formed toward the helix main axis. The latter is often considered more practical as a wider bandwidth and higher efficiency can be achieved. The *biconical antennas* (Fig. 4.20c) is a modified dipole antenna intended for a wider bandwidth. It is composed of two cones (solid or shell) with a gap near the center, across which an AC driving voltage is applied. It can be considered as a dipole antenna, in which each half wire is tapered to obtain broadband characteristics. The radiation beam is formed symmetrically on the side, similar to the case of dipole antennas. The *bow-tie antenna* (Fig. 4.20d) is a planar version of the biconical antenna. Although bow-tie antennas are more sensitive to frequency variation than biconical antennas, its simpler structure and compatibility with integrated circuit have made it a popular option for THz integrated circuits [13]. The *log-periodic antenna* (Fig. 4.20e) and *spiral antenna* (Fig. 4.20f) will be described in more detail in Sect. 4.2.5.

Aperture antennas are characterized with the presence of a physical aperture, through which the radiation or reception of electromagnetic waves are made. The most outstanding property of aperture antennas is the high gain. It can easily exceed

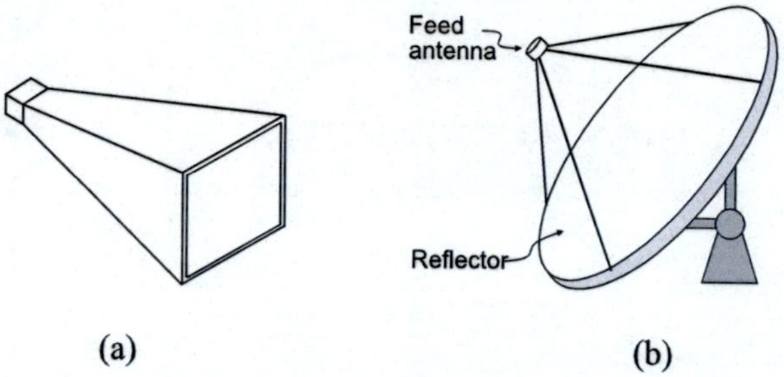

Fig. 4.21 Examples of aperture antennas: (**a**) Horn antenna, (**b**) Dish antenna

20 dBi and, if designed properly, it can reach a gain higher than ~50 dBi depending on the actual type. Horn antennas and reflector antennas are two widely employed examples of aperture antennas, which are sketched in Fig. 4.21. A *horn antenna* (Fig. 4.21a) can be considered as an open-ended rectangular waveguide with a tapered transition aperture attached at the end, which serves as a matching structure to the open space, suppressing reflection and increasing the bandwidth. The aperture is typically of rectangular or conical shapes in most practical cases. Horn antennas show a moderate bandwidth of 50% or larger. Although they are usually built as individual components, a horn antenna formed on a planar substrate for compatibility with integrated circuits has also been reported for THz operation [14]. *Reflector antennas* (Fig. 4.21b) are composed of a reflector and a feed antenna located at the focal point of the reflector. The reflector, typically with a parabolic surface, reflects and focuses the incident plane wave into the feed antenna (for receiver), or radiate a plane wave by reflecting the normally incident beam emitted from the feed antenna. For the feed antenna, horn antennas or open-ended waveguides are frequently employed, while dipole antennas can be also used at low frequency ranges such as UHF. As the bandwidth of a reflector itself is large, limited by the reflector size (lower boundary) and the surface distortion (upper boundary), the bandwidth of the entire antenna is usually determined by that of the feed antenna. Often called dish antennas, the reflector antennas are a kind of antenna that can achieve the highest gain and are deployed to various parts of private and public applications that require an extremely high gain including satellite and space communications.

4.2.3 Dipole Antennas

In the following sections, a few antenna types widely employed for planar THz systems will be reviewed, starting with dipole antennas. The operation of dipole antennas can be better understood if we first consider an open-ended two-wire transmission line as shown in Fig. 4.22. When a wave is guided by the open-ended transmission line, there will be a reflection at the end of the transmission line, establishing a standing wave with nodes formed at every $\lambda/2$ point from the end. This will cause a periodic distribution of the current on the wire as shown on the top

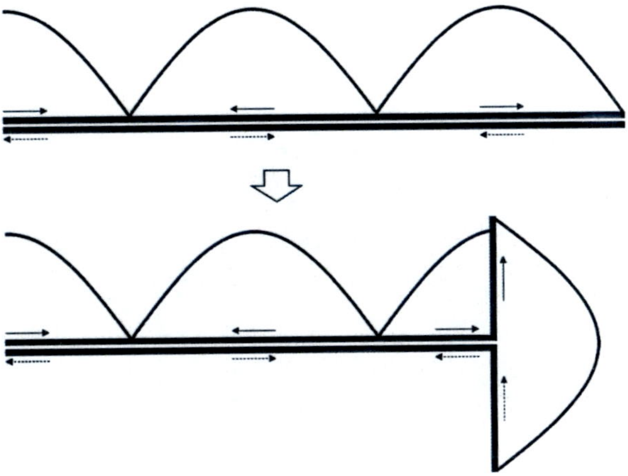

Fig. 4.22 Development of an open-ended two-wire transmission line into a dipole antenna. The curves show the standing wave formed along the wire, and the arrows indicate the direction of the AC current. (Adapted with permission from [6], © 2012 Wiley)

of Fig. 4.22. Also shown in the figure as arrows, the direction of the current on the lines will alternate as it passes every node. Notably, the current direction is the *opposite* for the two line sections facing each other, a characteristic of differential signal. The electric and magnetic fields will be confined close to the wires. Now, if a portion of the wires are bent outward as shown on the bottom of Fig. 4.22, the current will oscillate up and down in the *same direction* for the two wires, and the alternating fields will be exposed to the open space, resulting in radiation. This is the basic structure of a dipole antenna. Assumed in the figure is that the bent was made at $\lambda/4$ point from the end, resulting in a dipole with a total length of $\lambda/2$. A dipole antenna with this particular dimension is called the half-wave dipole antenna (also shown in Fig. 4.19a). As the length exactly matches a single period of the standing wave, a fundamental resonant mode is supported and the radiation efficiency will be maximized. This explains the popularity of the half-wave dipole antennas, and this subsection will also focus mostly on this type of dipole antenna.

The current distribution on the half-wave dipole will be sinusoidal of a half cycle, with a peak at the center and zeros on both ends. The corresponding voltage distribution will be also a sinusoidal half cycle, but with a peak at the ends and a zero at the center. This electrical configuration will result in a radiation pattern as follows:

$$F(\theta) = \frac{\cos\left((\pi/2)\cos\theta\right)}{\sin\theta}, \tag{4.60}$$

where θ is the angle from the axis of dipole. It shows only θ-dependence, as is verified with the 3D radiation pattern shown in Fig. 4.23a. 2D E-plane radiation pattern is shown in Fig. 4.23b, which is slightly elliptic with a HPBW of 78°. Naturally, dipole antennas show a linear polarization aligned along the direction of the dipole (called vertical linear polarization). H-plane radiation pattern is a simple circle. Equation (4.60) can be compared with the case for a very small dipole (see

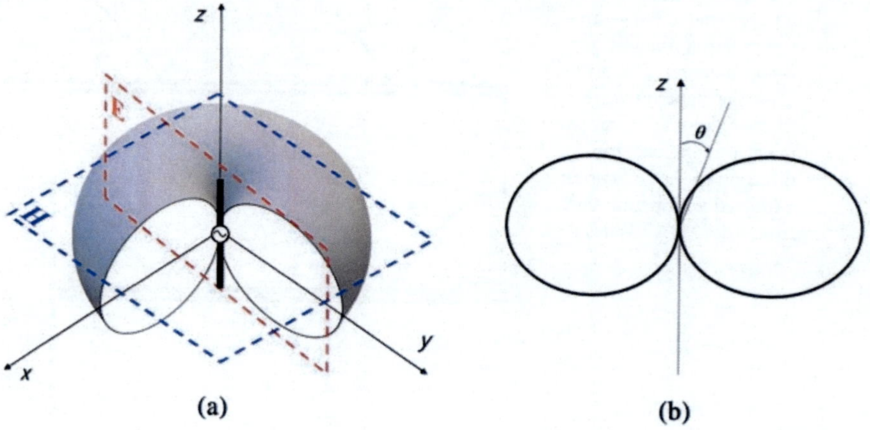

Fig. 4.23 Radiation pattern of a half-wavelength dipole antenna: (**a**) 3D pattern, (**b**) 2D *E*-plane pattern

Fig. 4.13), for which $F(\theta) = \sin\theta$. In this simple case, the *E*-plane radiation pattern will be a pair of perfect circles in contact, leading to a HPBW of 90°. The directivity of a half-wave dipole can be calculated from the total output power and the maximum radiation intensity, leading to a value of 1.643 or 2.15 dBi. This is slightly larger than that of a short dipole, which is 1.5. The directivity increases with the dipole length, which can be expected from the increased effective aperture with longer dipoles. Accordingly, the radiation pattern will be more elliptic for longer dipoles, further deviating from the circle of short dipoles.

The impedance of a half-wave dipole can be also calculated, which is $73 + j42.5\ \Omega$ assuming the wire of the dipole is very thin. As the resistance (real part of impedance) value is close to 50 Ω, half-wave dipoles are reasonably compatible with 50-Ω systems. For 75-Ω transmission lines, which are adopted for some applications, half-wave dipoles provide an almost ideal resistance value. Both resistance and reactance values show a strong dependence on the dipole length, both increasing (more positive) at least until the length becomes close to λ. A slight reduction of the dipole length leads to zero reactance, which is often practiced by design. In fact, this is the actual resonance point of the dipole with a finite wire thickness. Note that the characteristics and the numbers derived above are for the ideal dipole antenna placed in free space. When the antenna is built on top of a dielectric material, which is the case for the on-chip antennas that will be discussed shortly, they will surely deviate from the ideal states. Being a resonant antenna, half-wave dipole antennas exhibit relatively narrow bandwidth, roughly around in the order of 10% in terms of input matching (VSWR < 2). One effective way to improve the bandwidth is to increase the thickness of the dipole (or metal width of planar-type dipole).

One variance of the dipole antenna is the folded dipole antenna (see Fig. 4.19b). It can be understood as two dipole antennas placed in parallel and connected on both ends. The structure supports two modes depending on the current distribution, the transmission mode and the antenna mode, the latter being used for antenna operation. The radiation pattern is similar to that of regular dipole antennas, but the impedance is quadrupled, which is useful for some applications that involve high impedance transmission lines such as 300-Ω lines. Another advantage of the folded dipole antenna is the broader bandwidth, which makes them a popular choice as the feeding antenna for other types of antenna including Yagi-Uda antennas.

The on-chip integration of dipole antennas is a feasible option as dipole antennas can be configured compatible with integrated circuits based on standard semiconductor fabrication processes [15, 16]. There are many advantages for antennas if they are integrated with other circuit blocks on a single chip. For example, there will be no need for off-chip connection schemes such as wire-bonding, which limits the frequency response as well as the reliability of the whole system. Also, on-chip integration allows the co-design of antenna and adjacent circuits, which will significantly relax the requirement for 50-Ω matching, lifting one major constraint for antenna design and potentially removing some of matching circuits [15]. One prerequisite for on-chip integration, however, is the reasonable size of antennas, which has been a great barrier for low frequency operations. In this regard, THz integrated circuits are favorably positioned, as the size of half-wavelength falls on an acceptable range from a practical point of view for integration. Remember that $\lambda/2$ for 300 GHz is around 500 μm in free space, and is further reduced down by a factor of $\sqrt{\varepsilon_r}$ if embedded in a dielectric, where ε_r is the dielectric constant of the embedding dielectric material. For example, with the dielectric constant of ~3.9 of silicon dioxide (SiO_2), often used as inter-metal dielectric material for the standard Si processes, the reduction factor will be around 2, resulting in $\lambda/2$ of around 250 μm at 300 GHz. Note that if the field lines around the dipole conductor is not fully embedded in the dielectric material and partially exposed to the air (free space), which is typically the case for on-chip implementation of dipole antennas, the effective dielectric constant ε_{eff} should be used instead of ε_r. This will be discussed in more detail in Sect. 4.3.2.

There have been lots of reports that successfully employed dipole antennas, including folded variants, for THz integrated circuits including folded versions [17–19]. When a dipole antenna is built on top of the BEOL (back-end of the line) structure, the upper side of the antenna is (mostly) the air, while the back side is a BEOL dielectric layer (usually several μm), which is stacked on top of the semiconductor substrate (around 100–200 μm thick, depending on the extent of the wafer thinning). With this configuration, only a fraction of the total radiation will radiate upward, leading to a main beam formed toward the substrate due to the higher dielectric constant of the material beneath the antenna structure. In theory, when an antenna is placed on top of an infinitely large homogenous dielectric material filling the bottom half of the space, the portion of the air radiation will be roughly $1/\varepsilon_r^{3/2}$ of the total radiation, where ε_r is the dielectric constant of the dielectric material [20]. Substitution of $\varepsilon_r = 11.7$ of Si and 3.9 of SiO_2 into the relation will lead to 2.5% and 13%, respectively. Although the substrates of practical uses have a finite dimension and thickness, still most of the radiation from the on-chip integrated antennas without metal shield underneath will propagate downward through the substrate, until it reaches the bottom of the substrate and emerge in the air. A major issue here is that the substrate is naturally lossy and causes a degradation in the radiation efficiency, which is more serious with Si substrate as it is more lossy than other substrate options such as GaAs or InP. For higher frequency bands such as the THz band, the situation is aggravated with the excitation of the substrate modes due to the short wavelength that is comparable to the substrate thickness [21]. With the onset of the substrate mode, the substrate wave will propagate inside the substrate that serves as a waveguide retaining the wave, which will additionally degrade the radiation efficiency and also affect the radiation pattern. There are techniques proposed to suppress the substrate mode. The formation of randomly located cavities or thru-substrate vias will help to mitigate the mode generation [22], while mounting the chip on top of a silicon lens is also an effective and widely practiced technique [19, 23].

As a supplementary note, it can be mentioned that slot antennas are also widely employed as THz on-chip antennas [24, 25]. It is basically a conductor plane with a hole, or *slot*, carved into it, for which a signal feed is made at the two slot edges across the slot. In principle, a slot antenna can be considered as a complementary structure of a dipole antenna with the dimension of the slot opening equivalent to that of the dipole conductor. This mutual relation leads to the identical far-field radiation patterns for the two types of antennas with the same complementary dimension, except for the interchange of *E*-field and *H*-field. Also, there is a simple relation between the input impedances of the two complementary antennas: the product of the two impedances is $\eta^2/4$, where η is the intrinsic impedance of the medium in which the antennas are immersed. There are occasions where slot antennas are better suited than dipole counterparts, such as when the antenna is integrated with other slot-based on-chip planar structures [26]. For related details for slot antennas, the readers may refer to [27].

4.2.4 Patch Antennas

Patch antennas are a kind of planar antennas based on the microstrip structures [28], which is also called the microstrip antenna for this reason. It is composed of a thin metal conductor placed on top of a dielectric layer with a ground plane on the bottom as shown in Fig. 4.24a. The metal conductor, or patch, can be formed in various shapes, including square, rectangular, and circular among the most popular. There are also a few different options for signal feeding into the metal conductor, such as the edge-feed that accepts signal from a microstrip line on the patch edge, or the probe-feed in which signal is fed through a coaxial cable from the bottom. Also, the proximity coupling provides a signal feed from a microstrip line located *beneath* the patch without a direct contact, which helps to suppress spurious radiation and improve bandwidth, but requires an extra metal layer. In the discussion in this

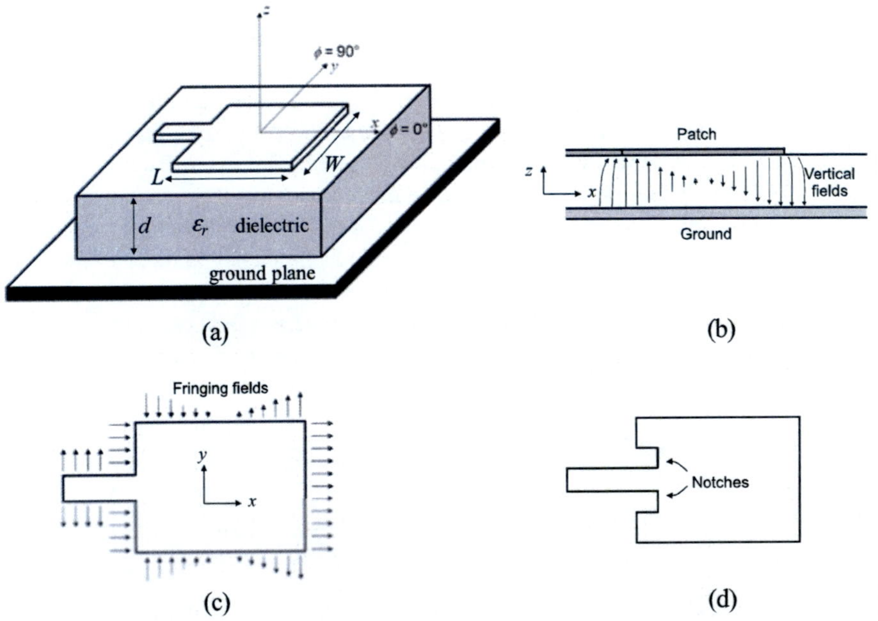

Fig. 4.24 (**a**) Structure of a typical patch antenna. (**b**) Side view showing the vertical components of the electric field. (**c**) Top view showing the horizontal components of the fringing electric field. (**d**) Patch antenna with notches. (Adapted with permission from [6], © 2012 Wiley)

subsection, we will assume an edge-fed rectangular patch antenna with a length *L* and width *W*, as depicted in Fig. 4.24a.

Patch antennas are usually operated as a resonant antenna, and the signal fed into the patch form a standing wave across the patch along the direction of the signal propagation. In the given structure, a node (the point where the electric field becomes zero) is formed in the mid-point, while two antinodes (the point where the field becomes maximum) are formed at the ends of the patch along the propagation direction (see Fig. 4.24b) with a half-wave standing wave. Another notable feature of patch antennas is the electric field mostly confined inside the space between the patch and the bottom ground plane, leaving only fringing field along the edges of the patch exposed to the air (see Fig. 4.24c). As a result, the radiation is made along the narrow gap (or "slot") between the patch and the ground plane at the edges, especially along the two *end* edges supporting the standing wave (i.e., edges along the *y* direction) where the field variation is maximum. Note that the horizontal (parallel to the patch plane) components of the fringing field on the two *side* edges of the patch (i.e., edges along the *x* direction) will have the opposite direction to each other across the patch (and side-by-side as well), and the radiation from these side edges will be canceled out at the far field making no contribution. So will the fringing field along the feeding microstrip lines. On the other hand, the horizontal component of the fringing field for the two end edges have the same direction, leading to reinforcement and broadside (perpendicular to the patch plane) radiation upward. This is a clear difference from the case of on-chip dipole antennas that radiate downward through the substrate. This upward radiation is favored in most practical cases as the issues related to substrate loss are absent and mounting on Si lenses is not needed. However, radiation through the narrow gap between the patch and the ground will limit the operation as will be discussed shortly.

The radiation properties of a patch antenna are mostly determined by the length L and width W of the patch, along with the thickness d and dielectric constant ε_r of the dielectric layer underneath. The operation frequency is dictated by L, a natural consequence of the standing wave being formed along the patch length direction. In this case, one would expect a relation of $L = \lambda/2$ is valid considering the half-wave standing wave. However, the actual physical length L required will be slightly smaller than this, because the fringing field protruding outside the patch also contributes to the total effective length. To be more quantitative, the physical length will be expressed as:

$$L = \frac{\lambda}{2} - 2\Delta L = \frac{\lambda_0}{2\sqrt{\varepsilon_{\mathrm{eff}}}} - 2\Delta L, \tag{4.61}$$

where λ_0 is the free space wavelength, $\varepsilon_{\mathrm{eff}}$ is the effective dielectric constant, and ΔL the fringing length that represents the effect of the fringing field on either side of the patch length, which is approximated as [5, 6]:

$$\Delta L = 0.412 \frac{(\varepsilon_{\mathrm{eff}} + 0.3)(W/d + 0.264)}{(\varepsilon_{\mathrm{eff}} - 0.258)(W/d + 0.8)} d. \tag{4.62}$$

As for $\varepsilon_{\mathrm{eff}}$ for microstrip structures including patches, the following approximate expression is widely accepted [29]:

$$\varepsilon_{\mathrm{eff}} = \frac{\varepsilon_r + 1}{2} + \frac{\varepsilon_r - 1}{2} \left(1 + 12\frac{d}{W}\right)^{-1/2}, \quad \text{if } W \geq d. \tag{4.63}$$

Beware that this formula is based on a quasi-static analysis and does not explain the frequency dependence of $\varepsilon_{\mathrm{eff}}$. We will come back to this issue in Sect. 4.3.2. While the patch length L dominantly influences the wavelength and thus the operation frequency of the antenna, the width W is a parameter that can be used to tune the input impedance of the antenna for a fixed L. For an edge-fed half-wave rectangular patch antenna, the input impedance can be approximately expressed as [6, 30]:

$$Z_A = R_A = 90 \frac{\varepsilon_r^2}{\varepsilon_r - 1} \left(\frac{L}{W}\right)^2 (\Omega). \tag{4.64}$$

Note that the reactance is zero as the resonant condition is assumed. As an additional method to control the input impedance (and thus the antenna matching), the formation of a recess notch on both sides of the microstrip feedline is also widely adopted (see Fig. 4.24d). The input resistance decreases with the recess depth (along x direction) until it reaches $L/2$ at which the resistance becomes zero, then it bounces back eventually to the original value.

As briefly mentioned, the main lobe of the radiation is made on the broadside of the patch antenna, a direction perpendicular to the patch plane. With the coordinate denoted in Fig. 4.24a, the electric field that contributes to the far-field radiation is immersed in the x–z plane. Hence, in terms of the spherical coordinate, the E-plane corresponds to the $\phi = 0°$ plane, while the H-plane is for $\phi = 90°$. The radiation patterns for the two primary planes are:

$$F_E(\theta) = \cos\left(\frac{\beta L}{2}\sin\theta\right) \quad \text{for } E\text{-plane}, \tag{4.65}$$

$$F_H(\theta) = \cos\theta \frac{\sin\left(\frac{\beta W}{2}\sin\theta\right)}{\frac{\beta W}{2}\sin\theta} \quad \text{for } H\text{-plane}, \tag{4.66}$$

where β is the phase constant in free space. With the pattern, a linear polarization along the feed direction (x-direction in Fig. 4.24a) is obtained for rectangular patch antennas. The directivity of a rectangular patch antennas depends on the patch width and shows a larger value than dipole antennas. If we assume $d \ll \lambda_0/2\pi$, the directivity is around 6.6 or 8.2 dBi when $W \ll \lambda_0$, whereas it asymptotically approaches $8(W/\lambda_0)$ for $W \gg \lambda_0$ [5].

One major downside of half-wave patch antennas is the narrow bandwidth, lower than that of half-wave dipole antennas, typically falling below 10% for on-chip cases. An empirical expression for the (fractional) bandwidth of rectangular patch antennas is given as:

$$B = 3.77\frac{\varepsilon_r - 1}{\varepsilon_r^2}\frac{W}{L}\frac{d}{\lambda_0} \quad \text{if } d \ll \lambda_0. \tag{4.67}$$

It is apparent from the relation that large W/L and d/λ_0 ratios help to improve the bandwidth. The dependence on d, in particular, would favor higher frequency operation, as it will correspond to a longer electrical dimension. For instance, for on-chip patch antennas with typical parameter values of $d = 10$ μm, $\varepsilon_r = 3.9$ (SiO_2), and $W/L = 1$, the estimated bandwidth for 100 GHz is 2.4%, but it will increase up to 7.2% at 300 GHz. The radiation efficiency also increases with dielectric thickness d, an obvious result of the radiation barely making its way out through the gap (slot) between the patch and the ground plane along the patch edge.

The inherent planar configuration of patch antennas has made them one of the most popular choices for integrated on-chip antennas, especially for high-frequency applications that allow a reasonable size for half-wave structures. With standard Si technologies, the patch is typically formed with the top metal layer while the ground plane is made of one of the bottom metal layers, between which is located the BEOL dielectric layer. Hence, no extra layer or lithographic patterning step is required to build patch antennas, indicating the full compatibility of the antenna with the standard fabrication process. One major challenge with this configuration arises from the fact that the maximum available dielectric layer thickness is predefined by the fabrication process, typically around 5–10 μm, limiting the performance, especially efficiency and bandwidth. A report shows that the increase of dielectric thickness from 3 to 6 μm improves the simulated efficiency from 12% to 43% near 400 GHz [31]. One may consider drastically increasing *d* by utilizing a backside metal as the ground plane, in which case the Si substrate thickness will be added for the total thickness. However, for standard Si technologies, backside metallization is not offered as a standard process step. Even if in-house post-process backside metallization were carried out, the resultant exposure to the lossy Si substrate will degrade radiation properties. Regarding this issue, the situation for III–V technologies based on GaAs or InP substrates is much more favorable, since backside metallization is often available *and* the substrate loss is significantly smaller than in Si wafers, making it a viable option to use the backside metal as the ground for patch antennas. In this case, however, the generation of the substrate mode may degrade the efficiency, partially negating the advantage of increased dielectric thickness.

Despite these issues, lots of application cases of on-chip patch antennas for THz operation have been reported, many of them integrated with functional THz circuits [32–36]. Various modifications have also been made for those on-chip THz patch antennas to improve the performance or meet the specific applications, some of which are briefly described below. Differential patch antennas, as opposed to the standard patch antennas in the single-ended configuration, have been adopted for some THz applications to facilitate a direct connection to the adjacent circuit block in the differential configurations [32, 33]. In such a differential patch antenna, which is driven by a pair of signal feeds carrying a differential signal, the orientation of the standing wave is dictated by the location of the feed point, not necessarily along the feed direction as is the case for regular single-ended patch antennas. It was reported that stacking a dielectric resonator on top of a CMOS on-chip patch antenna resulted in an antenna gain improvement by more than 5 dB around 340 GHz [37], which is a similar approach to the antenna with superstrate that was demonstrated for on-chip slot ring structures [38]. It was also reported that a segmented metallic resonator, formed with the top metal layer, was stacked on top of a CMOS on-chip patch antenna and resulted in a significantly improved efficiency and bandwidth around 300 GHz [39]. To mitigate the generation of the substrate mode in patch antennas with backside ground metal plane as mentioned above, a cavity antenna operating near 300 GHz has been developed in which via holes are formed around the patch, effectively suppressing the substrate mode in the InP substrate [40].

4.2.5 Broadband Antennas

As mentioned earlier, broadband antennas refer to a group of antennas that exhibit a wide bandwidth characteristic. Especially, those with exceptionally wide bandwidth (10:1 or more) are called the frequency-independent antennas [41, 42], which will be the main focus of this subsection. It is known that frequency-independent antennas are based on a design that is mainly characterized by the angle rather than the length, since a finite length typically involves frequency-selective resonances while an angle does not. Also, frequency-independent antennas are self-complementary, meaning that they internally include mutually complementary structures at the same time, so that a shift or rotation of one structure may exactly overlay on the complementary counterpart. These descriptions may sound too conceptual to the readers, but they will make better sense as examples of this type of antennas are introduced shortly below. For now, in a plain language, it can be said that a frequency-independent antenna includes patterns with a wide range of width so that the resonance can be obtained for a wide range of frequency. Two types of frequency-independent antennas will be briefly reviewed: spiral antennas and log-periodic antennas.

4.2.5.1 Spiral Antennas

It is known that an antenna with a pattern described by the following equiangular spiral curve is practically frequency-independent beyond a certain value [41]:

$$r = r_0 e^{a\phi}, \tag{4.68}$$

where ϕ is the angle from a reference axis, r_0 is the initial radius at $\phi = 0$, a is the flare rate that indicates the rate of radius increase. The actual shape of the curve is depicted in Fig. 4.25a. The spiral antenna with shapes based on an equiangular spiral

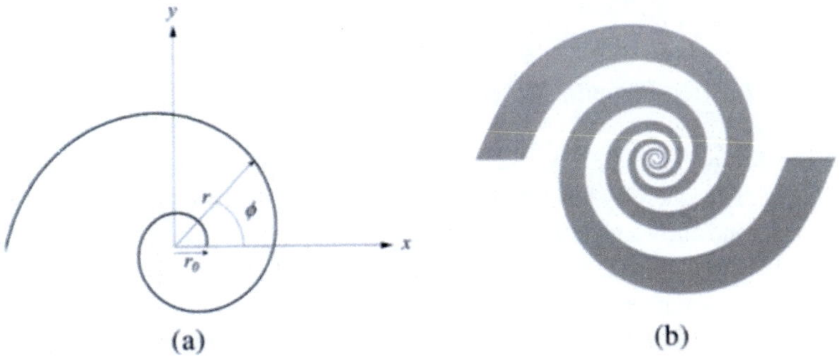

Fig. 4.25 (**a**) Equiangular spiral curve with key structure parameters (Reproduced with permission from [6], © 2012 Wiley). (**b**) The pattern of a typical of equiangular spiral antenna (log-spiral antenna)

curve is called the *equiangular spiral antenna* [43], which is the most popular type of spiral antennas. It is also widely referred to as a log-spiral antenna. As shown in Fig. 4.25b, it is composed of two symmetric interwound metallic patterns, the edges of which are all based on an equiangular spiral curve, except for the outmost portion that is truncated for a finite dimension. The inner minimum radius and the outer maximum radius of the pattern determine the upper and the lower boundary of the operation frequency band, respectively, which roughly corresponds to a quarter-wavelength of the associated frequency. The signal is fed across the inner parts of the two spiral arms. It can be seen from the figure that the antenna is self-complementary, as the shape of the area covered by the metal pattern is basically identical to the shape of the unpatterned area. Hence, in the given configuration, a spiral dipole antenna and a complementary slot antenna coexist. Also, the pattern is dictated by the angle as can be seen from Eq. (4.68), with the length being a dependent parameter of the angle. With these features, the equiangular spiral antenna is a good example of the frequency-independent antennas.

The radiation is made on the broadside of the antenna along vertical directions perpendicular to the spiral plane, upwards and downwards, with an approximate radiation pattern of cosθ, leading to a pair of symmetric main lobes if the antenna structure is placed in the middle of a homogeneous medium (such as the air). However, for the case of on-chip implementation, which is frequently found for THz applications, the radiation will be mostly toward the substrate as was the case for on-chip dipole antennas due to the higher dielectric constant of the substrate than that of the air. The radiation is circularly polarized, with the sense (left- or right-handed) opposite for the two beams upwards and downwards. The antenna impedance is theoretically $\eta/2$ in the uniform medium (188.5 Ω in the air), but the actual value is lower than this due to the finite arm length, non-zero thickness, etc. A slot version of the equiangular spiral antenna is also available, in which the spiral arm patterns are made of slots carved into a metal plane. There are other versions of spiral antennas that are not based on Eq. (4.68). Archimedean spiral antennas maintain the same metal width and gap as the spiral patterns flare out. A 3D version of the equiangular spiral antenna is the conical spiral antenna.

4.2.5.2 Log-Periodic Antennas

The log-periodic antenna is another widely deployed antenna that belongs to the class of frequency-dependent antennas [44]. In this type of antenna, the geometric variation follows a periodic function governed by the logarithm of the distance. This geometric feature is translated into the electrical characteristics of the antenna, such as the impedance and radiation parameters, varying periodically over the logarithm of the wavelength and thus the frequency. One well-known example is the TV antennas that used to cover the roofs of many house buildings but are now fading away, officially called the log-periodic dipole array (LPDA). For this antenna, an array of dipole wire antennas is placed with the spacing varying exponentially with distance (thus varying linearly with its logarithm). Here, we will focus on a different

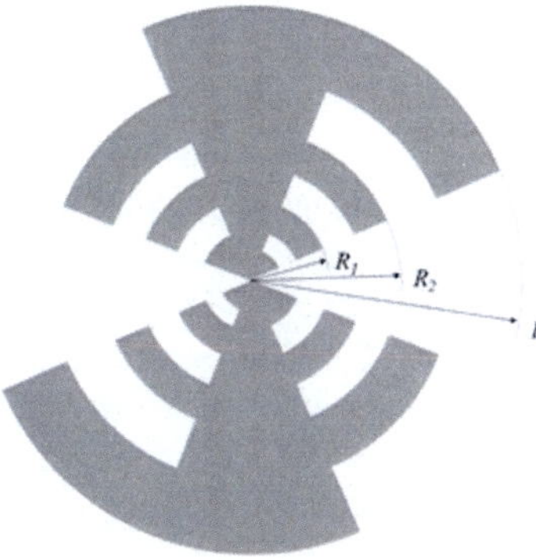

Fig. 4.26 Pattern of a typical log-periodic toothed planar antenna

configuration, which is called the *log-periodic toothed planar antenna*, as shown in Fig. 4.26. Being based on a planar structure, it is widely adopted for compact planar applications including on-chip antennas on semiconductor substrates.

It can be considered as a planar bow-tie antenna added with tooth-shaped periodic patterns protruding outward from the bow-tie. As the distance of each tooth from the center (R) increases exponentially, its ratio between adjacent teeth (R_n/R_{n+1}) will remain constant, a similar situation to the equiangular spiral antennas. Hence, it is expected that the antenna performance will vary periodically over the logarithm of frequency, and if the variation is not significant, the antenna can be considered frequency independent. The bandwidth will be determined by the dimension of the smallest and the largest tooth, which correspond to a quarter-wavelength of the boundary frequencies. The signal feed is made between the inner edges of the two conductor patterns near the center. With a proper adjustment of the tooth length, the patterns can be made self-complementary, a signature of frequency-independent antennas. The radiation pattern is on the broadside of the antenna, with a pair of symmetric main lobes formed normal to the pattern plane if the structure is immersed in a homogeneous medium. If built on a substrate, it will have a downward lobe prevailing over the other, as was the case for other dipole-based antennas discussed before. Its polarization is linear, oriented along the edges of the teeth. The impedance will be again expected to be $\eta/2$ assuming homogeneous medium, although the actual value may vary. There are whole different variations of log-periodic antennas depending on the detailed geometric structure, such as log-periodic toothed wedge antenna, log-periodic toothed trapezoid antenna, log-periodic zig-zag antenna, and so forth.

Both equiangular spiral antennas and log-periodic toothed planar antenna have been widely employed for broadband THz applications, often integrated with thermal detectors such as bolometers that are known for a wide bandwidth [45]. They have also been popular choices for optics-based photoconductive THz emitters and detectors for both pulses [46, 47] and photomixer-based CW waves [48, 49]. When these antennas are used as a photoconductive antenna, the driving laser beam is incident on the inner gap between the two conductor arms, which obviates the need for an electric signal feed. In this case, this active region around the gap is often patterned with interdigitated or other similar structures for improved conversion efficiency [49, 50]. The details of optical methods will be provided in Chap. 5. Currently, the adoption of the frequency-independent antennas has been limited to the integration with single components such as individual detectors. The integration with large-scale THz circuits based on mainstream semiconductor technologies is yet to come. This may be partly due to the moderate bandwidths currently exhibited by those highly integrated THz circuits, which would not require such a wide bandwidth as is provided by the frequency-independent antennas at this point. However, we can expect that this type of antennas will soon respond to the future growing needs from various types of THz circuits for a wideband operation.

4.3 Waveguides

In the sections above, we went through the propagation of electromagnetic waves in free space or homogenous media, and the antenna as a device that generates such electromagnetic waves from the guided waves or vice versa. To complete the series of discussion, it is natural to bring our attention to the waveguides that physically confine and guide electromagnetic waves. For this purpose, in this section, we will review three types of waveguides: metallic rectangular waveguides, planar waveguides built on dielectric substrates, and dielectric waveguides with various configurations. Additionally, transition between different waveguide types, especially between rectangular waveguides and two popular planar waveguide, microstrip lines and CPW, will be described.

Before moving onto these topics, it will help to remind the readers of the types of wave propagation and modes inside the waveguides. There can be three types of waves inside waveguides depending on the propagation scheme: transverse electromagnetic (TEM), transverse electric (TE), and transverse magnetic (TM) waves. TEM waves carry only transverse components for both electric and magnetic fields, without any longitudinal field component. TE (TM) waves include a longitudinal component for magnetic (electric) field while electric (magnetic) field is still purely transverse. A common confusion that perplexes some students in accepting this definition is a simple fact they should have learned in the electromagnetics classes: the propagation direction of electromagnetic waves is always perpendicular to both electric and magnetic fields. This makes them wonder how there can be a *longitudinal* field component at all for a propagating wave in TE or TM modes. This confusion can be resolved by clarifying the reference of the *longitudinal* (and transverse) direction: The longitudinal direction here is defined with reference to the waveguide axis direction, not the physical propagation direction of electromagnetic waves inside the waveguide. The waveguide axis direction and the wave propagation direction are aligned only for TEM mode. Hence, in either TE or TM modes, electromagnetic waves pass through the waveguide as they continuously bump on the waveguide boundaries.

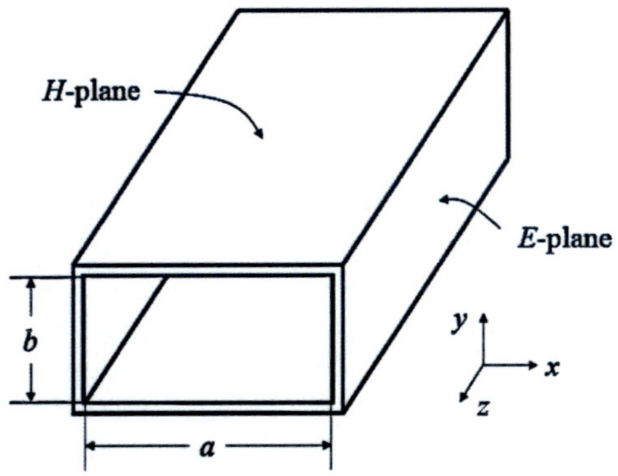

Fig. 4.27 Structure and dimension of a rectangular waveguide

4.3.1 Rectangular Waveguides

Rectangular waveguides have long been used from the early stage of microwave engineering history, and their use has been recently extended for THz band applications. While planar waveguides, which will be discussed in Sect. 4.3.2, are increasingly popular for high-frequency applications owing to their compatibility with integrated circuits, rectangular waveguides are still widely employed in many practical situations, such as packaging and inter-package links. A rectangular waveguide is basically a hollow pipe with a rectangular cross section. The structure is typically made of metal, although they can also be built based on dielectric, silicon for instance, if the interior wall surface is properly coated with a conducting film. The inner space is usually left empty, or filled with the air, while in principle any dielectric can occupy the space. Figure 4.27 shows a typical rectangular waveguide, with a cross-section width and height of a and b along x- and y-axis, respectively. The aspect ratio of the cross section for standard rectangular waveguides is 2:1, or $a = 2b$. The planes parallel to y–z plane (including the narrow-side walls) and x–z plane (including the broad-side walls) are E-plane and H-plane, respectively, as they contain the corresponding field in the basic operation mode, or the dominant mode as will be described later.

The nomenclature of standard rectangular waveguides, as briefly mentioned in Chap. 1, is WRx, where x is the width of the broad side (a in Fig. 4.27) shown in the unit of 10 mils. “WR” stands for Waveguide Rectangular. The dimensions of standard rectangular waveguides are summarized in Table 4.1, along with the lower and upper boundaries of the operation frequency range that will be discussed later in this section. One can easily notice that the dimensions of the rectangular waveguides operating in the THz band are very tiny, mostly below millimeter range. In addition to the raised precision level required for the fabrication, the small dimension would make the connection between waveguide parts quite challenging, as even a slight misalignment would lead to a significant loss. One unique feature of rectangular waveguides is that the body (or the interior surface) is composed of one piece of conductor. A main consequence of such a configuration is that TEM mode is

Table 4.1 Waveguide standard dimensions and frequency boundaries

Standard	*a* (in.)	*b* (in.)	*a* (mm)	*b* (mm)	f_{down} (GHz) Lower boundary (cutoff frequency)	f_{up} (GHz) Upper boundary
WR284	2.840	1.420	72.1	36.1	2.1	4.2
WR137	1.372	0.686	34.8	17.4	4.3	8.6
WR90	0.900	0.450	22.9	11.4	6.6	13.1
WR62	0.622	0.311	15.8	7.90	9.5	19.0
WR42	0.420	0.210	10.7	5.33	14	28
WR28	0.280	0.140	7.11	3.56	21	42
WR22	0.224	0.112	5.69	2.84	26	53
WR19	0.188	0.094	4.78	2.39	31	63
WR15	0.148	0.074	3.76	1.88	40	80
WR12	0.122	0.061	3.10	1.55	48	97
WR10	0.100	0.050	2.54	1.27	59	118
WR8.0	0.080	0.040	2.03	1.02	74	148
WR6.5	0.065	0.033	1.65	0.83	91	182
WR5.1	0.051	0.026	1.30	0.65	116	231
WR3.4	0.034	0.017	0.86	0.43	174	347
WR2.0	0.020	0.010	0.51	0.25	295	590
WR1.5	0.015	0.008	0.38	0.19	393	787
WR1.0	0.010	0.005	0.25	0.13	590	1180

not allowed inside the waveguide, leaving TE and TM modes only available states for propagation. More detailed properties of rectangular waveguides will be reviewed below mainly with TE modes, with a brief comparison with TM modes.

4.3.1.1 Propagation in TE Modes

For TE modes, the magnetic field is composed of both longitudinal (H_z) and transverse components (H_x and H_y), while the electric field has only the transverse components (E_x and E_y). The non-zero field components can be obtained from the wave equation, on which the boundary condition (the tangential components of *E*-field are zero at the conductor inner surfaces) are applied, leading to the following expressions [51]:

$$H_z = A_{mn} \cos\left(\frac{m\pi x}{a}\right) \cos\left(\frac{n\pi y}{b}\right) \mathrm{e}^{-j\beta z}, \tag{4.69}$$

$$H_x = \frac{j\beta m\pi}{k_c^2 a} A_{mn} \sin\left(\frac{m\pi x}{a}\right) \cos\left(\frac{n\pi y}{b}\right) \mathrm{e}^{-j\beta z}, \tag{4.70}$$

$$H_y = \frac{j\beta n\pi}{k_c^2 b} A_{mn} \cos\left(\frac{m\pi x}{a}\right) \sin\left(\frac{n\pi y}{b}\right) \mathrm{e}^{-j\beta z}, \tag{4.71}$$

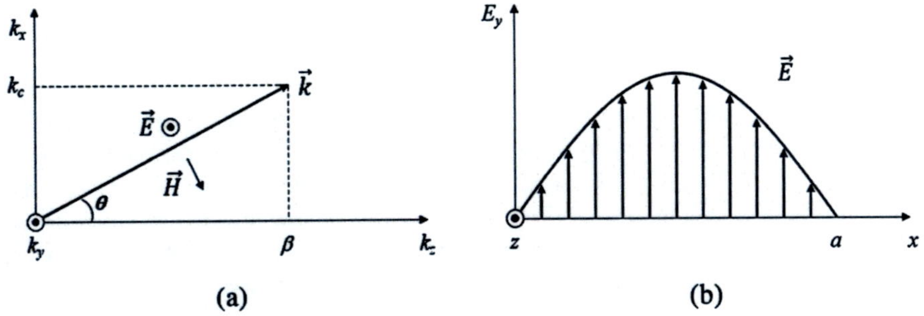

Fig. 4.28 (**a**) Components of a wave vector **k** assuming k_c has only x component, which corresponds to TE_{m0} modes. (**b**) Profile of E-field for TE_{10} mode

$$E_x = \frac{j\omega\mu n\pi}{k_c^2 b} A_{mn} \cos\left(\frac{m\pi x}{a}\right) \sin\left(\frac{n\pi y}{b}\right) e^{-j\beta z}, \tag{4.72}$$

$$E_y = -\frac{j\omega\mu m\pi}{k_c^2 a} A_{mn} \sin\left(\frac{m\pi x}{a}\right) \cos\left(\frac{n\pi y}{b}\right) e^{-j\beta z}, \tag{4.73}$$

where A_{mn} is an arbitrary coefficient, k_c is the cutoff wave number, β is the propagation constant, and $m = 0, 1, 2\ldots$ and $n = 0, 1, 2\ldots$ Note that the full expressions for the fields can be obtained by multiplying a factor of $e^{j\omega t}$ to account for the sinusoidal time dependence.

Understanding the behavior of the waves with the fields given above should start with reviewing the definitions of k_c and β. Simply speaking, k_c and β are the transverse and longitudinal components of the wave vector **k**, whose direction aligns with the wave propagation with a magnitude of $2\pi/\lambda$. This is depicted in Fig. 4.28a. From the definitions, the following relation holds:

$$k^2 = k_c^2 + \beta^2, \tag{4.74}$$

where k is the magnitude of **k**, also known as the wave number. A key point here is that k and β are not identical, as opposed to the case with the plane waves, or TEM mode propagation. This is a direct consequence of the fact that the wave propagation inside a rectangular waveguide is not aligned with the waveguide axis along the z-axis, or the longitudinal direction, due to the existence of the non-zero z-component of H-field. In a sense, the wave is composed of the transverse and longitudinal components, each corresponding to wave number of k_c and β, respectively. We can also define corresponding wavelengths, the cutoff wavelength λ_c and the guide wavelength λ_g, each satisfying the relation $k_c = 2\pi/\lambda_c$ and $\beta = 2\pi/\lambda_g$. They can be understood as effective wavelengths seen along the transverse and longitudinal directions, respectively, while λ is the true wavelength that can be observed along the propagation direction.

If we consider the two-dimensional cross-section of rectangular waveguides normal to the longitudinal direction, the transverse direction can be further sub-divided into two components, each corresponding to x and y direction, such that:

$$k_c^2 = k_{cx}^2 + k_{cy}^2, \tag{4.75}$$

where k_{cx} and k_{cy}, are x and y component of k_c, respectively. Again, we can define corresponding wavelengths for each component of k_c, as denoted by λ_{cx} and λ_{cy}, so that $k_{cx} = 2\pi/\lambda_{cx}$, and $k_{cy} = 2\pi/\lambda_{cy}$. Now, with this picture in mind, one can easily see from Eqs. (4.70)–(4.73) that standing waves are effectively formed along the *transverse direction*, both x and y directions. In this view, m and n represent the number of nodes along x- and y-axis, respectively, satisfying the relations $a = m\lambda_{cx}/2$ and $b = n\lambda_{cy}/2$, or equivalently, $k_{cx} = m\pi/a$ and $k_{cy} = n\pi/b$. For the *longitudinal direction*, wave propagation is maintained as implied by the propagation factor $e^{-j\beta z}$. This overall configuration is referred to as TE_{mn} mode. It should be noted that the solution for $m = 0$ and $n = 0$ is trivial as it corresponds to $\mathbf{E} = \mathbf{H} = 0$. That is, there is no TE_{00} mode allowed inside a rectangular waveguide.

4.3.1.2 Cutoff Frequency

One may wonder why the transverse component k_c is called the "cutoff" wave number. To understand this, consider a modified expression of Eq. (4.74): $\beta = \sqrt{k^2 - k_c^2}$. If k is smaller than k_c, β will be imaginary. With an imaginary β, the wave will decay exponentially and does not propagate any more. Hence, k_c is the smallest allowed value for k to maintain the wave propagation, leading to the name "cutoff". Now, one can easily infer that if there is a "cutoff wave number", there should be a "cutoff frequency" as well. Yes, this is certainly true, as manifested by the simple one-to-one relation between k and f, or $ck = 2\pi f$, where c is the speed of light. To arrive at the expression for the cutoff frequency f_c, first consider the expression for k_c with the boundary condition mentioned earlier:

$$k_c = \sqrt{k_{cx}^2 + k_{cy}^2} = \sqrt{\left(\frac{m\pi}{a}\right)^2 + \left(\frac{n\pi}{b}\right)^2}. \tag{4.76}$$

To explicitly show the dependence on m and n, k_c may be expressed as $k_{c,mn}$. Based on this relation, the expression for f_c becomes:

$$f_{c,mn} = \frac{ck_{c,mn}}{2\pi} = \frac{c}{2}\sqrt{\left(\frac{m}{a}\right)^2 + \left(\frac{n}{b}\right)^2}. \tag{4.77}$$

This is the well-known cutoff frequency formula for TE_{mn} modes, below which the mode ceases to exist (beware that a and b are the waveguide dimensions, whereas c is the speed of light in this equation and the description below). Each mode has its own cutoff frequency, and the mode with the lowest f_c is called the *dominant mode*. For rectangular waveguides with $a > b$, the dominant mode is TE_{10}. In this case, $f_{c,10}$ is simply $c/2a$. The mode with the second lowest cutoff frequency will depend on the

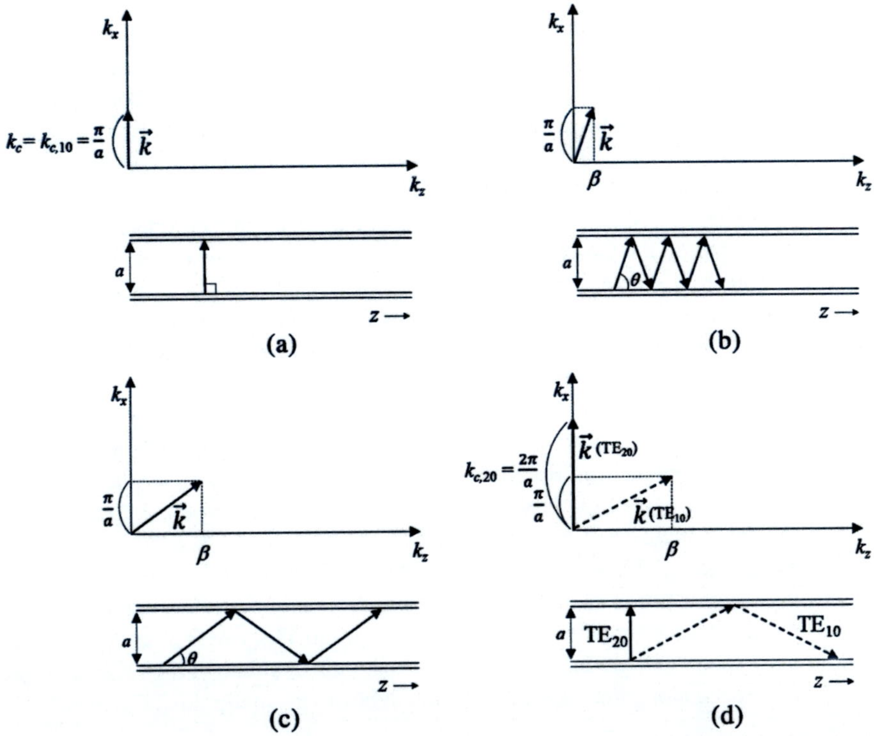

Fig. 4.29 Decomposition of a wave vector **k** (top figures), shown along with the direction of the corresponding wave propagation inside a rectangular waveguide (bottom figures) for various frequency points: **(a)** At $f_{c,10}$ (f_c of TE_{10}), **(b)** Near $f_{c,10}$, **(c)** Away from $f_{c,10}$, **(d)** At $f_{c,20}$ (f_c of TE_{20})

ratio between a and b. For standard rectangular waveguides with $a = 2b$, this mode is TE_{20}, which will be excited if frequency is increased beyond $f_{c,20}$, or c/a. Hence, standard rectangular waveguides will maintain a single-mode operation for frequency from $c/2a$ up to c/a, beyond which the waveguide will be *overmoded*. The operation frequency range of standard rectangular waveguides is typically given as the range for the single-mode operation, or between $c/2a$ and c/a, as listed in Table 4.1, although wave propagation is allowed even in the overmoded condition (but not allowed below f_c as mentioned earlier). The "recommended operation frequency" is narrower than this range to allow margins for practical reasons. In fact, the frequency ranges for selected waveguide standards as shown in Table 1.3 are based on this recommended operation frequency.

It would be useful to observe what actually happens inside a standard rectangular waveguide around the dominant TE_{10} mode including the behavior near the cutoff frequencies. The readers should remember that, in this mode, E-field has only y component and k_c has only x component (i.e., $k_c = k_{cx}$). Now, take a look at Fig. 4.29, which illustrates the various cases for the wave propagation path inside the waveguide with gradually increasing frequency from $f_{c,10}$ up to $f_{c,20}$. First consider the case of $f = f_{c,10}$, in which $k = k_{c,10}$ and $\beta = 0$ as shown in Fig. 4.29a. This condition indicates that there exists only the transverse component of the wave vector and the wave does not make propagation along the longitudinal direction. If f is slightly pushed beyond $f_{c,10}$, as is shown in Fig. 4.29b, k becomes larger than $k_{c,10}$, and β takes on a non-zero real value. This triggers wave propagation, which will progress along the waveguide following a zig-zagged path inside the waveguide as depicted in the bottom of Fig. 4.29b. Note that, in reality, a pair of waves with opposite transverse component will propagate in a symmetric fashion, although it is not shown in the figure for simplicity. The angle θ between k and β, or the angle the propagating wave forms against the waveguide wall, satisfies a relation of $\sin\theta = k_{c,10}/k = f_{c,10}/f = \lambda/\lambda_{c,10}$. With a further increase of f, k grows larger and so does β (see Fig. 4.29c), which would lead to a more straightened path of the wave propagation due to reduced θ, as described in the bottom of Fig. 4.29c. As f continues to climb up and eventually reaches $f_{c,20}$, k will become large enough to excite TE_{20} mode with $k = k_{c,20} = 2\pi/a$. At this point (and beyond), the decomposition of k into k_c and β will have two scenarios depending on the mode, which is depicted in Fig. 4.29d: First, if the wave stays with TE_{10} mode, β continues to absorb the entire increase in k, while k_c remains at $k_{c,10} = \pi/a$. Second, if the wave enters TE_{20} mode, k_c will satisfy a new boundary condition of $k_c = k_{c,20} = 2\pi/a$, which will take up most of k, leading to an abrupt reduction of β. As a result, there will be two modes available inside the waveguide, TE_{10} and TE_{20}. With a further push of f, β for TE_{20} mode will gradually recover the magnitude, although the (averaged) propagation speed along the waveguide axis will be much smaller for TE_{20} than TE_{10} mode.

4.3.1.3 Characteristic Impedance

The characteristic impedance of TE modes, Z_{TE}, is given as the ratio of E-field and H-field normal to each other, or

$$Z_{TE} = \frac{E_x}{H_y} = -\frac{E_y}{H_x} = \frac{\omega\mu}{\beta} = \frac{k}{\beta}\eta, \tag{4.78}$$

where η is the intrinsic impedance of the dielectric inside the waveguide (in free space, $\eta = \eta_0 = 376.7\ \Omega$). This impedance can be compared with the TEM case (although not allowed in rectangular waveguide), where $Z_{TEM} = \eta$. This is as expected from the fact that $k/\beta = 1$ for TEM, since there is no longitudinal component for the field and thus no transverse component of the wave vector in TEM mode. Equation (4.78) is valid for any TE mode, but the actual value will vary over the mode since the ratio of k/β will be different for each mode. A better insight can be obtained if the expression for Z_{TE} is modified as follows based on the relation $\beta^2 = k^2 - {k_c}^2$:

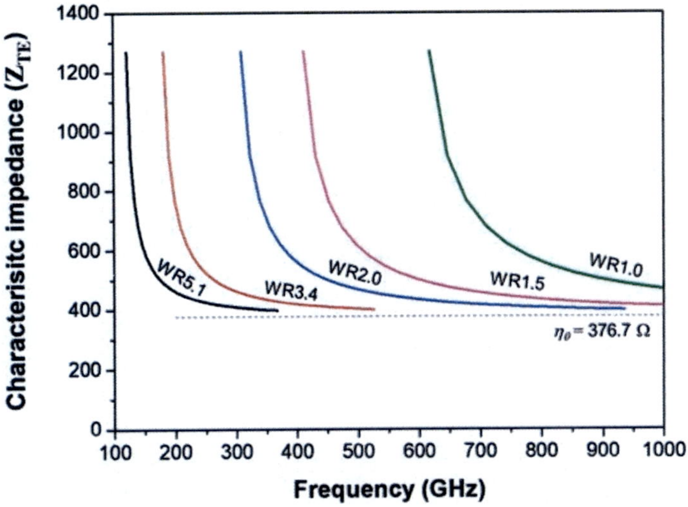

Fig. 4.30 Characteristic impedance Z_{TE} of TE_{10} mode for various waveguide standards at the THz band. It is assumed that the waveguide interior is filled with the air. Note that the values approach the intrinsic impedance η_0 (shown as dotted line) as frequency increases

$$Z_{TE} = \frac{k\eta}{\beta} = \frac{k\eta}{\sqrt{k^2 - k_c^2}} = \frac{\eta}{\sqrt{1 - (k_c/k)^2}} = \frac{\eta}{\sqrt{1 - (f_c/f)^2}}, \tag{4.79}$$

It is clear from Eq. (4.79) that Z_{TE} is a function of frequency for each mode and will show a sharp increase near the corresponding f_c. Also, η will serve as the lower boundary for Z_{TE}, which will be approached with a very high f, provided the mode is maintained. Examples are shown in Fig. 4.30 for various waveguide standards operating at the THz band assuming the interior is filled with the air.

4.3.1.4 Loss

The loss of metallic rectangular waveguides predominantly arises from the finite conductivity of the waveguide walls if dielectric loss can be neglected, which is true for waveguides filled with the air. The corresponding attenuation due to the conductor loss for TE_{10} can be expressed in terms of the waveguide dimension as follows [51]:

$$\alpha_c = \frac{R_s}{a^3 b\beta k\eta}\left(2b\pi^2 + a^3k^2\right)\ (\text{Np/m}), \tag{4.80}$$

where R_s is the sheet resistance of the waveguide wall. R_s can be obtained by $1/\sigma\delta$, where σ is the conductivity and δ is the skin depth given by $1/\sqrt{\pi f\mu\sigma}$, μ being the

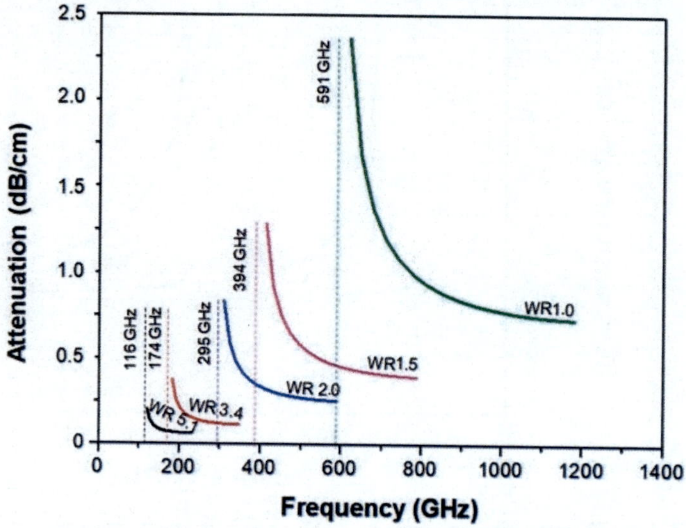

Fig. 4.31 Attenuation of TE_{10} mode for various waveguide standards at the THz band. It is assumed that the interior is filled with the air and copper is used for the waveguide wall structure. Also shown in dotted lines are f_c for each waveguide standard

permeability of the wall conductor. Some readers may not be familiar with the unit Np (Neper). It is simply a ratio that is shown in its natural logarithm. For a linear ratio of x_1/x_0, for instance, its expression in Np is $\ln(x_1/x_0)$. Then, one can easily show that the conversion from Np to dB can be made by multiplying a factor of $10\log(e)$ (or $20\log(e)$ if the ratio is for *E*- or *H*-field) to the value expressed in terms of Np. Hence, the attenuation α_c shown above can be expressed in terms of dB/m by multiplying a factor of $20\log(e)$ to the value given Eq. (4.80). The attenuation in TE_{10} mode for various waveguide standards in the THz band is plotted in Fig. 4.31. It is assumed that the interior is filled with the air and copper is used for the wall structure. When compared with f_c, also shown in the plot for reference, it is clear that attenuation sharply increases as the frequency approaches f_c.

4.3.1.5 TM Mode Basics

Up to this point, the properties for TE modes have been discussed. For TM modes, the fields are expressed as follows [51]:

$$E_z = B_{mn} \sin\left(\frac{m\pi x}{a}\right) \sin\left(\frac{n\pi y}{b}\right) e^{-j\beta z}, \tag{4.81}$$

$$E_x = -\frac{j\beta m\pi}{k_c^2 a} B_{mn} \cos\left(\frac{m\pi x}{a}\right) \sin\left(\frac{n\pi y}{b}\right) e^{-j\beta z}, \tag{4.82}$$

$$E_y = -\frac{j\beta n\pi}{k_c^2 b} B_{mn} \sin\left(\frac{m\pi x}{a}\right) \cos\left(\frac{n\pi y}{b}\right) e^{-j\beta z}, \tag{4.83}$$

$$H_x = \frac{j\omega\varepsilon n\pi}{k_c^2 b} B_{mn} \sin\left(\frac{m\pi x}{a}\right) \cos\left(\frac{n\pi y}{b}\right) e^{-j\beta z}, \tag{4.84}$$

$$H_y = -\frac{j\omega\varepsilon m\pi}{k_c^2 a} B_{mn} \cos\left(\frac{m\pi x}{a}\right) \sin\left(\frac{n\pi y}{b}\right) e^{-j\beta z}, \tag{4.85}$$

where B_{mn} is another arbitrary coefficient. The fields given above may look similar to the corresponding fields in TE modes at first glance. However, a careful inspection reveals that all the field components for TM modes will be zero if either m or n is zero, which is one outstanding difference from TE modes. This means that m and n should be positive integers for TM modes for non-trivial solutions, not allowing TM_{00}, TM_{01}, and TM_{10}. Hence, TM_{11} is the lowest-order TM mode. Another difference can be found in the characteristic impedance Z_{TM}, which is given as

$$Z_{TM} = \frac{E_x}{H_y} = -\frac{E_y}{H_x} = \frac{\beta}{k}\eta = \eta\sqrt{1 - (f_c/f)^2}. \tag{4.86}$$

Hence, η will serve as the *upper* boundary of Z_{TM}, as opposed to the case of Z_{TE}, which will be approached with a very large f. Near f_c, Z_{TM} will show a value close to zero.

Other than these, the properties of TM_{mn} modes will be largely similar to those of TE_{mn} modes. Notably, the relation $k^2 = k_c{}^2 + \beta^2$ still holds for TM modes, leading to the identical expression for the cutoff frequency $f_{c,mn}$ for TM and TE modes. If we consider a particular case of $a = 2b$, which is valid for standard rectangular waveguides, the expression for $f_{c,mn}$ will be reduced to the following for both TE and TM modes:

$$f_{c,mn} = \frac{c}{2a}\sqrt{m^2 + 4n^2} \quad \text{if } a = 2b. \tag{4.87}$$

This expression predicts that, as the frequency is increased from the dominant mode TE_{10}, we will experience in sequence the onset of the modes TE_{20}/TE_{01}, TE_{11}/TM_{11}, TE_{21}/TM_{21}, TE_{30}, and so on. Several modes are listed in Table 4.2 along with the cutoff frequency values for various waveguides at the THz band. In most practical applications of rectangular waveguides, a single-mode operation is preferred. Therefore, the cutoff frequencies of TE_{10} and TE_{20} serve as the lower and upper boundaries of operation frequency, respectively, for a given waveguide, which were assumed in Table 4.1.

Table 4.2 Cutoff frequency of the selected modes for various waveguides

Mode	Cutoff frequency f_c (GHz)					
	WR6.5	WR5.1	WR3.4	WR2.0	WR1.5	WR1.0
TE_{10}	91	116	174	295	393	590
TE_{20}	182	231	347	590	787	1180
TE_{11}	203	259	388	660	880	1320
TM_{11}	203	259	388	660	880	1320
TE_{21}	257	327	491	835	1113	1669
TM_{21}	257	327	491	835	1113	1669
TE_{30}	272	347	521	885	1180	1770
TE_{31}	327	417	626	1064	1419	2128
TM_{31}	327	417	626	1064	1419	2128

4.3.2 *Planar Waveguides*

Planar waveguides, or planar transmission lines, can be easily fabricated by placing patterned conductor films on top of a substrate. This configuration is compatible with integrated circuits based on standard semiconductor processes, and thus has made planar waveguides a popular choice for various high-frequency integrated circuit applications. The compatibility with integrated circuits is enhanced with increasing frequency due to the reduced size, a favored feature for THz applications. While there are various types of planar waveguides in practical uses, the most popular one is the microstrip line. Also widely used are coplanar waveguide (CPW) and slot lines. In this subsection, each of these three types will be briefly discussed. For more details, the readers are referred to [52].

4.3.2.1 Microstrip Lines

The *microstrip line* is made of a single conductor strip placed on top of a dielectric substrate with grounded backside. Figure 4.32 shows the structure of a typical microstrip line and the field lines in normal operation. Its simple structure and compatibility with interconnect lines of other electronic components explain its popularity with a practical basis. As shown in Fig. 4.32b, the field lines are spread over both air and dielectric regions, which complicates the propagation properties of microstrip lines. If there were wave propagating in TEM mode along the microstrip line with the given structure, the fields in the air and in the dielectric will show a phase velocity difference by a factor of $\sqrt{\varepsilon_r}$ of the dielectric, leading to a phase mismatch at the boundary. This would be inconsistent with the boundary condition that enforces the continuity of the fields across the boundary. This expected inconsistency indicates that microstrip lines cannot support TEM mode, which would have been supported if the line were fully immersed in a homogeneous medium. However, the longitudinal components of *E*- and *H*-fields are relatively small, leading to the wave behavior quite close to that of TEM mode. Hence, microstrip

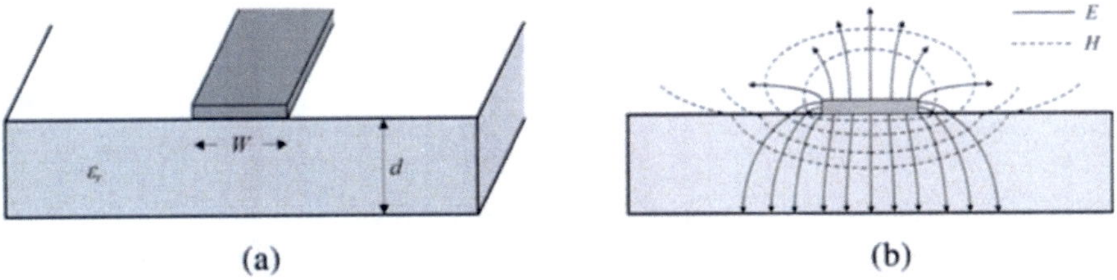

Fig. 4.32 (**a**) Structure of a microstrip line. (**b**) Cross-sectional field lines

lines are often regarded as a quasi-TEM transmission line. In this case, the signal line can be assumed to be immersed in a homogeneous medium with the effective dielectric constant $\varepsilon_{\mathrm{eff}}$, which was given earlier in Eq. (4.63). Again, the expression in Eq. (4.63) is valid only at low frequencies and will increase and approach ε_{r} as frequency increases as will be discussed shortly.

The characteristic impedance Z_0 of microstrip lines can be approximated as follows [53]:

$$\begin{aligned} Z_0 &= \frac{60}{\sqrt{\varepsilon_{\mathrm{eff}}}} \ln\left(\frac{8d}{W} + \frac{W}{4d}\right)\ (\Omega) && \text{for } \frac{W}{d} \le 1 \\ Z_0 &= \frac{120\pi}{\sqrt{\varepsilon_{\mathrm{eff}}}\left[\frac{W}{d} + 1.393 + 0.667 \ln\left(\frac{W}{d} + 1.444\right)\right]}\ (\Omega) && \text{for } \frac{W}{d} \ge 1 \end{aligned} \qquad (4.88)$$

According to these expressions, Z_0 decreases with increasing W/d ratio. A larger $\varepsilon_{\mathrm{eff}}$ also reduces the impedance. This behavior is somewhat expected if we recall the characteristic impedance of the ideal transmission line, which is given as $\sqrt{L/C}$, where L and C represent the inductance and capacitance per unit length, respectively. A wider width will decrease L but increase C, while a smaller d will also contribute to increase C. A larger $\varepsilon_{\mathrm{eff}}$ will raise C, of course, all consistent with the behavior predicted by Eq. (4.88). The approximated expressions, however, are based on quasi-static analysis. Hence, their accuracy is limited, and, more importantly, they do not reflect the frequency dependence of the line characteristics that actually exist as will be shortly discussed below. Nevertheless, these empirical formulas will be still useful for the initial design of microstrip lines, although the final design should be polished with the help of more rigorous analyses now readily available with various EM simulation tools. A quick estimation with Eq. (4.88) shows that an on-chip microstrip line with 50 Ω will require W/d of roughly 1.6 with $\varepsilon_{\mathrm{eff}}$ of 4, close to that of SiO_2. This translates to W of 16 μm, if d of 10 μm is assumed for a typical Si technology. A design based on EM simulation suggests a slightly smaller W, but not very far from the quick estimation.

The loss is another property of waveguides that affects the performance of the circuit or system when employed [54]. Three types of losses can be considered in microstrip lines: conductor loss, dielectric loss, and radiation loss [55]. The

Table 4.3 Skin depth of selected metals at various frequencies

	Skin depth δ (μm)		Sheet resistance R_s (mΩ)	
Frequency (GHz)	Aluminum	Copper	Aluminum	Copper
30	0.473	0.376	56.1	44.6
60	0.335	0.266	79.3	63.1
100	0.259	0.206	102.3	81.4
300	0.150	0.119	177.3	141.0
600	0.106	0.084	250.7	199.4
1000	0.082	0.065	323.6	257.4

Following conductivity (bulk) was assumed: $\sigma = 3.77 \times 10^7$ S/m for aluminum, 5.96×10^7 S/m for copper

conductor loss arises from the finite conductivity of the metallic strip and the ground plane, in conjunction with the skin effect and surface roughness. For a small conductor loss, it certainly helps to employ metals with a lower resistivity, although in many cases, especially when using standard semiconductor technologies, the option is limited to copper or aluminum. As the field is confined approximately within a skin depth δ below the surface, a larger skin depth will result in a lower conductor loss. A concern for THz application is that the skin depth decreases with increasing frequency as indicated by the relation ($\delta = 1/\sqrt{\pi f \mu \sigma}$), leading to a higher conductor loss. This is true to any metal-based waveguides and emerges as a major challenge for their adoption to THz applications. For reference, the skin depths of aluminum and copper calculated based on the bulk conductivity are shown in Table 4.3 for various frequency points, along with the sheet resistance R_s ($=1/\sigma\delta$) calculated from these values. The surface roughness also affects the conductor loss, as the effective current path length increases with the roughness. The *dielectric loss* mainly stems from the lossy substrate of the microstrip lines. It shows a larger value with a larger loss tangent of the dielectric, as expected, and also with a larger "filling factor", which indicates the extent of the field lines immersed inside the dielectric instead of the air. The *radiation loss* is due to the microstrip line acting as an antenna [56]. It is more pronounced when the substrate dielectric constant is small so that the fields are only weakly coupled to the dielectric. Although it is the conductor loss that dominates the total loss of microstrip lines in most cases, if lines are formed directly on semiconductor substrates, especially on Si, the dielectric loss become significant due to their lossy characteristics and may exceed the conductor loss. Rather fortunately, most of on-chip microstrip lines of today are built on top of the BEOL dielectric layer, which is electrically separated from the lossy semiconductor substrate by a metal ground plane, leading to relatively small dielectric losses.

Conventional analyses on microstrip lines are quasi-static and assumes near-DC operation condition. However, microstrip lines are dispersive and indeed show frequency-dependent behaviors, partly due to the propagation that is not strictly TEM. The effective dielectric constant ε_{eff} is one example, and its value will deviate

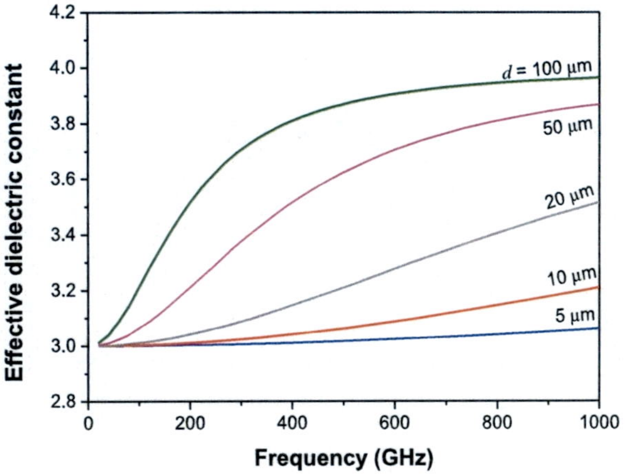

Fig. 4.33 Effective dielectric constant $\varepsilon_{eff}(f)$ shown as a function of frequency for various dielectric thickness d values, assuming $W/d = 1.5$, $\varepsilon_r = 4$, $\varepsilon_{eff}(0) = 3$, and $Z_0 = 50\ \Omega$

from the quasi-static expression given in Eq. (4.63) as frequency increases. The frequency-dependent ε_{eff} can be approximated as follows [53]:

$$\varepsilon_{eff}(f) = \varepsilon_r - \frac{\varepsilon_r - \varepsilon_{eff}(0)}{1 + G(f/f_p)^2}, \tag{4.89}$$

where $G = 0.6 + 0.009Z_0$ and $f_p = Z_0/(8\pi d)$ with Z_0 in Ω, d in cm, and f in GHz. From the formula, we can see that $\varepsilon_{eff}(f)$ will increase and eventually converge to ε_r with increasing frequency, an important fact that affects the dispersion properties of the waveguides that employ dielectric materials. Assuming its accuracy extends to the THz band, Eq. (4.89) is plotted as a function of frequency up to 1 THz in Fig. 4.33 for various d values, with W/d fixed at 1.5 and ε_r of 4 (thus $\varepsilon_{eff}(0) = 3$ from Eq. (4.63)) as well as $Z_0 = 50\ \Omega$. It shows that $\varepsilon_{eff}(f)$ does deviate from the quasi-static value (i.e., $\varepsilon_{eff}(0)$) and converge to ε_r at high frequencies. It is noted, though, that the deviation trend strongly depends on d as well, indicating that the driving factor is λ/d rather than the frequency itself. In fact, for d of 5–10 μm, typical values for on-chip microstrip lines in Si technologies, $\varepsilon_{eff}(f)$ remains reasonably close to $\varepsilon_{eff}(0)$ up to a few hundred GHz. The onset frequencies of various higher order modes in microstrip lines, which may arise from the surface wave inside the dielectric or the finite cross section of the strip conductor, also show an inverse relation to d [51], indicating λ/d as a dominant factor. Many of these onset frequencies occur at around a few THz if d is limited to 5–10 μm, a typical range for microstrip lines implemented in Si technologies. These observations indicate that microstrip lines can be applied to THz application in a way similar to the lower frequency cases unless the dielectric thickness is excessively large in terms of the wavelength.

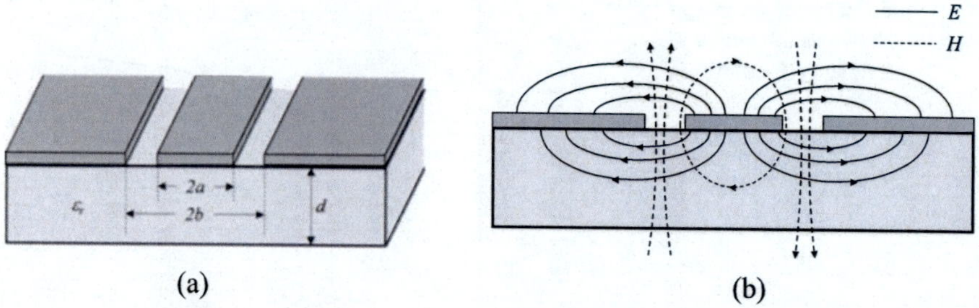

Fig. 4.34 (a) Structure of a coplanar waveguide. (b) Cross-sectional field lines

4.3.2.2 Coplanar Waveguides

The *coplanar waveguide* (CPW) is made of a conductor strip on a dielectric substrate with two ground planes placed on the sides in parallel with a gap [57]. Sketched in Fig. 4.34 is the structure of a typical CPW shown together with the field lines. CPW can be formed on a substrate with a single metal layer, as opposed to microstrip lines that need two metal layers. Such a simple fabrication process is one of the main reasons for its wide adoption in many practical applications. Also, the fully planar configuration does not need vertical via holes that typically involve undesired parasitics as well as the difficulty in the fabrication. This feature simplifies the implementation of the shunts between signal and ground lines. Besides, CPW is known for low dispersion, a major advantage over microstrip lines. On a practical note, it is suggested for CPWs that the ground edge distance b is larger than a half-wavelength at the operation frequency and the ground plane width is larger than $5b$ [55].

One issue with CPW is the parasitic slotline mode (odd mode) that appears together with the desired coplanar mode (even mode). To suppress the unwanted slotline mode, airbridges are usually employed to connect the two ground planes across the signal line, preferably with an interval less than a quarter wavelength and particularly near line discontinuities. This will require one more level of conductor layer, though, possibly with via holes across the airbridge and ground levels. Although the inclusion of airbridges complicates the fabrication process and increases the line loss, it is still often practiced for CPWs. Another issue, especially for on-chip CPWs, is the field lines penetrating into the lossy semiconductor substrate, leading to significantly increased total loss. To avoid this, a modified CPW structure that employs an additional ground plane at the bottom of the dielectric, called the conductor-backed CPW (CBCWP) or the guided microstrip lines, is often employed to electrically separate the CPW structure from the substrate [58]. Another challenge from a practical point of view arises from the complexity in the layout, as CPWs require both signal line and ground plane on the same metal level. This is more problematic for operation at high frequencies including the THz band, as the CPW layout becomes more compact with reduced line lengths.

As in the case of microstrip lines, the field lines in CPW are distributed over both dielectric and air (see Fig. 4.34), leading to quasi-TEM mode propagation. As for the effective dielectric constant that is employed to account for the given configuration, the situation is simpler than the microstrip lines, since the field distribution is symmetric across the air–dielectric boundary if an infinitely thick dielectric substrate is assumed. With this assumption, the quasi-static effective dielectric constant ε_{eff} is simply given as the average of the dielectric constant of the air and the substrate such that:

$$\varepsilon_{\text{eff}} = \frac{\varepsilon_{\text{r}} + 1}{2}. \tag{4.90}$$

With a finite dielectric thickness, which is more practical, ε_{eff} is reduced, apparently due to the field lines exposed to the air in the space below the dielectric [59]. For stacked dielectric layers, as in the case of on-chip CPWs formed on top of a BEOL dielectric layer deposited on a semiconductor substrate, the dielectric constant of both layers will affect ε_{eff}, while the effect of the top layer will be more dominant due to its denser field lines [60].

The characteristic impedance of CPW depends on the properties of the dielectric as well as the geometry of the conductor structures, the latter being represented by three parameters, a, b, and d, as denoted in Fig. 4.34 (assuming the ground plane is infinitely wide). The compact forms proposed in the literature for Z_0 with CPW mostly involve the elliptic integrals [52, 59] and are not introduced here. The general trend predicted by the formulas is a smaller Z_0 with a larger a/b and larger ε_{eff}, again predictable from the dependence of Z_0 on the unit length inductance and capacitance as discussed above. d is a less critical parameter for CPW than in the case of microstrip lines, as it does not intimately affect the field configuration in CPW, except its effect on ε_{eff}. The loss of CPW comprises the conductor loss, dielectric loss, and radiation loss, as was the case for microstrip lines. The conductor loss is more affected by the lateral dimension than in microstrip lines. The a/b ratio of around 0.5–0.7 will show the smallest conductor loss, while a/b close to 0 or 1 will lead to a sharp increase in the conductor loss [52]. The dielectric loss of CPW will be affected by the loss tangent and the filling factor in a similar way as in the microstrip line case. As for the radiation loss, it is noted that the parasitic slotmode, often excited near the line discontinuities, is particularly responsible for the radiation from CPW structure.

4.3.2.3 Slot Lines

The *slot line* is composed a conductor plane on a dielectric substrate with a gap, or slot, that separates the conductor into two parts as shown in Fig. 4.35 [61]. As was the case for CPW, slot lines can be built with a single-level metallization process, an attractive feature for many planar applications. The slot width W serves as the main geometrical parameter, which determines the overall line characteristics together

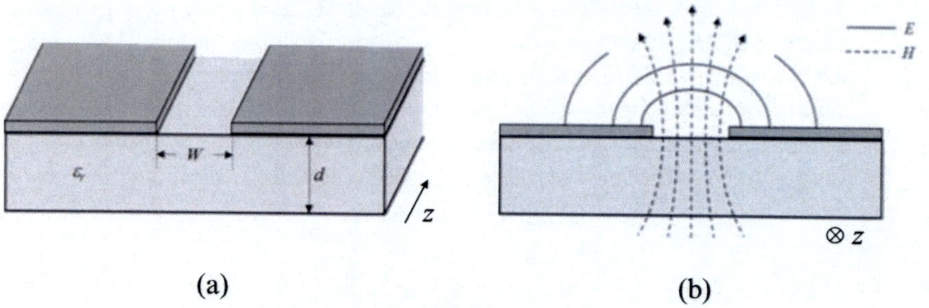

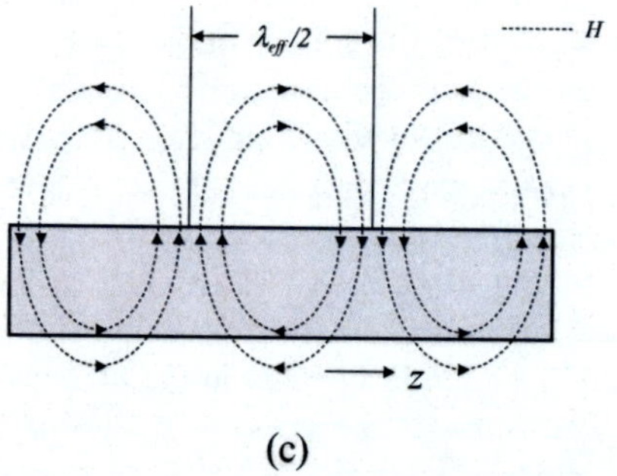

Fig. 4.35 (**a**) Structure of a slot line. (**b**) Field lines in the transverse section. (**c**) *H*-field lines in the longitudinal section

with the dielectric constant of the substrate. The dielectric thickness *d* plays a minor role as in CPW. The effective dielectric constant is identical to that of CPW, the average of the dielectric constant of the air and the substrate, which is reduced if a finite *d* is employed. As slot lines are formed on a large conductor plane with a simple slot formation, it is possible for them to be built on the ground plane of microstrip lines, allowing co-design with microstrip lines for various passive components.

The propagation mode significantly deviates from TEM and is rather close to TE, which results in the highly dispersive behavior of slot lines, limiting its broadband applications [52, 62]. Also, the presence of a significant longitudinal *H*-field component (see Fig. 4.35c) tends to promote radiation off the slot line surface, leading to a significant radiation loss. In fact, slot line structures are occasionally employed for antennas, resonant or non-resonant, to make use of this tendency for radiation. To suppress the radiation, dielectric substrates with a high ε_r are preferred, which helps to confine the field inside the dielectric. Or, its radiating property can be exploited to use the slot line for a dual purpose of waveguide and radiating element [26].

The characteristic impedance Z_0 increases with a wider gap, which is expected with a smaller unit length capacitance for a large W. A larger ε_{eff} reduces Z_0 as was the case for other waveguides discussed above, which is affected by both ε_r and d. The frequency dependence of Z_0 is stronger than in the cases with microstrip lines and CPW, a result from the non-TEM propagation as mentioned above. The loss of slot lines is again composed of the conductor, dielectric, and radiation losses. One notable behavior is a significant increase in the conductor loss for a narrow slot width. In this case, a large ε_{eff}, available with controlling either ε_r or d, would help to allow a wider slot width for a given Z_0. This strategy, however, will not be readily available for on-chip realization of slot lines, where the vertical dimension and the material parameters including dielectric constant are largely predetermined by the fabrication process. The radiation loss is, as just mentioned above, significant for slot lines. One may want to lower the radiation loss with the adoption of substrates with high ε_r, which is, again, not a practical option for on-chip slot-lines.

4.3.3 Dielectric Waveguides

As mentioned earlier, the conductor loss in metal-based waveguides becomes increasingly problematic at higher frequencies, emerging as a major issue in their THz applications. An effective and natural way to cope with this challenge is to fully eliminate the conductor from the waveguides. A free space propagation may be an option as mentioned at the beginning of this chapter, but the path loss will be excessively high if there is no confinement with optical components such as lenses. A favorable alternative will be the dielectric waveguides. As electromagnetic waves tend to reside inside a region with a higher dielectric constant, in principle, the waves can be guided with a dielectric material with a higher dielectric constant than surrounding materials without conductor-based confinement. Also, the presence of the wave inside the dielectric with a higher dielectric constant helps the total reflection at the boundary if the propagating electromagnetic waves form a shallow angle against the boundary. In fact, the optical fibers, serving as the heart for optical communication links, are a form of the dielectric waveguide that applies to the optical spectrum of electromagnetic waves. In this regard, the application of the dielectric waveguides to the high-frequency bands including the THz band seems highly feasible.

There can be various structural options for dielectric waveguides. A dielectric rod may be the simplest realization, as shown in Fig. 4.36a. Although a rectangular cross section is assumed in the figure, the cross section may take on various shapes: It may be a circle, as is for the case of optical fibers, or may include a hollow inside. There can be other types that are more compatible with planar structures as well. Figure 4.36b shows the dielectric slab waveguide, where ε_{r2} is larger than both ε_{r1} and ε_{r3} for field confinement, as is required for waveguides. The structure shown in Fig. 4.36c, called the image line [63], includes a ground plane, which enables a convenient integration with active devices that need DC biasing and also serves as a

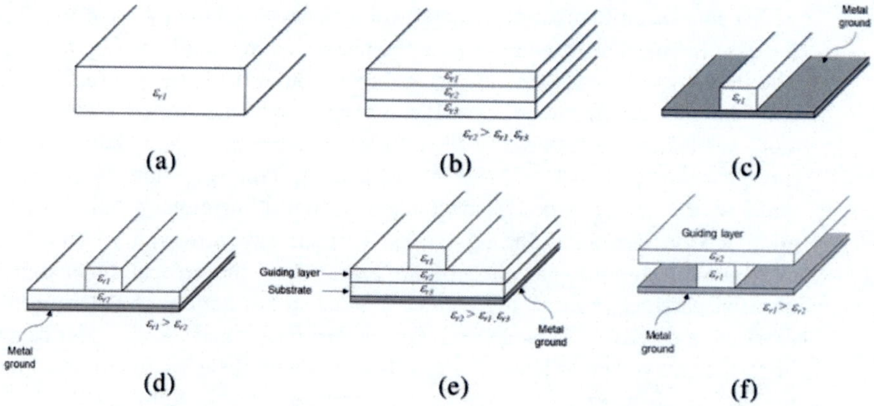

Fig. 4.36 Various dielectric waveguides: (**a**) Dielectric rod, (**b**) Dielectric slab waveguide, (**c**) Image line, (**d**) Insulated image line, (**e**) Strip dielectric waveguide, (**f**) Inverted strip dielectric waveguide

heat sink. However, the metallic plane may cause a conductor loss, the elimination of which was the main motivation of adopting dielectric waveguides. A compromised solution is the insulated image line shown in Fig. 4.36d, in which an insulating dielectric layer is inserted below the guiding dielectric strip [64]. A main idea here is to maintain the dielectric constant of the insulating layer (ε_{r2}) smaller than that of the guiding strip (ε_{r1}), so that the field is mostly confined in the upper dielectric, avoiding the conductor loss. The opposite choice for ε_{r1} and ε_{r2} (i.e., $\varepsilon_{r1} < \varepsilon_{r2}$) is also possible for the given structure, though, if one favors the reduction of the radiation loss rather than the conduction loss. There have been other variants as well, such as the strip dielectric waveguide (Fig. 4.36e) [65], in which both radiation and conduction losses are suppressed with ε_{r2} larger than both ε_{r1} and ε_{r3}, and the inverted strip dielectric waveguide with $\varepsilon_{r1} > \varepsilon_{r2}$ (Fig. 4.36f), which inherits the advantages of the strip dielectric waveguide while additionally eliminating the dielectric loss from the bottom layer [66].

As for the properties of the wave propagating inside dielectric waveguides, it should be noted that TEM mode is not supported by dielectric waveguides. They support TE and TM modes, as well as the hybrid modes in which longitudinal field components exist for both *E*-field and *H*-field. The hybrid modes are available only for non-planar waveguides with two-dimensional field confinements, not available for fully planar waveguides such as the slab waveguide shown in Fig. 4.36b. One characteristic feature of the dielectric waveguides is that the fields in dielectric waveguides are only loosely confined, and they can partially extend outside the dielectric boundary of the waveguide. This is true even for the cases with the total internal reflection mentioned above. The loose field confinement is likely to cause radiation loss, particularly at the curved section, junction, and discontinuities. Also, such a field configuration leads to the absence of analytic solutions to fully describe the fields. Even for the simple rectangular dielectric waveguide structure shown in

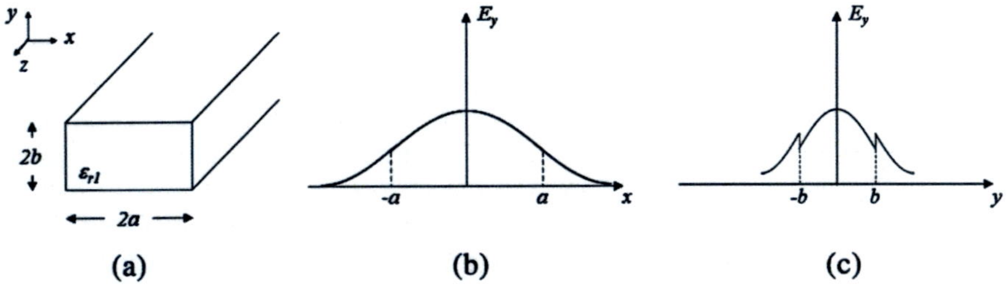

Fig. 4.37 (**a**) Structure of a rectangular dielectric slab used for Eqs. (4.91)–(4.94). (**b**) Profile of E_y along x-axis, (**c**) Profile of E_y along y-axis. The outside of the slab is the air

Fig. 4.36a, the exact solution is not available, which is in contrast to the metallic rectangular waveguides where compact analytical solutions can be readily obtained with the help of well-defined boundary conditions as discussed earlier.

Consider rectangular dielectric waveguides as an example for a little bit of details. Inside rectangular dielectric waveguides, all the modes are hybrid as mentioned above, and they can be grouped into two families, E^x_{mn} and E^y_{mn} [67]. For E^x_{mn} modes, which are obtained with x-direction polarization, the largest field components are E_x and H_y, while for the E^y_{mn} modes that are available with y-direction polarization, E_y and H_x are the strongest. The dominant mode is obtained for $m = n = 1$ for both cases. Let us take a brief look at the case of E^y_{11} based on the structure depicted in Fig. 4.37a, which is a simple dielectric slab surrounded by the air. Omitting the time factor $e^{j\omega t}$, approximate expressions for y-component of the E-field, E_y, are given as follows for various regions [67, 68]:

$$|x| \le a, |y| \le b: \quad E_y = E_0 \cos(k_x x) \cos(k_y y) e^{-j\beta z}, \tag{4.91}$$

$$|x| > a, |y| \le b: \quad E_y = E_1 \cos(k_x a) \cos(k_y y) e^{-k_{x0}(|x|-a)} e^{-j\beta z}, \tag{4.92}$$

$$|x| \le a, |y| > b: \quad E_y = E_2 \cos(k_x x) \cos(k_y b) e^{-k_{y0}(|y|-b)} e^{-j\beta z}, \tag{4.93}$$

$$|x| > a, |y| > b: \quad E_y = 0, \tag{4.94}$$

where β is the propagation constant, k_x, k_y and k_{x0}, k_{y0} are the x- and y-direction propagation constants inside and outside the waveguide, respectively. A quick inspection of the expression reveals that a sinusoidal profile is obtained for the E-field inside the waveguide, while it exponentially decays outside the boundary. We have $E_1 = E_0$ due to the continuity required for the *tangential* components of electric field at the dielectric boundary. On the other hand, for the *normal* components, $E_2 = \varepsilon_r E_0$ as the normal component of the electric flux density (**D**) should maintain continuity at the charge-free dielectric boundary. Based on these properties, the actual field distribution will appear as depicted in Fig. 4.37b, c. One can clearly observe a finite portion of the field extending outside the waveguide boundary as mentioned above. Consistent with this field extension, k_x and k_y are slightly smaller than $\pi/2a$ and $\pi/2b$, respectively, meaning that the half-wavelength is slightly larger than the physical dimension of the given waveguide for both x and y directions. This can be compared to the case of the metallic rectangular waveguide with the same dimension, where k_x and k_y exactly match $\pi/2a$ and $\pi/2b$, respectively, for the corresponding mode. H-field components of the dielectric waveguide can be calculated based on the E-field given above. For E_{11}^{x} mode, the roles of E-field and H-field will be reversed. As we move up with increasing frequency beyond the dominant mode, odd- and even-modes will be alternately show up for the higher modes, as are characterized with sine and cosine profiles, respectively. The details of the fields and modes inside dielectric waveguides can be found in other sources [67–69]. The loss and dispersion characteristics will be dependent on the specific dielectric material used in the guiding region.

There have been recent increasing efforts to apply the dielectric waveguides for communication links based on millimeter-wave and THz carrier frequencies. In [70], a rectangular dielectric waveguide based on polystyrene, with a cross-section dimension of 8 mm × 1.1 mm, was successfully employed for a link between transmitter and receiver carrying a 12.5 Gbps AKS modulated signal at 57 and 80 GHz. A similar data link experiment through a dielectric waveguide was reported in [71], for which a circular hollow polymer fiber tube was employed as a waveguide for a 2.5 Gbps CPFSK (continuous-phase frequency shift keying) modulated signal transmission at 120 GHz over a distance up to 7 m. The waveguide with an outer and inner diameter of 2 mm and 1 mm, respectively, exhibited a loss of 2–4 dB/m for a frequency range of 110–140 GHz.

4.3.4 Waveguide Transitions

Various types of waveguides have been so far described. The best-suited waveguide will vary from application to application, and the optimum waveguide type will be selected based on the details of the needs from the given application. Obviously, there may arise situations where two different types of waveguides need to be connected. The most common situation is the connection between a planar waveguide and a rectangular waveguide for packaging purposes, in which *waveguide transition* structures need to be employed. There have been a variety of transition structures developed for this purpose in microwave and mm-wave bands, many of which can be adopted for the THz region as well. In this subsection, a brief overview of those transition structures will be presented with focus on two cases: transition between a microstrip line and a rectangular waveguide, and transition between a CPW and a rectangular waveguide.

The challenge in the transition between a planar waveguide to a rectangular waveguide stems from the fact that planar waveguides are composed of two conductors, signal and ground, while rectangular waveguides are made of one single conductor body. How can those two highly heterogeneous structures be linked together, preferably with a low insertion loss and broadband characteristics? A clue can be found if we recall the transverse electric field distribution in the rectangular waveguides. When the dominant TE_{10} mode is assumed, E_y will be given as a sine function of x with a peak located at the middle of the broad-side of the rectangular cross section and zeros at both edges, as indicated by Eq. (4.73). E_x will be non-existent. This implies that the largest potential difference is established between the centers of the two broad-side walls of the waveguide. Consequently, the best approach is to link the signal and ground conductors of the planar waveguides to the two broad-side centers of the rectangular waveguide. Two traditionally popular transition structures, ridge waveguide transitions and fin-line transitions, are both based on this approach. Another approach is to radiate (or detect) electromagnetic waves with the signal line placed inside the rectangular waveguide, which is the basis for the E-plane probe transition. These three types of transitions are discussed below.

4.3.4.1 Ridge Waveguide Transition

To make a transition from a microstrip line to a rectangular waveguide based on the first approach mentioned above, it is needed to place the ground plane of the microstrip line on the bottom broad-side wall of the waveguide and stretch the signal line upward to reach the top broad-side wall. One structural solution is to form a "ridge" on the top broad-side wall along the waveguide axis, which drops down to touch the microstrip signal line on one side and gradually tapers off on the other side. This ridge waveguide transition structure is illustrated in Fig. 4.38a. As the transition structure needs to provide impedance transformation from a microstrip line (tens of Ω's) to a rectangular waveguide (hundreds of Ω's) for a wide frequency range, a

Fig. 4.38 Structures of ridge waveguide transitions: (**a**) Waveguide-to-microstrip line, (**b**) Waveguide-to-CPW

careful structural optimization needs to be made for matching. It can be achieved by a stepped structure with each step made of roughly $\lambda/4$ length [72]. However, as frequency increases, a gradually tapered structure is preferred from a manufacturing viewpoint, typically with a cosine profile that may extends a few guided wavelengths long [52]. A similar structure has been also applied to CPW to rectangular waveguide transitions [73]. There is only a minor difference from the microstrip line case, which is related to the ground plane being formed on the top side for CPW, as opposed to the microstrip lines. Figure 4.38b shows a ridge waveguide transition for CPWs. Ridge waveguide transitions have been a popular choice for various microwave applications partly owing to its broadband characteristics, but they require a precise machining process, which becomes increasingly challenging with increasing frequency. The cross-sectional dimension of rectangular waveguides for the THz band already enters the sub-mm range (see Table 4.1), and the ridges are of a fractional size of this cross section, prohibitively small for precision machining. In view of this limitation, it may be fair to say that ridge waveguide transitions are not the best option for THz applications.

4.3.4.2 Fin-Line Transition

Another way to make transition is to employ the fin-line structure [74]. For the transition with a microstrip line as shown in Fig. 4.39, the signal line and the ground plane are patterned, so that they are gradually curved into the side edges in the opposite direction, forming fin-shaped transitional regions on top and bottom sides of the substrate, respectively. The patterned structure is inserted into the rectangular waveguide on the E-plane, or with its main plane in parallel with the narrow-side walls of the waveguide (see Fig. 4.39b). In the completed structure, the signal line and ground plane of a microstrip line will be in contact with the central regions of the two broad-side walls of the waveguide, which is desired as mentioned above. As the two fin patterns are situated in the antipodal position to each other, it is often called the antipodal fin-line transition. Also, it can be viewed as a double-ridge waveguide transition in a sense, as we may regard the two metal patterns on the substrate as variant types of ridge. CPW to rectangular waveguide transition can also be implemented with fin-line structures. However, as the ground planes are placed on both sides of the signal line on the same metal level for CPW, the formation of the two fin patterns on the (laterally) opposite side is not straightforward. There have been various structural solutions reported, a couple of which are mentioned here. Figure 4.39c shows a structure where two fins on the opposite sides are formed, one with the signal line and the other with one of the ground planes [75]. The other ground plane is linked to the other ground with air bridges. In the case shown in Fig. 4.39d, one fin is patterned with one of the ground planes, which is regular. The other fin, however, is formed with the signal line and the other ground plane combined together, the gap between which is terminated with a slot-line radial stub that serves as a CPW-to-slot line transition [76]. The fin-line transition is compatible with high-frequency bands as the fin structures are realized with planar

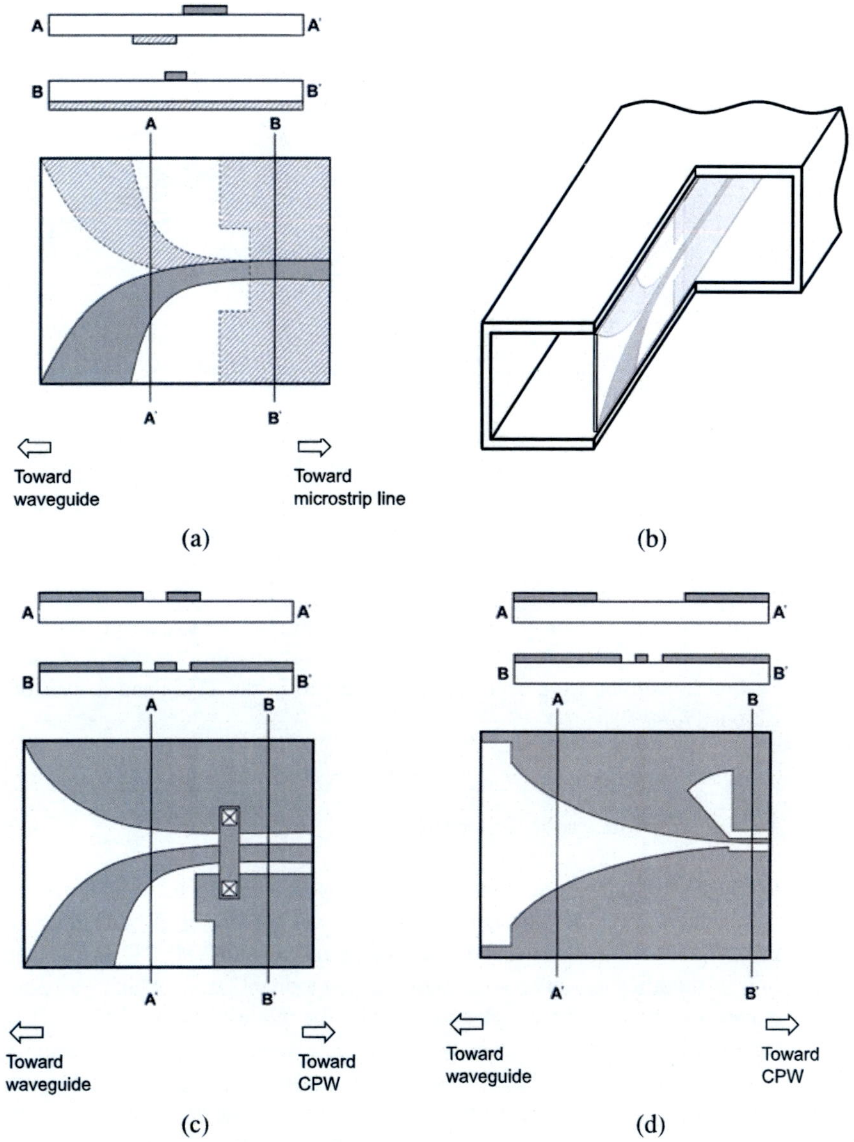

Fig. 4.39 Fin-line transitions: (**a**) Waveguide-to-microstrip line, (**b**) Waveguide-to-microstrip line fixed inside the waveguide, (**c**) Waveguide-to-CPW with airbridge links, (**d**) Waveguide-to-CPW with slot line radial stub

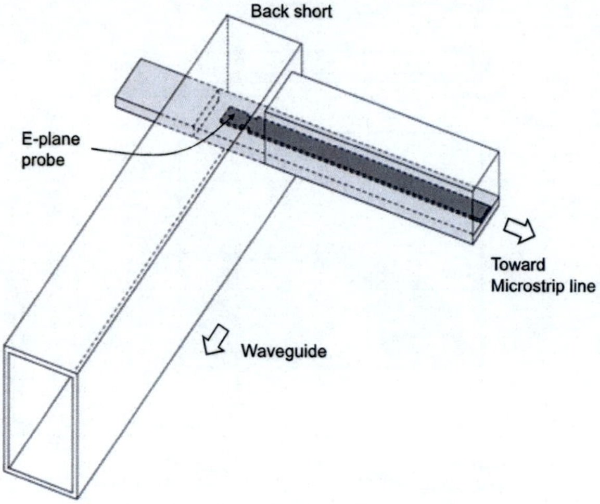

Fig. 4.40 *E*-plane probe transition to microstrip line

metal patterning that easily scales with increasing frequency, as opposed to the ridge waveguide structures that requires a precision machining process. For this reason, fin-line transitions have been widely employed in the THz band for various applications [77–79].

4.3.4.3 *E*-plane Probe Transition

While the two transitions described above rely on a contact between the signal line of the planar waveguides and the rectangular waveguide wall, *E*-plane probe does not involve such a direct contact between the two conductors [80, 81]. Instead, the signal line is modified to serve as a radiating component, from which electromagnetic waves are radiated and coupled into one of the modes supported by the rectangular waveguide at the given frequency. In a reciprocal way, of course, the same structure can be used to detect electromagnetic waves propagating inside the rectangular waveguide. A typical structure of *E*-plane probe is shown in Fig. 4.40. The signal line is inserted and suspended inside the rectangular waveguide, lying on the *E*-plane of the waveguide (hence the name "*E*-plane probe"), typically with the line width modulated for impedance transformation. Although rare, it is possible for the surface of the signal line to face the propagation direction of the waveguide [82], which is not an *E*-plane probe in a strict sense as the probe lies on the *H*-plane in this case. In most cases, dielectric substrate is also stretched into the waveguide together with the signal line to provide a mechanical support to the thin metallic "probe". Because it is typical for the signal line to protrude into the waveguide from the side, the waveguide is terminated with a backshort that serves to maximize the conversion. The *E*-plane probe can be readily applied to the transition between CPW and rectangular waveguide as well with a similar structure [83], as the treatment of the two-sided ground plane is not an issue for the configuration needed. As *E*-plane probes have an excellent scalability in size, they are well suited for THz applications. There have been many reports with successful adoption of *E*-plane probes for various applications, up to beyond 2 THz [84]. There have also been reports on THz on-chip dipole antennas that serve as a probe for transition with the entire chip placed inside the waveguide structure [85, 86].

References

1. P.F. Goldsmith, *Quasioptical Systems* (IEEE Press, Piscataway, NJ, 1998)
2. O. Svelto, D.C. Hanna, *Principles of Lasers*, 5th edn. (Springer, Berlin, 1998)
3. A.E. Siegman, *Lasers* (University Science Books, Sausalito, CA, 1986)
4. H.T. Yura, S.G. Hanson, Optical beam wave propagation through complex optical systems. J. Opt. Soc. Am. **4**(10), 1931–1948 (1987)
5. C.A. Balanis, *Antenna Theory: Analysis and Design* (John Wiley & Sons, Hoboken, NJ, 2016)
6. W.L. Stutzman, G.A. Thiele, *Antenna Theory and Design* (John Wiley & Sons, Hoboken, NJ, 2012)
7. G.M. Rebeiz, Millimeter-wave and terahertz integrated circuit antennas. Proc. IEEE **80**(11), 1748–1770 (1992)
8. I. A. P. Society, *IEEE Standard for Definitions of Terms for Antennas* (IEEE, New York, 1993). 9780738189277, https://books.google.com/books?id=dc17swEACAAJ
9. J.S. McLean, A re-examination of the fundamental limits on the radiation Q of electrically small antennas. IEEE Trans. Antennas Propag. **44**(5), 672 (1996). https://doi.org/10.1109/8.496253
10. R.C. Hansen, Fundamental limitations in antennas. Proc. IEEE **69**(2), 170–182 (1981). https://doi.org/10.1109/PROC.1981.11950
11. Y.P. Zhang, M. Sun, L.H. Guo, On-chip antennas for 60-GHz radios in silicon technology. IEEE Trans. Electron Devices **52**(7), 1664–1668 (2005). https://doi.org/10.1109/TED.2005.850628
12. S.-S. Hsu, K.-C. Wei, C.-Y. Hsu, H.-R. Chuang, A 60-GHz millimeter-wave CPW-fed Yagi antenna fabricated by using 0.18-μm CMOS technology. IEEE Electron Device Lett. **29**(6), 625–627 (2008). https://doi.org/10.1109/LED.2008.920852
13. F. Schuster et al., A broadband THz imager in a low-cost CMOS technology, in *IEEE International Solid-State Circuits Conference Digest of Technical Papers* (2011), pp. 42–43. https://doi.org/10.1109/isscc.2011.5746211
14. T. Tajima, H. Song, K. Ajito, M. Yaita, N. Kukutsu, 300-GHz step-profiled corrugated horn antennas integrated in LTCC. IEEE Trans. Antennas Propag. **62**(11), 5437–5444 (2014). https://doi.org/10.1109/TAP.2014.2350520
15. H.M. Cheema, A. Shamim, The last barrier: On-chip antennas. IEEE Microw. Mag. **14**(1), 79–91 (2013). https://doi.org/10.1109/MMM.2012.2226542
16. K.O. Kenneth et al., On-chip antennas in silicon ICs and their application. IEEE Trans. Electron Devices **52**(7), 1312–1323 (2005). https://doi.org/10.1109/TED.2005.850668
17. E. Ojefors, U.R. Pfeiffer, A 650GHz SiGe receiver front-end for terahertz imaging arrays, in *IEEE International Solid-State Circuits Conference Digest of Technical Papers* (2010), pp. 430–431
18. U.R. Pfeiffer, E. Ojefors, A 600-GHz CMOS focal-plane array for terahertz imaging applications, in *34th European Solid-State Circuits Conference* (2008), pp. 110–113

19. D. Yoon, J. Kim, J. Yun, M. Kaynak, B. Tillack, J.-S. Rieh, 300-GHz direct and heterodyne active imagers based on 0.13-μm SiGe HBT technology. IEEE Trans. Terahertz Sci. Technol. **7** (5), 536–545 (2017)
20. A. Babakhani, X. Guan, A. Komijani, A. Natarajan, A. Hajimiri, A 77-GHz phased-array transceiver with on-Chip antennas in silicon: Receiver and antennas. IEEE J. Solid State Circuits **41**(12), 2795–2806 (2006). https://doi.org/10.1109/JSSC.2006.884811
21. N.G. Alexopoulos, P.B. Katehi, D.B. Rutledge, Substrate optimization for integrated circuit antennas. IEEE Trans. Microw. Theory Tech. **31**(7), 550–557 (1983). https://doi.org/10.1109/TMTT.1983.1131544
22. H.-J. Song, J.-Y. Kim, K. Ajito, N. Kukutsu, M. Yaita, 50-Gb/s direct conversion QPSK modulator and demodulator MMICs for terahertz communications at 300 GHz. IEEE Trans. Microw. Theory Tech. **62**(3), 600–609 (2014)
23. J. Kim et al., Three-dimensional terahertz tomography with transistor-based signal source and detector circuits operating near 300 GHz. IEEE Trans. Terahertz Sci. Technol. **8**(5), 482–491 (2018). https://doi.org/10.1109/TTHZ.2018.2851542
24. M. Asada, S. Suzuki, N. Kishimoto, Resonant tunneling diodes for sub-terahertz and terahertz oscillators. Jpn. J. Appl. Phys. **47**(6), 4375–4384 (2008). https://doi.org/10.1143/jjap.47.4375
25. R. Han, E. Afshari, A CMOS high-power broadband 260-GHz radiator array for spectroscopy. IEEE J. Solid State Circuits **48**(12), 3090–3104 (2013). https://doi.org/10.1109/JSSC.2013.2272864
26. Z. Hu, M. Kaynak, R. Han, High-power radiation at 1 THz in silicon: A fully scalable array using a multi-functional radiating mesh structure. IEEE J. Solid State Circuits **53**(5), 1313–1327 (2018). https://doi.org/10.1109/JSSC.2017.2786682
27. H.G. Booker, Slot aerials and their relation to complementary wire aerials (Babinet's principle). J. Inst. Electric. Eng. IIIA Radiolocation **93**(4), 620–626 (1946)
28. K. Carver, J. Mink, Microstrip antenna technology. IEEE Trans. Antennas Propag. **29**(1), 2–24 (1981)
29. E.O. Hammerstad, Equations for microstrip circuit design, in *1975 5th European Microwave Conference*, (IEEE, New York, 1975), pp. 268–272
30. D.R. Jackson, N.G. Alexopoulos, Simple approximate formulas for input resistance, bandwidth, and efficiency of a resonant rectangular patch. IEEE Trans. Antennas Propag. **39**(3), 407–410 (1991). https://doi.org/10.1109/8.76341
31. E. Seok et al., Progress and challenges towards terahertz CMOS integrated circuits. IEEE J. Solid State Circuits **45**(8), 1554–1564 (2010). https://doi.org/10.1109/JSSC.2010.2049793
32. E. Ojefors, J. Grzyb, Y. Zhao, B. Heinemann, B. Tillack, U.R. Pfeiffer, A 820GHz SiGe chipset for terahertz active imaging applications, in *IEEE International Solid-State Circuits Conference* (2011), pp. 224–226
33. D. Yoon, K. Song, J. Kim, M. Kaynak, B. Tillack, J.-S. Rieh, A D-band active imager in a SiGe HBT technology. J. Infrared Millimeter Terahertz Waves **36**(4), 335–349 (2015)
34. E. Seok et al., A 410GHz CMOS push-push oscillator with an on-chip patch antenna, in *IEEE International Solid-State Circuits Conference (ISSCC)*, (IEEE, New York, 2008), pp. 472–629
35. R. Han et al., A 280-GHz Schottky diode detector in 130-nm digital CMOS. IEEE J. Solid State Circuits **46**(11), 2602–2612 (2011). https://doi.org/10.1109/JSSC.2011.2165234
36. S. Chai, S. Lim, S. Hong, THz detector with an antenna coupled stacked CMOS plasma-wave FET. IEEE Microw. Wirel. Compon. Lett. **24**(12), 869–871 (2014). https://doi.org/10.1109/LMWC.2014.2353211
37. C. Li, T. Chiu, 340-GHz low-cost and high-gain on-chip higher order mode dielectric resonator antenna for THz applications. IEEE Trans. Terahertz Sci. Technol. **7**(3), 284–294 (2017). https://doi.org/10.1109/TTHZ.2017.2670234
38. J.M. Edwards, G.M. Rebeiz, High-efficiency elliptical slot antennas with quartz superstrates for silicon RFICs. IEEE Trans. Antennas Propag. **60**(11), 5010–5020 (2012). https://doi.org/10.1109/TAP.2012.2207353

39. W. Choe, J. Jeong, Broadband THz CMOS on-chip antenna using stacked resonators, in *2017 IEEE International Symposium on Radio-Frequency Integration Technology* (2017), pp. 208–210. https://doi.org/10.1109/RFIT.2017.8048251
40. K.M. Lee, I.J. Lee, S. Jeon, M. Kim, J.S. You, 300 GHz InP rectangular cavity antenna, in *2015 IEEE International Symposium on Antennas and Propagation* (2015), pp. 2105–2106. https://doi.org/10.1109/APS.2015.7305442
41. V. Rumsey, Frequency independent antennas. IRE Int. Conven. Record **5**, 114–118 (1957). https://doi.org/10.1109/IRECON.1957.1150565
42. P.E. Mayes, Frequency-independent antennas and broad-band derivatives thereof. Proc. IEEE **80**(1), 103–112 (1992). https://doi.org/10.1109/5.119570
43. J. Dyson, The equiangular spiral antenna. IRE Trans. Antennas Propag. **7**(2), 181–187 (1959)
44. R. DuHamel, D. Isbell, Broadband logarithmically periodic antenna structures. IRE Int. Conven. Record **5**, 119–128 (1957). https://doi.org/10.1109/IRECON.1957.1150566
45. A.D. Semenov et al., Terahertz performance of integrated lens antennas with a hot-electron bolometer. IEEE Trans. Microw. Theory Tech. **55**(2), 239–247 (2007). https://doi.org/10.1109/TMTT.2006.889153
46. S.-P. Han et al., Compact fiber-pigtailed InGaAs photoconductive antenna module for terahertz-wave generation and detection. Opt. Express **20**(16), 18432–18439 (2012)
47. S. Lepeshov et al., Boosting terahertz photoconductive antenna performance with optimised plasmonic nanostructures. Sci. Rep. **8** (2018)
48. S. Verghese, K.A. McIntosh, E.R. Brown, Highly tunable fiber-coupled photomixers with coherent terahertz output power. IEEE Trans. Microw. Theory Tech. **45**(8), 1301–1309 (1997). https://doi.org/10.1109/22.618428
49. R. Mendis, C. Sydlo, J. Sigmund, M. Feiginov, P. Meissner, H.L. Hartnagel, Tunable CW-THz system with a log-periodic photoconductive emitter. Solid State Electron. **48**(10–11), 2041–2045 (2004)
50. S. Yang, M.R. Hashemi, C.W. Berry, M. Jarrahi, 7.5% optical-to-terahertz conversion efficiency offered by photoconductive emitters with three-dimensional plasmonic contact electrodes. IEEE Trans. Terahertz Sci. Technol. **4**(5), 575–581 (2014). https://doi.org/10.1109/TTHZ.2014.2342505
51. D.M. Pozar, *Microwave Engineering* (John Wiley & Sons, Hoboken, NJ, 2009)
52. R. Garg, I. Bahl, M. Bozzi, *Microstrip Lines and Slotlines* (Artech house, Boston, MA, 2013)
53. I.J. Bahl, D.K. Trivedi, A designer's guide to microstrip line. Microwaves **16**(5), 174–182 (1977)
54. R.A. Pucel, D.J. Masse, C.P. Hartwig, Losses in microstrip. IEEE Trans. Microw. Theory Tech. **16**(6), 342–350 (1968). https://doi.org/10.1109/TMTT.1968.1126691
55. L. Maloratsky, *Passive RF and Microwave Integrated Circuits* (Elsevier, Amsterdam, 2003)
56. L.J. van der Pauw, The radiation of electromagnetic power by microstrip configurations. IEEE Trans. Microw. Theory Tech. **25**(9), 719–725 (1977). https://doi.org/10.1109/TMTT.1977.1129201
57. C.P. Wen, Coplanar waveguide: A surface strip transmission line suitable for nonreciprocal gyromagnetic device applications. IEEE Trans. Microw. Theory Tech. **17**(12), 1087–1090 (1969). https://doi.org/10.1109/TMTT.1969.1127105
58. A. Komijani, A. Hajimiri, A wideband 77-GHz, 17.5-dBm fully integrated power amplifier in silicon. IEEE J. Solid State Circuits **41**(8), 1749–1756 (2006)
59. G. Ghione, C. Naldi, Analytical formulas for coplanar lines in hybrid and monolithic MICs. Electron. Lett. **20**(4), 179–181 (1984). https://doi.org/10.1049/el:19840120
60. S.S. Gevorgian, Basic characteristics of two layered substrate coplanar waveguides. Electron. Lett. **30**(15), 1236–1237 (1994). https://doi.org/10.1049/el:19940861
61. S.B. Cohn, Slot line on a dielectric substrate. IEEE Trans. Microw. Theory Tech. **17**(10), 768–778 (1969). https://doi.org/10.1109/TMTT.1969.1127058

62. E.A. Mariani, C.P. Heinzman, J.P. Agrios, S.B. Cohn, Slot line characteristics. IEEE Trans. Microw. Theory Tech. **17**(12), 1091–1096 (1969). https://doi.org/10.1109/TMTT.1969.1127106
63. P. Toulios, R. Knox, Rectangular dielectric image lines for millimeter integrated circuits, in *Western Electronic Show and Convention* (1970), pp. 25–28
64. W.V. McLevige, T. Itoh, R. Mittra, New waveguide structures for millimeter-wave and optical integrated circuits. IEEE Trans. Microw. Theory Tech. **23**(10), 788–794 (1975). https://doi.org/10.1109/TMTT.1975.1128684
65. T. Itoh, R. Mittra, New waveguide structures for millimeter-wave integrated circuits, in *IEEE-MTT-S International Microwave Symposium* (1975), pp. 277–280. https://doi.org/10.1109/MWSYM.1975.1123359
66. T. Itoh, Inverted strip dielectric waveguide for millimeter-wave integrated circuits. IEEE Trans. Microw. Theory Tech. **24**(11), 821–827 (1976). https://doi.org/10.1109/TMTT.1976.1128967
67. E.A.J. Marcatili, Dielectric rectangular waveguide and directional coupler for integrated optics. Bell Syst. Tech. J. **48**(7), 2071–2102 (1969). https://doi.org/10.1002/j.1538-7305.1969.tb01166.x
68. P. Bhartia, I.J. Bahl, *Millimeter Wave Engineering and Applications* (Wiley, New York, 1984)
69. D. Marcuse, *Theory of Dielectric Optical Waveguides* (Elsevier, Amsterdam, 2013)
70. S. Fukuda et al., A 12.5+12.5 Gb/s full-duplex plastic waveguide interconnect. IEEE J. Solid State Circuits **46**(12), 3113–3125 (2011). https://doi.org/10.1109/JSSC.2011.2168870
71. N.V. Thienen, W. Volkaerts, P. Reynaert, A multi-gigabit CPFSK polymer microwave fiber communication link in 40 nm CMOS. IEEE J. Solid State Circuits **51**(8), 1952–1958 (2016). https://doi.org/10.1109/JSSC.2016.2580605
72. M.V. Schneider, B. Glance, W.F. Bodtmann, Microwave and millimeter wave hybrid integrated circuits for radio systems. Bell Syst. Tech. J. **48**(6), 1703–1726 (1969). https://doi.org/10.1002/j.1538-7305.1969.tb01147.x
73. G.E. Ponchak, R.N. Simons, A new rectangular waveguide to coplanar waveguide transition, in *IEEE MTT-S International Microwave Symposium Digest* (1990), pp. 491–492. https://doi.org/10.1109/mwsym.1990.99626
74. L.J. Lavedan, Design of waveguide-to-microstrip transitions specially suited to millimetre-wave applications. Electron. Lett. **13**(20), 604–605 (1977). https://doi.org/10.1049/el:19770434
75. J.V. Bellantoni, R.C. Compton, H.M. Levy, A new W-band coplanar waveguide test fixture, in *IEEE MTT-S International Microwave Symposium Digest* (1989), pp. 1203–1204. https://doi.org/10.1109/mwsym.1989.38940
76. V.S. Mottonen, Wideband coplanar waveguide-to-rectangular waveguide transition using fin-line taper. IEEE Microw. Wirel. Compon. Lett. **15**(2), 119–121 (2005). https://doi.org/10.1109/LMWC.2004.842855
77. C. Groppi, C.Y.D. d'Aubigny, A.W. Lichtenberger, C.M. Lyons, C.K. Walker, Broadband finline ortho-mode transducer for the 750–1150 GHz band, in *Proc. 16th Int. Symp. Space Terahertz Technol.* (2005), pp. 513–518
78. L.A. Samoska, An overview of solid-state integrated circuit amplifiers in the submillimeter-wave and THz regime. IEEE Trans. Terahertz Sci. Technol. **1**(1), 9–24 (2011). https://doi.org/10.1109/TTHZ.2011.2159558
79. P. Kittara, P. Grimes, G. Yassin, S. Withington, K. Jacobs, S. Wulff, A 700-GHz SIS antipodal finline mixer fed by a Pickett-Potter horn-reflector antenna. IEEE Trans. Microw. Theory Tech. **52**(10), 2352–2360 (2004). https://doi.org/10.1109/TMTT.2004.835976
80. T.Q. Ho, Y.-C. Shih, Y.-C. Shih, Spectral-domain analysis of E-plane waveguide to microstrip transitions. IEEE Trans. Microw. Theory Tech. **37**(2), 388–392 (1989). https://doi.org/10.1109/22.20065
81. Y.-C. Leong, S. Weinreb, Full band waveguide-to-microstrip probe transitions, in *IEEE MTT-S International Microwave Symposium Digest*, vol. 4 (1999), pp. 1435–1438. https://doi.org/10.1109/MWSYM.1999.780219

82. Y. Shih, T. Ton, L.Q. Bui, Waveguide-to-microstrip transitions for millimeter-wave applications, in *IEEE MTT-S International Microwave Symposium Digest* (1988), pp. 473–475. https://doi.org/10.1109/MWSYM.1988.22077
83. L. Samoska, A. Peralta, H. Ming, M. Micovic, A. Schmitz, A 20 mW, 150 GHz InP HEMT MMIC power amplifier module. IEEE Microw. Wirel. Compon. Lett. **14**(2), 56–58 (2004). https://doi.org/10.1109/LMWC.2003.822575
84. P.H. Siegel, R.P. Smith, M.C. Graidis, S.C. Martin, 2.5-THz GaAs monolithic membrane-diode mixer. IEEE Trans. Microw. Theory Tech. **47**(5), 596–604 (1999). https://doi.org/10.1109/22.763161
85. K.M.K.H. Leong et al., A 340–380 GHz integrated CB-CPW-to-waveguide transition for sub millimeter-wave MMIC packaging. IEEE Microw. Wirel. Compon. Lett. **19**(6), 413–415 (2009). https://doi.org/10.1109/LMWC.2009.2020043
86. W. Choe, J. Jeong, A broadband THz on-chip transition using a dipole antenna with integrated balun. Electronics **7**(10), 236 (2018)

Chapter 5
THz Optical Methods

It was mentioned earlier in this book that there can be two approaches to the THz research, one upward approach from the electronics and the other downward from the optics. While the purpose of this book is to cover the topics relevant to the electronic approach, it is true that the THz research has long been dominated by the approach based on the optics, which has inevitably influenced THz electronics. Therefore, it is desired for the intended readers of this book, who are expected to have their backgrounds in the electronics, to be aware of the basic concepts commonly shared in the optics-driven THz territory. It will promote the interaction with the researchers from the other side of the THz spectrum with a lowered "language barrier" and also help to import the ideas on that side for a productive blending. When those with electronics background attempt to learn the basics of optics-based THz, however, they are often overwhelmed by the amount and details carried by the references and books targeted at the optics majors. This chapter is prepared to serve as a quick guide to optics-based THz for the electronics-based THz researchers. Those who are seeking for a comprehensive coverage of optics-based THz are referred to other references intended mostly for such purposes [1–3]. In this chapter, two major topics will be discussed, which are, naturally, the THz wave generation and detection. The discussion on generation is divided into two sections, one on the pulse and the other on the continuous wave (CW) generation. The section on detection is described mainly for the pulse detection, as the CW detection is commonly performed by electrical means even in the optical environment.

5.1 THz Pulse Generation

One outstanding characteristic of optics-based THz research is the heavy reliance on ultra-short pulses, which contrasts with the electronics-based THz that mostly involves continuous waves. The readers should be aware that pulses in the time domain occupy a wide bandwidth in the frequency domain if translated by Fourier

© Springer Nature Switzerland AG 2021
J.-S. Rieh, *Introduction to Terahertz Electronics*,
https://doi.org/10.1007/978-3-030-51842-4_5

transform. When the pulses are tailored to have a very short time period, in the order of roughly below 1 picosecond (ps), their corresponding band coverage in the frequency domain will be extended to reach the THz spectrum, enabling various exciting applications. Such extremely short pulses became readily available with optical methods, in large part owing to the advent of femtosecond lasers [4]. Among the various approaches developed for THz pulse generation, the two most widely adopted techniques, which are based on photoconductive (PC) material and optical rectification with nonlinear electro-optic (EO) materials, will be introduced in this section. The former technique involves free electrons while the latter is based on bound electrons for THz pulse generation, although they are both driven by femtosecond lasers. Although less popular, THz pulses can be generated based on the semiconductor surface field, which also rely on femtosecond lasers. All these techniques will be briefly discussed in this section.

5.1.1 Generation with Photoconductive Antennas

In semiconductors, when an electron is excited with energy large enough to enable the transition from the valence to the conduction band, an electron-hole pair is generated. This process can be triggered by various causes such as thermal, electrical, and optical processes. Once generated, they become free carriers, allowed to travel around inside the crystal. Their lifespan, however, is finite, as those of humans. After a finite time, the generated electrons and holes recombine with each other and become bound carriers again. The statistical average time for survival is called the carrier lifetime. It strongly depends on the quality of the semiconductor crystal, as defects create traps to serve as stepping stones for inter-band transition, effectively promoting the recombination. In most semiconductor applications, low defect density is preferred for various reasons including high mobility and low leakage current. One rare exception is the application for THz pulse generation with photoconductive antennas, as is described below.

When a semiconductor crystal is irradiated by a very short optical pulse with a high intensity, a huge number of electron–hole pairs are immediately generated and the material becomes conductive (photoconductive) temporarily. The generated carriers will eventually vanish by recombination after an average time that equals the lifetime of the given semiconductor material. The transient change in the carrier density can be converted into a form of current (photocurrent) with a proper bias scheme, and the generated transient current can result in radiation with a proper radiating structure, or antenna. The radiated electromagnetic wave takes a form of pulse, and its time duration will be dictated by the time derivative of the transient photocurrent, leading to a narrower pulse with a faster change over time. As our purpose is to attain a very short pulse for THz operation, the surge and decay of the photocurrent should be very fast. Toward this end, two critical components are needed: a very short optical pulse and a semiconductor material with a very short

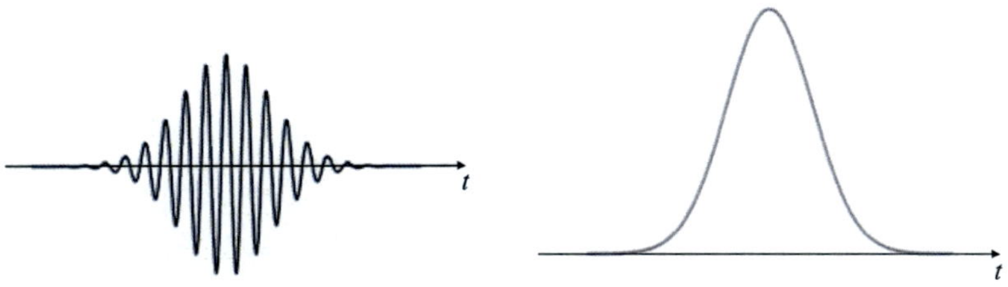

$E_{opt}(t) \propto e^{-at^2} \cos \omega t$

(a)

$I(t) \propto e^{-2at^2}$

(b)

Fig. 5.1 Profiles of an ultrashort optical pulse in the time domain: (**a**) Electric field (**b**) Averaged intensity

carrier lifetime. The former is achieved with a femtosecond laser while the latter is typically realized with low-temperature (LT)-GaAs.

A femtosecond laser is a pulse laser that generates ultrashort optical pulses whose temporal duration is well below 1 ps, typically in the range of 10–100 fs (1 fs $= 10^{-15}$ s $= 10^{-3}$ ps). Femtosecond lasers are usually based on solid-state gain media, the most popular one being Ti:sapphire (Ti-doped Al_2O_3). Ti:sapphire lasers can be tuned over a wavelength of 650–1100 nm, while the most efficient operation is obtained near 800 nm, for which a separate pump laser with ~520 nm wavelength is used for optical pumping. Considering that a wavelength of 800 nm corresponds to a time period of ~2.7 fs, we can see that a typical optical pulse from a femtosecond laser contains several to a few tens of cycles for the alternating electric (or magnetic) field. However, it will be the averaged intensity, which shows a Gaussian profile in the time domain, that will affect the THz (electrical) pulse generation. Figure 5.1 shows the profiles of electric field and the averaged intensity of an ultrashort optical pulse in the time domain. Although a relatively recent invention with its first appearance around early 1980s, femtosecond lasers have been rapidly matured and are now readily available as commercial products, which can be conveniently purchased to serve as a key component in ultrafast optics setups.

LT-GaAs commonly refers to a GaAs epitaxial layer grown at a low temperature. In the epitaxial growth of semiconductor layers, typically with an MBE or MOCVD, the crystal quality is critically affected by the growth temperature. A higher temperature will result in a higher quality crystal as there is a higher chance for atoms to find the right site on the crystal surface. As our purpose is to obtain a *low-quality* crystal with lots of defects for this particular application, a semiconductor layer grown at a low temperature is desired, the opposite direction from the conventional approaches. GaAs is the most popular and readily available compound semiconductor with a photoconductive characteristic, and thus widely adopted for this purpose. Typical growth temperature of LT-GaAs with MBE is around 200 °C [5–7], which can be compared to the conventional high-quality GaAs growth temperature of around 500 °C. As-grown LT GaAs will have too many defects to achieve a reasonable

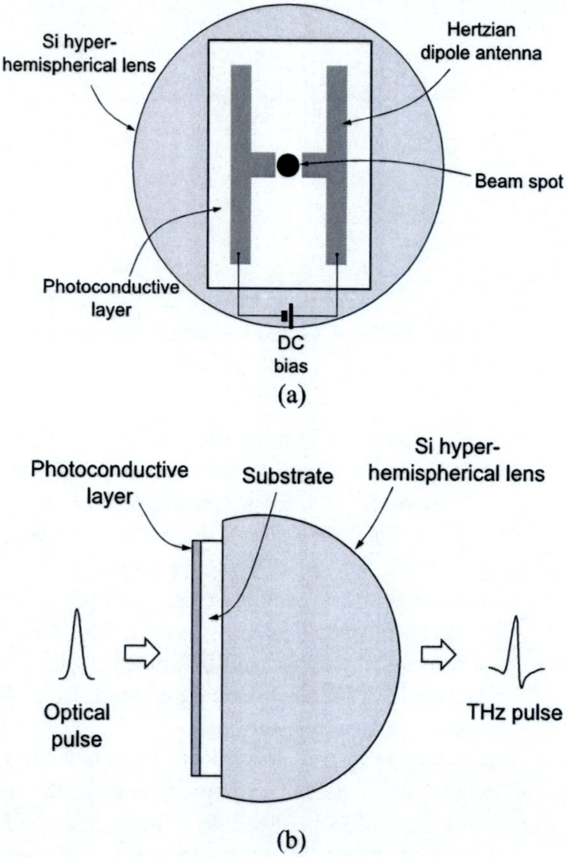

Fig. 5.2 Structure of a typical photoconductive antenna mounted on a Si hyper-hemispherical lens: (**a**) Front view (**b**) Side view

electron mobility that is required for THz generation with photoconductive material. Hence, a post-growth annealing is routinely carried out, after which a typical mobility of 200–300 cm^2/Vs is achieved with a carrier lifetime of around 0.2–0.5 ps [5–7]. While RD-SOS (Radiation-Damaged Si on Sapphire) is also employed for the application, in which a Si layer grown on Sapphire is intentionally damaged by a high-energy ion implantation of O^+ ions [8], the mobility by this method is lower and the carrier lifetime is longer than the LT-GaAs case. This makes LT-GaAs a more popular choice. The following discussion is made assuming an LT-GaAs layer is employed as the photoconductive material needed for the application.

With a femtosecond laser and an LT-GaAs layer available at hand, THz (electrical) pulses can now be obtained. For the actual generation of THz pulses, a more sophisticated structure than simply an LT-GaAs layer is needed, which is called the photoconductive antenna. A schematic of a typical photoconductive antenna is illustrated in Fig. 5.2. An LT-GaAs layer, which is inherently photoconductive, is grown on a substrate, and a planar metallic electrode is patterned on top of the

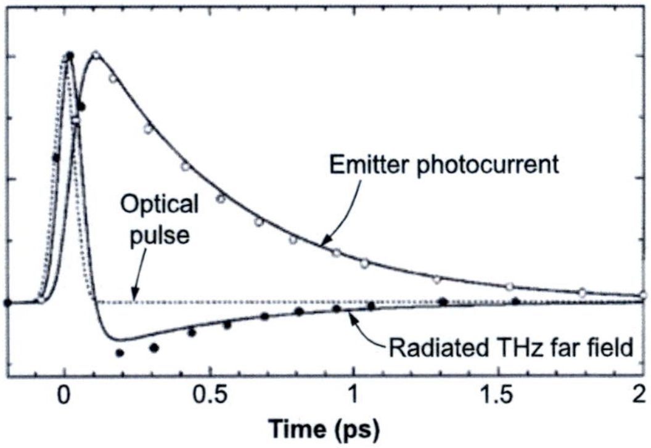

Fig. 5.3 Calculated temporal profiles of optical pulse, photocurrent, and radiated field (lines). Also shown are data points (symbols). (© 2001 IEEE [10])

photoconductive layer. The H-shaped metallic electrode, which is in fact composed of two conductors separated by a small gap in the middle, serves as a dipole antenna as well as a current path. The photoconductive antenna is often mounted on a hyper-hemispherical dielectric lens (typically Si lens) for an improved radiation efficiency as well as enhanced directivity. The patterned electrode will be DC-biased across the two conductors, which will induce an electric field across the gap.

When ultrashort optical pulses from the femtosecond laser is applied on the photoconductive antenna, with the beam spot landing on the middle gap of the antenna pattern, a sudden surge in the carrier concentration will occur due to the electron–hole pair generation triggered by the photo excitation. The generated carriers will form a photocurrent, which runs along the metallic electrode path as driven by the applied DC bias. However, the very short lifetime of the photoconductive layer will dictate a rapid recombination of the generated carriers, leading to a fast reduction of the carrier concentration and thus of the current level. Hence, throughout this progress of events, the photocurrent goes through an abrupt increase followed by a rapid reduction. Such a transient behavior of the photocurrent will induce the radiation of electromagnetic waves from the antenna through the dielectric lens. The radiated field will be proportional to the time derivative of the photocurrent as follows [9]:

$$E_{\mathrm{THz}}(t) \propto \frac{\partial I_{\mathrm{photo}}(t)}{\partial t}. \tag{5.1}$$

Consequently, the field will show a transient profile, as characterized by a positive cycle followed by a negative one. This temporal progress of the field is shown in Fig. 5.3 along with those of the photocurrent and the optical pulse [10]. The duration of the field transition will be typically around 1 ps, similar to that of the photocurrent. This is longer than the time span of the applied optical pulses, but short enough to reach the THz regime in the frequency domain. This is clearly seen in Fig. 5.4, which compares the profile of a generated THz pulse in the time domain and its corresponding spectrum in the frequency domain [11], exhibiting a pulse of about a half ps reaching well beyond 2 THz. The readers are cautioned not to confuse the

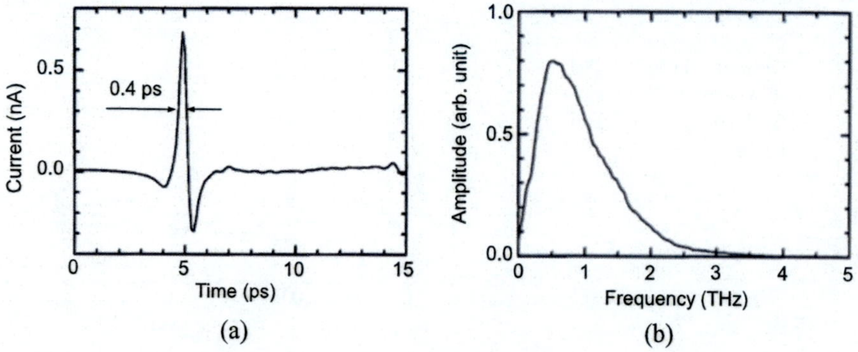

Fig. 5.4 (**a**) THz pulse in the time domain. (**b**) Corresponding profile in the frequency domain. (Reprinted with permission from [11] © The Optical Society)

electric field of the radiated THz pulse (E_{THz}) with the electric field of the incident optical pulse (E_{opt}). The latter is based on a sinusoidal waveform with a period of a few fs, or hundreds of THz in frequency, originated from a femtosecond laser. An optical pulse would comprise several to a few tens of oscillation cycles of the optical field as mentioned earlier (see Fig. 5.1), while the transient field of the generated THz pulse covers a time span in the same order of (but longer than) the optical pulse duration (see Fig. 5.3). The efficiency of THz pulse generation with the photoconductive antenna is known to be higher than with the electro-optic method to be described below, which can be further improved by employing an optimized antenna structure [12].

5.1.2 *Generation with Optical Rectification*

Let us recall one of the basic electromagnetic properties of dielectric materials. When there is an external electric field applied to a dielectric, the electrons will be slightly displaced from the original equilibrium position. The displacement of an electron will form an electric dipole and consequently a non-zero electric dipole moment, making the dielectric polarized. Note that the electrons are still bound electrons as they do not escape the host atoms. Now, let us assume the applied electric field is oscillating, as is the case for electromagnetic waves, including light. In this situation, the electrons will also be oscillating around the original equilibrium point at the same rate as the oscillating field. What happens when the magnitude of the oscillating field increases? Naturally, the maximum displacement of the electrons will also be increased. However, the details may differ depending on the actual intensity of the applied field. If the field intensity is not excessive, the relation between the field magnitude and the electron maximum displacement will remain in the linear regime. This means that when the field magnitude is doubled, for instance, the electron displacement will also be doubled. When the field magnitude

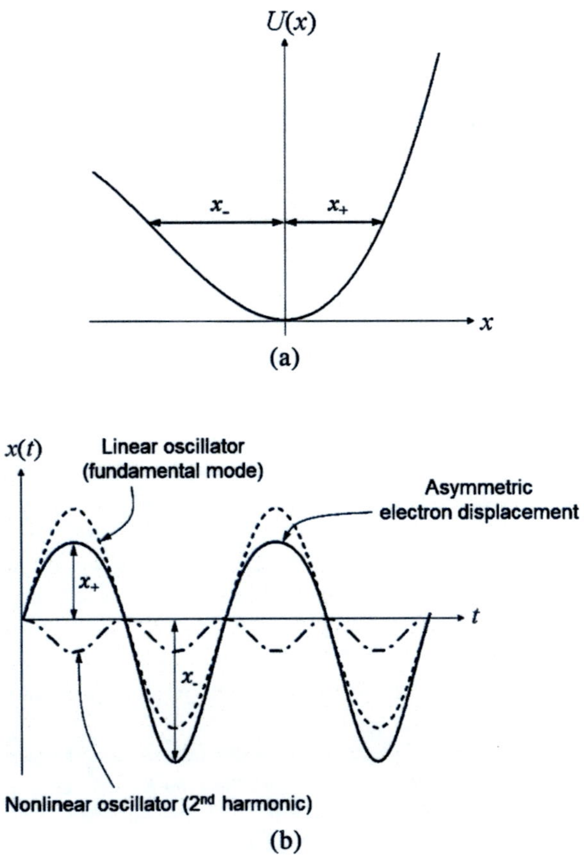

Fig. 5.5 (**a**) Asymmetric potential profile experienced by electrons in ZnTe lattice, which arises from the stronger electronegativity of Te than that of Zn. (**b**) Displacement of electrons under a strong optical illumination. (Adapted by permission from Springer Nature: [1] © 2009)

is tripled, so does the electron displacement. However, when the applied field has a very strong intensity, the linear relation does not hold any more and the response of the electrons becomes nonlinear. Consequently, various nonlinear optical phenomena begin to take place, including *optical rectification*.

Optical rectification is typically observed with electro-optic (EO) crystals, in which the optical properties are affected by the presence of electric field. The most popular electro-optic crystal is ZnTe (zinc telluride) with a zincblende crystal structure [13, 14], which is taken as an example for the description below [1]. In this EO crystal, the electrons that form bonds between Zn and Te atoms experience an asymmetric potential profile as the electronegativity of Te is stronger than that of Zn, as is depicted in Fig. 5.5a. The potential profile dictates the pattern of the electron oscillation in response to the illumination with light. If the illumination is weak and the magnitude of the oscillating electric field is moderate, the electrons will show a nearly symmetric harmonic oscillator-like behavior despite the asymmetric potential profile. However, when the illumination is strong, the displacement of the electrons will exhibit an obvious skew toward the negative direction, as is depicted in

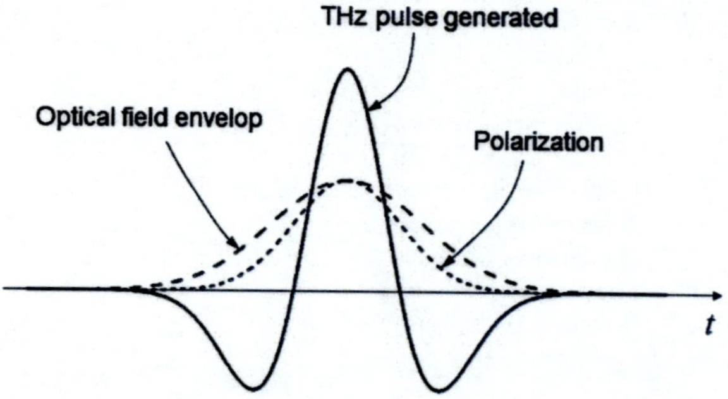

Fig. 5.6 Profiles of the optical field Gaussian envelop, the resultant polarization, and the generated THz pulse by optical rectification

Fig. 5.5b. A closer look reveals that the asymmetric displacement of the electrons can be decomposed into two parts: a linear fundamental-mode oscillation (at the frequency identical to that of the applied electric field) and a nonlinear second harmonic oscillation that also includes a DC component. While the availability of the second harmonic oscillation is one thing useful for some applications by itself, the major consequence that interests us here is the existence of the DC component in it. The DC component in the electron displacement gives rise to a DC or *rectified* polarization. This process is called the optical rectification. Here, the polarization is a quantity defined as the volume density of the electric dipole moments, which is different from the polarization for electromagnetic waves that was discussed in Sect. 4.2.

Let us now consider a special case that is more relevant to our current interest, in which the optical illumination on the EO crystal is based on an ultrashort optical pulse from a femtosecond laser instead of continuous wave (CW). In a single optical pulse, the electric field will oscillate with a magnitude confined by a Gaussian envelop (see Fig. 5.6). If the intensity is strong enough, a shift in polarization will occur due to optical rectification as expected with the electron behaviors described above. However, differently from the CW case we assumed in the previous discussion, the rectified polarization induced by the pulse is not a constant but has a temporal variation. Its profile will also be Gaussian as illustrated in Fig. 5.6, since the induced rectified polarization (second-order nonlinear by its nature) is known to be proportional to the averaged incident field intensity [15]. So, what will be the major consequence of this very fast time-domain variation in the rectified polarization? The variation in the rectified polarization means the variation in the average electron displacement. Therefore, due to the electron movement, a pulse radiation will occur, and it is short enough to be a THz pulse. The radiated field will show a following relation with the time-dependent rectified polarization [2]:

$$E_{\mathrm{THz}}(t) \propto \frac{\partial^2 P(t)}{\partial t^2}. \tag{5.2}$$

With a rectified polarization with an ideal Gaussian profile $\left(P(t) \propto \mathrm{e}^{-2at^2}\right)$, the generated THz pulse will take on a similar pulse shape but flanked with a shallow negative swing on both sides as depicted in Fig. 5.6. Because the electro-optical properties of EO materials including ZnTe depend on the orientation of the fields, an optimal incident angle against the crystal axis needs to be maintained for the incoming optical field to achieve the best efficiency. While the description above is based on a time-domain picture, the THz pulse generation with optical rectification can be equivalently explained in the frequency domain as a difference-frequency mixing between two frequency components within the broad frequency spectrum of the incident optical pulse.

5.1.3 Generation with Surge Current at Semiconductor Surface

When semiconductor surface is illuminated by an ultrashort optical pulse by a femtosecond laser, a surge current will develop due to the carrier generation and the surface field, eventually leading to THz pulse generation. There are two distinct mechanisms for the generation, both of which will be in effect to a certain level, although the relative dominance will be dependent on the specific condition. They are described below.

5.1.3.1 Generation by Surface Field

At the surface of semiconductors, where the lattice periodicity stops and lots of dangling bonds are formed, energy levels are not prohibited anymore in the middle of the energy bandgap, resulting in almost continuous surface states distributed across the bandgap. This situation gives rise to the Fermi energy pinning at the surface, and this in turn leads to a band bending near the surface, resulting in the surface depletion region with a built-in field normal to the surface. When the semiconductor surface is illuminated with an ultrashort optical pulse from a femtosecond laser, lots of electrons and holes will be generated near the surface, and they will be accelerated by the built-in field inside the surface depletion region, building up a surge current normal to the surface. Note that the surge current is the *drift* current by nature. This process is described in Fig. 5.7 for both *n*-type and *p*-type semiconductors. As the current will last for a very short time with the ultrashort optical pulse, in part helped by the rapid recombination via the abundant surface states, the radiation will be a pulse in the THz regime, again dictated by the relation given in Eq. (5.1). In a sense, this situation is similar to the case for THz pulse generation with photoconductive antennas. The difference is that, for this surge current case, the generated carriers are accelerated by the *intrinsic* surface field instead of the *external* DC bias, and the rapid recombination is due to the *surface states* rather than the *bulk states* due to defects. This technique can be applied to both *n*-type and *p*-type semiconductors, although the time-domain profile of the generated

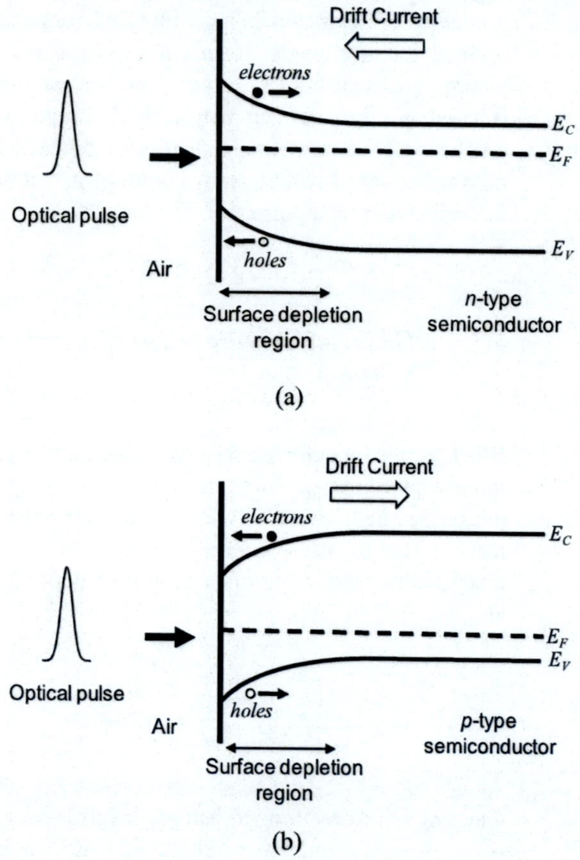

Fig. 5.7 Drift current established at semiconductor surface due to built-in depletion field: (**a**) *n*-type semiconductor (**b**) *p*-type semiconductor

THz pulses will be slightly different (vertically flipped when shown along the time axis), which is due to the opposite direction of the surge current for the two types of semiconductor as shown in Fig. 5.7.

5.1.3.2 Generation by Photo-Dember Effect

Let us maintain our focus on semiconductor surface, but with a selected group of semiconductors, which have a narrow bandgap and a large mobility difference between electrons and holes. InAs (indium arsenide) and InSb (indium antimonide) belong to this category. For InAs, which has a bandgap around 0.35 eV, electron mobility is ~40,000 cm^2/Vs, but hole mobility is only ~500 cm^2/Vs at room temperature. For InSb with a bandgap of ~0.17 eV, electron and hole mobilities are ~77,000 and ~850 cm^2/Vs, respectively. When these semiconductors are illuminated, the drift current due to the built-in field as discussed above, for which electron and hole current components have the same direction, will exist but only with a small

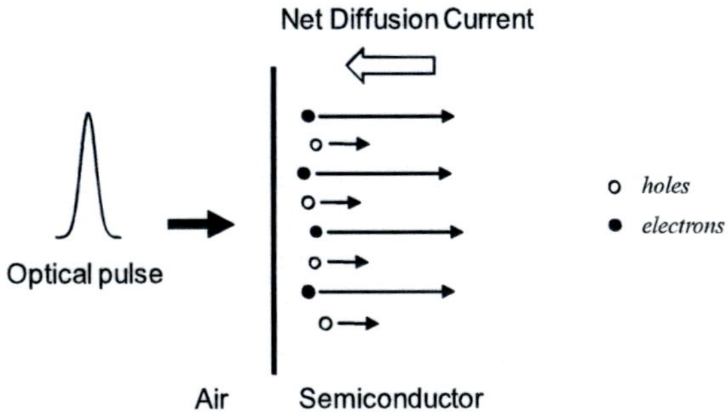

Fig. 5.8 Diffusion current established at semiconductor surface due to carrier concentration gradient

amount. This is because the built-in potential in the depletion region is weak for the narrow-bandgap materials as a result of the small band bending. Now recall another current component in semiconductors, the *diffusion* current. It is driven by the gradient of carrier concentration, either of electrons or holes. When electrons and holes are generated near the surface by photo excitation caused by pulse illumination, both electrons and holes will diffuse toward the bulk due to the established concentration gradient, building up diffusion currents for both carrier polarities. Although the direction of the electron and hole fluxes should be the same due to the same gradient vector, the direction of the two diffusion currents will be opposite because of the different charge polarities, as shown in Fig. 5.8. In such a situation, if there is no difference between the mobilities (thus the diffusion constants) of electrons and holes, the two diffusion currents would cancel each other, assuming an identical concentration gradient for the two carriers. However, when they are different, there will exist a finite net diffusion current, which will be greater with the larger mobility difference. Hence, for those semiconductors with highly imbalanced mobilities for electrons and holes, a strong net diffusion current will develop upon the illumination, which will dominate over the surface field-driven drift current if band bending is small. A dipole will also be established normal to the surface due to the separation of the distributions of electrons and holes. This is called the photo-Dember effect. The fast surge and relaxation of the developed diffusion current will end up generating a THz pulse when illuminated by an ultrashort optical pulse from a femtosecond laser.

5.2 THz CW Generation

Optical methods can also be applied to generate continuous wave (CW) signal that falls within the THz spectrum. In Chap. 2, we have overviewed various methods to generate CW signals based on electronic approaches. In a sense, we can say that the two different approaches, electronic and optical, are competing for the same technical solution. After all, however, we benefit from having multiple options for THz CW generation that have their own pros and cons. Most optics-based THz CW generation techniques rely on the frequency down-conversion by mixing two CW signals in the optical spectrum with wavelengths located close to each other. For the down-conversion, basically the same techniques employed for THz pulse generation, i.e., photoconductive antenna and optical rectification, can be employed, which are widely referred to as photomixing and frequency difference mixing (or generation), respectively. Photomixing can also be achieved with *p-i-n* diodes, including the uni-traveling-carrier photodiode (UTC-PD), which will also be treated in this section. Lasers, a prevailing solution for CW coherent signal generation at the optical spectrum, can also be used for THz CW generation by extending the output wavelength to reach the THz band without help of mixing-based frequency down-conversion. Such an extremely long-wavelength (from the optics viewpoint) radiation is available with some types of lasers, as will also be described in this section.

5.2.1 Generation with Photomixing

When there exist two periodic optical fields with slightly different frequencies, an optical beat is created with a frequency that corresponds to the difference of the two original frequencies. This is very similar to the acoustic beat generated from two acoustic waves with slightly different pitches, as is often utilized for tuning of musical instruments. For practical purposes, the optical beats are usually obtained with two aligned beams from a pair of CW lasers with slightly different wavelengths. When such optical beats are incident on a photoconductive device, carriers will be generated to form a photocurrent, in a similar way as described in Sect. 5.1.1. With the optical beat, which can be tuned for THz frequency range by adjusting the wavelengths of the pair of lasers, the photocurrent will also be modulated at the same frequency. This will result in a CW THz signal radiated through a properly designed radiating structure, typically a wideband antenna. This process is known as photomixing or optical heterodyne down-conversion [16]. The photoconductive device that generates THz CW signals in response to the optical beat is called the photomixer. The most common type of photomixers is based on the photoconductive material deposited on top of a semiconductor substrate, as was employed for THz pulse generation discussed before, which will be the main subject of this section. Additionally, photomixers based on *p-i-n* diode will also be described with focus on UTC-PD as an example.

5.2.1.1 Photomixing Based on Photoconductive Material

The process for CW THz wave generation with photoconductive material can be described as follows [1]. For two interfering optical fields with frequencies of ω_1 and ω_2, the total field applied to the photomixer can be expressed as:

$$E_i(t) = E_1(t)e^{-j\omega_1 t} + E_2(t)e^{-j\omega_2 t} \tag{5.3}$$

where E_1 and E_2 are the amplitudes of the fields corresponding to optical frequencies of ω_1 and ω_2, respectively. The associated intensity of the total field is given as:

$$I_i(t) = \frac{1}{2}c\varepsilon_0|E_i(t)|^2 = I_0 + 2\sqrt{I_1 I_2}(\cos(\omega_1 + \omega_2)t + \cos(\omega_1 - \omega_2)t) \tag{5.4}$$

where c is the speed of light and ε_0 is the permittivity in vacuum. $I_0 = I_1 + I_2$ is the average intensity of the incident total optical fields, with I_1 and I_2 representing the intensities for individual field components. The photocurrent induced inside the photomixer with an incident beam with the intensity given above will show a modulation frequency equal to the beat frequency of $\omega = \omega_1 - \omega_2$, as given by:

$$I_{\text{photo}}(\omega) = I_{\text{photo,DC}} + \frac{I_{\text{photo,AC}}}{\sqrt{1 + \omega^2\tau^2}}\cos(\omega t - \phi), \tag{5.5}$$

where $I_{\text{photo,DC}}$ and $I_{\text{photo,AC}}$ represent DC and AC components of the photocurrent, respectively, ϕ is the phase delay given by $\tan^{-1}(\omega\tau)$, and τ is the carrier lifetime. Note that the other component of the incident intensity with the sum frequency of $\omega_1 + \omega_2$, which should belong to the optical band, will survive the process. However, as Eq. (5.5) implies, the photomixer behaves as a low-pass filter with $1/\tau$ as the pole frequency, which would eliminate the sum frequency component. The THz field radiated by the modulated photocurrent is governed by the time-derivative of the photocurrent, as was the case for the THz pulse generation with Eq. (5.1). This will lead to $E_{\text{THz}}(t)$ proportional to $\sin(\omega t - \phi)$, manifesting its oscillation also at the beat frequency of ω. Therefore, the heterodyne frequency down-conversion can be obviously achieved with a photomixer. The power of the THz radiation from the photomixer will be eventually given as follows:

$$P_{\text{THz}}(\omega) = \frac{1}{2}\frac{I^2_{\text{photo,AC}}}{\left(1 + (\omega\tau)^2\right)}\frac{R_A}{\left(1 + (\omega R_A C_A)^2\right)}, \tag{5.6}$$

where R_A is the antenna radiation resistance and C_A is the device capacitance. The power roll-off with increasing frequency, as is expected from this expression, includes the effect of the finite carrier lifetime as well as the RC time constant. The former reflects the material property that is also considered in Eq. (5.5), while the latter is rather an extrinsic property that stems from the device structure [16, 17].

The actual structure of the photomixers based on photoconductive material very much resembles the photoconductive antenna described earlier for THz pulse generation, which is not surprising as the radiation mechanisms are virtually identical. These photomixers also employ an LT GaAs layer grown on a semi-insulating GaAs substrate, which is mounted on a hyper-hemispherical Si lens. When it comes to the antenna structure, it is typical for the photomixers to employ broadband antennas

Fig. 5.9 Log-spiral antenna with an internal interdigitated structure used for photomixing

such as log-spiral or bowtie antennas, rather than the simpler dipole-based structures common for THz pulse generation. The gap at the center between the two conductor patterns in the broadband antennas, where the incoming beams are focused, usually takes on an interdigitated shape to enhance the efficiency of the conversion. This can be seen in an example case shown in Fig. 5.9. In some cases, photomixers without an antenna structure are employed. For them, the operation relies upon the direct THz radiation from the generated carriers, which are accelerated by the incoming beams with modulated intensity. As the dimension of the illuminated area is large for these cases, comparable to or even larger than the wavelength of the radiated THz wave, they are called large area emitters (LAEs) [17].

As for the lasers that generate the beams to pump the photomixers, there is an obvious difference from the case for THz pulse generation. While femtosecond lasers are used for pulse generation, it should be CW lasers that are employed for CW THz signal generation. As a fractional change in the wavelength in the optical band will result in a significant variation in the generated THz wave frequency, tunable lasers are preferred to allow broadband THz wave generation. In most practical cases, one of the two lasers is tuned, while the other is fixed at a reference wavelength. The coupling of a pair of beams from the two lasers can be made through a proper combination of a mirror and a beam splitter. As the requirements for CW lasers are much relaxed compared to the case of femtosecond lasers, commercially available diode lasers can be readily employed, whose wavelength typically ranges over 800–850 nm [2]. Note that 1% change in the wavelength in this region (~8 nm), which is readily available in typical cases, results in nearly 4 THz variation in the down-converted frequency. Of course, it is not this entire frequency range that is converted to the tuning bandwidth of the generated THz radiation, due to the internal limiting factors of the photomixers such as the carrier lifetime and *RC* time constant as discussed above with Eq. (5.6). Still, the tuning bandwidth of photomixers is far wider than those of electrical sources, a major advantage of photomixers.

The radiation power of the THz waves from the photomixers based on photoconductive material currently reaches up to ~10 μW for emission frequencies larger than 100 GHz [18, 19]. This can be compared with the output power levels available with electrical signal sources, which exceed ~10 mW and ~1 mW at 100 GHz and 300 GHz, respectively, with IMPATT or Gunn diodes. RTDs also exhibit output power larger than ~0.1 mW at these frequencies (for the output power of the diode-based sources, see Fig. 2.9). The output power from semiconductor integrated circuits based on oscillators and power amplifiers now reaches beyond 10 mW around 300 GHz (for the output power of the transistor-based sources, see Fig. 2.33). Hence, it is fair to say that the photomixers fall behind the electrical sources in terms of the generated output power at the moment, while they provide much wider tunable bandwidths. The output power of the photomixers can be improved with increased beam intensity from the lasers, but an excessive optical power may end up damaging the LT GaAs structure with increased device temperature. Substrates with a higher thermal conductivity may allow higher incident beam power, for which Si substrate may be used in place of GaAs as the thermal conductivity of Si is higher than that of GaAs (1.41 W/cmK for Si vs 0.46 W/cmK for GaAs at $T = 300$ K) [20]. LAE mentioned above can partially address this issue by spreading out the beam, and thus the beam power, over a wide area. The optimization of the antenna structure can enhance the output power as well. For example, it was shown that a dual dipole structure slows down the roll-off by resonating out the device capacitance with inductive tuning, as was achieved by controlled dipole length [21]. A structure employing a Fabry–Pérot cavity for photomixers was also reported, which resulted in an output power of 0.35 mW around 300 GHz [22].

5.2.1.2 Photomixing Based on *p-i-n* Diodes

While the majority of photomixers employ photoconductive material for carrier generation, photomixing can also be attained by illuminating a *p-i-n* diode with a beating optical beam. A *p-i-n* diode is based on a similar structure as the standard *p-n* diodes, the main difference being an undoped intrinsic layer inserted between the *p*-type and *n*-type regions. With the inclusion of the intrinsic layer, which is fully depleted under the reverse bias, a thick depleted drift region will be formed, which can serve as a photon absorption layer. This feature makes the diode an attractive candidate for photo-sensing devices. The photomixing mechanism with *p-i-n* diodes employed as a photomixer is largely similar to the photomixing case with photoconductive material. When a reverse-biased *p-i-n* diode is illuminated with an optical beam, electrons and holes will be generated in the depleted intrinsic region. They will drift through the intrinsic region before they arrive at the *n*-type and *p*-type regions, respectively, building up a photocurrent. This is shown in Fig. 5.10a, which depicts the structure of a conventional *p-i-n* diode under illumination. With a fast response of the photocurrent, CW THz signal can be generated when the incident beam has an intensity modulation at THz frequency, which stems from the beating between the two optical beams with a slight wavelength difference as described earlier. In a sense, *p-i-n* diodes belong to the category of the large area emitters mentioned above.

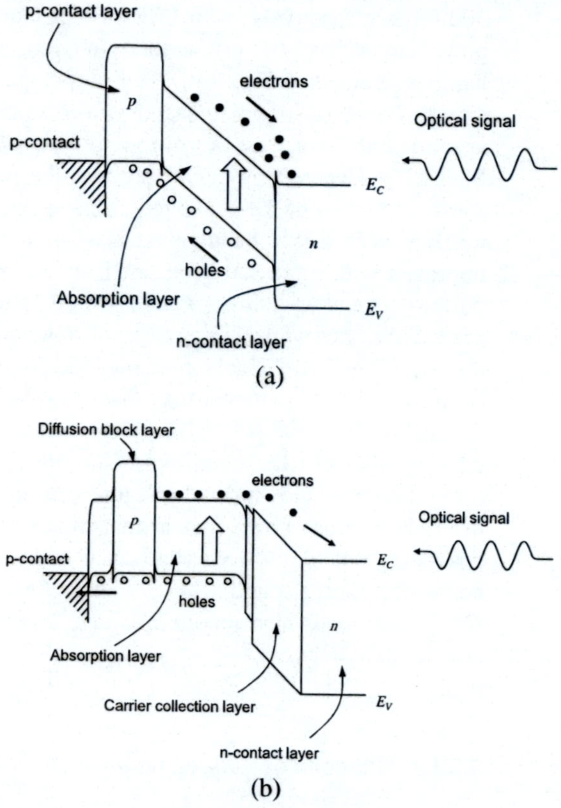

Fig. 5.10 Structure of diodes used for photomixing under illumination: (**a**) Conventional *p-i-n* diode (**b**) UTC-PD (© 2011 Nagatsuma T, Ito H. Published in [24] under CC BY-NC-SA 3.0 license. Available from: https://doi.org/10.5772/14800)

In typical *p-i-n* diodes, the velocity of the holes is much smaller than that of the electrons. Hence, the slow hole transport will impose a negative impact on the photo-response speed and thus the bandwidth of the *p-i-n* diode photomixer. Additionally, the slow drift of the holes will effectively promote the accumulation of holes in the drift region, resulting in substantial space charge density formed in the intrinsic region. The space charge will cause a field modulation and, consequently, a band bending, which will induce a local reduction of the field in the intrinsic region near the *n*-type region boundary. The reduced electric field will further slow down the carrier motion, eventually limiting the maximum current allowed in the diode. This will serve as a main limiting factor of the output power available with conventional *p-i-n* diode photomixers.

In view of this situation, if only the electrons are selected to contribute to the operation as the active carrier, while the effect of the hole transport is suppressed, one would expect improved photomixer bandwidth as well as higher saturation current. The UTC-PD was invented to serve this purpose [23]. In an UTC-PD, the absorption layer is *p*-type doped and separated from the intrinsic layer. Besides, it has a narrower bandgap than the intrinsic region as shown in Fig. 5.10b [24]. Therefore, when the absorption layer is illuminated with incoming optical beam, the photo-generated holes are rather confined in the *p*-type region and quickly relaxed as the majority carrier. On the other hand, the generated electrons enter the intrinsic region and are collected by the *n*-type region after drifting through the depleted intrinsic layer. Hence, the photocurrent is mostly composed of electrons, and its speed is governed by the electron velocity, leading to an improved operation speed and bandwidth. Also, a higher current level can be achieved compared to the conventional *p-i-n* diode case, since it does not suffer from the adverse band bending effect that limits the current saturation of the photomixer. The separation of the absorption and drift regions also helps to optimize the structure to reduce the device *RC* time constant that affects the bandwidth. It is noted that the band bending does occur in UTC-PD as well (in the opposite direction), but its effect is not as outstanding as the *p-i-n* diode case owing to the higher velocity of electrons, which is helped by the electron velocity overshoot. In fact, the average traveling distance for generated electrons to reach the *n*-type region is longer for UTC-PD due to the separate absorption layer, but it is compensated for by other merits including the higher electron velocity. The overall advantage of UTC-PDs leads to a wider bandwidth and, probably more importantly, higher output power compared to the conventional *p-i-n* diode photomixers.

A more detailed look into UTC-PD structure would help understand the operation. The absorption and intrinsic regions are usually composed of InGaAs and InP layers, respectively. The *p*-type absorption layer can be modified to create a quasi-electric field to help the electron acceleration, which can be achieved by bandgap grading or doping modulation. On top of the *p*-type absorption layer, another layer with a wider bandgap is routinely placed, typically an InGaAs layer with a higher Ga composition, which would block the diffusion of electrons toward the anode (top contact). Between the absorption and the intrinsic regions, an n^+ InP layer can be inserted to suppress the bandgap spike, facilitating a smooth launching of electrons into the drift region. This feature is similar to the layer structure in typical III-V HBTs, which are designed to help electron acceleration at the base-collector boundary. The ballistic transport of electrons with velocity overshoot in the intrinsic region in UTC-PDs as mentioned above also resembles the situation at the base-collector depletion region in HBTs. The difference is that the electrons are photo-generated in UTC-PDs, while they are injected from the *n*-type emitter in HBTs.

The output power of UTC-PD is higher than that of the conventional *p-i-n* diodes by a factor of an order or two, as indicated by the results reporting ~20 mW near 100 GHz [25] and ~100 μW close to 300 GHz [26]. It is notable that the large output power can be achieved simultaneously with a large bandwidth, decoupling the trade-off between output power and bandwidth in the conventional *p-i-n* diodes. There are some variants of UTC-PDs, including those integrated with a broadband or narrow-band antenna for a more efficient THz wave radiation [27].

5.2.2 *Generation with Frequency Difference Mixing by Optical Rectification*

The mixing between two beams with slightly different frequencies and the consequent THz wave generation can be obtained with optical rectification based on electro-optic materials as well. In Chap. 2, we observed that a mixing between two signals can be obtained as they go through a nonlinear (electrical) device, and, as a result, an output signal with a frequency that corresponds to the difference between the frequencies of the two incoming signals can be generated. The principles of the generation of THz CW signal by frequency difference mixing is conceptually identical to such an electrical mixing with a nonlinear device. The only difference for the frequency difference mixing is the fact that the operation frequency is much higher, around the optical band, and the nonlinear "device" is in fact a plain optical medium, rather than established structure such as diode or transistor, with a preferred electro-optical property.

Consider two beams propagating together that are assumed to be linearly polarized. For the two beams, the total optical field will be expressed by Eq. (5.3), as was the case for photomixing case. When an electro-optic material with a second-order nonlinearity is illuminated with the combined beam, a polarization will be established due to optical rectification. The second-order nonlinear polarization will be proportional to the square of the incident total optical field, as is given below:

$$P(t) \propto E_1^2 + E_2^2 + 2E_1E_2(\cos(\omega_1 + \omega_2)t + \cos(\omega_1 - \omega_2)t). \tag{5.7}$$

Note that the expression is basically identical to the intensity of the incoming coupled beams as was shown in Eq. (5.4). Hence, it can be equally stated that the achieved polarization is proportional to the intensity of the incident optical field. Assuming the sum frequency component is suppressed due to the low-pass filtering in the process of optical rectification, as was the case for the photomixing, the time-dependence of the polarization will be simply given by $\cos(\omega_1 t - \omega_2 t)$, or $\cos(\omega t)$ with $\omega = \omega_1 - \omega_2$. If we recall Eq. (5.2), which dictates that the generated THz field is proportional to the second-order time-derivative of the polarization, we have $E_{THz}(t)$ again proportional to $\cos(\omega t)$. Hence, the generation of a CW THz signal with a frequency equal to the difference of the two input optical frequencies is achieved.

For difference frequency generation, various nonlinear materials have been employed, such as GaSe, GaP, quartz, and $LiNbO_3$. Among these, GaSe has been found most effective owing to its strong second-order nonlinearity as well as the low absorption rate at the THz frequency range [28]. As for the beam sources, conventional CW lasers can serve the purpose, similar to the photomixing case, which enables a compact setup for THz generation. The generation of CW THz signal from difference frequency method also provides a very wide tuning range as is the case for photomixing. However, the output power level is again a major issue for this technique, smaller than those from photomixing.

5.2.3 *Generation with Lasers*

Laser (Light Amplification by Stimulated Emission of Radiation) is a device that radiates coherent light based on stimulated emission. Before directly jumping into CW THz signal generation with lasers, it would be helpful to review the very basics of the laser operation principle.

5.2.3.1 Lasers (Very) Basics

Consider a simplified picture of a system composed of two electron energy levels, denoted as E_1 and E_2 in Fig. 5.11. There can be three scenarios for the interaction between electrons and photons. First, an electron in E_1 can absorb energy from an incident photon and make transition up to E_2, which is called the absorption (Fig. 5.11a). Second, an electron in E_2 can spontaneously make transition down to E_1 and create a photon, which is called the spontaneous emission (Fig. 5.11b). Third, an electron in E_2 can make transition down to E_1 stimulated by an incident photon and create an extra photon, which is called the *stimulated emission* (Fig. 5.11c). The laser operation is based on the stimulated emission. As two photons emerge from the system with a single incident photon, we can say *light amplification* is attained with the stimulated emission, a process behind the full name of "laser." The stimulated emission is obtained only when a certain condition is satisfied, namely the *population inversion*. That is, the electron concentration at E_2 needs to be larger than the concentration at E_1 for stimulated emission to occur. To achieve the population inversion, various methods to "pump" the electrons from E_1 to E_2 can be employed. In real systems, there can be more than two energy levels, although it is still a pair of levels that contribute to the photon generation by the stimulated emission. The medium with the desired energy levels, or the gain medium, is physically situated internal to an optical cavity, inside which created photons will be accumulated as they are reflected back and forth repeatedly by the partially transparent cavity

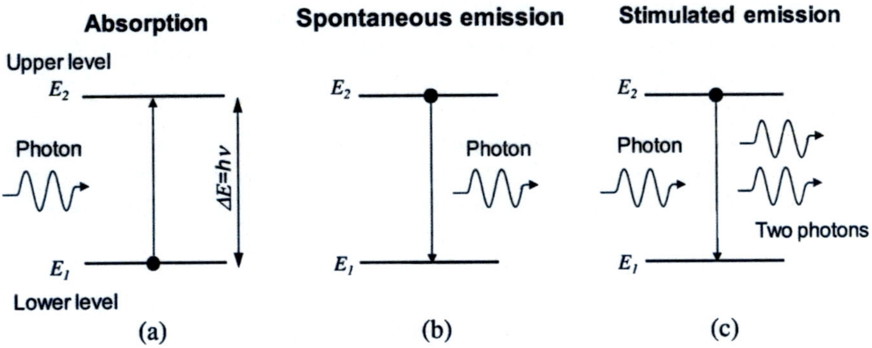

Fig. 5.11 Three scenarios of the interaction between electrons and photons: (**a**) Absorption (**b**) Spontaneous emission (**c**) Stimulated emission

boundaries. This will further enhance the stimulated emission, leading to a rapid growth in the photon density before reaching a saturation, which is responsible for the high radiation power available with a laser. A key point here in the context of *terahertz* lasers is the frequency of the radiation from the lasers. It is dictated by $\Delta E = E_2 - E_1$, with a relation $f = \Delta E/h$, where h is Planck constant. A major challenge for THz lasers is to obtain ΔE that corresponds to the THz spectrum, which is only a few meV, far smaller than what is usually available with typical laser systems. However, it is still achievable with some selected types of lasers, which are described below in this subsection.

5.2.3.2 *p*-type Ge Lasers

It is possible to obtain CW THz radiation with a *solid-state* laser. For most solid-state crystals, the energy bandgap between the conduction and valence band typically ranges in the order of ~eV, which is far larger than the energy levels corresponding to the THz spectrum. Therefore, the *inter-band transition* in solid-state materials cannot be used for THz radiation. On the other hand, there are chances for the *intra-band transition* between subbands internal to the valence or conduction bands to be exploited for lasing, if the transition energy falls within a range of a few meV. Such a transition can be achieved with *p*-type doped germanium (Ge) lasers [29, 30].

For Ge, the valence band is composed of the heavy hole (with a large effective mass) band and the light hole (with a small effective mass) band, the former showing a larger radius of curvature than the latter in the $E - k$ relation. There exists yet another band, namely the spin-orbit split-off band, in the valence band, but it hardly contributes to the conduction due to the negligibly small hole population, as it is located significantly away (~290 meV at $T = 300$ K) from the valence band edge. Figure 5.12a shows the hole bands of Ge in the $E - k$ space on the bottom half. It would be noteworthy that the maxima of these hole bands are all located at Γ point, whereas the conduction band for electrons (shown on the upper half of the figure) has its minimum at L point in the crystal Ge. Such an indirect bandgap structure, however, does not preclude Ge for laser operation, since it is not the inter-band transition that leads to its laser operation. For *p*-type Ge lasers, the intra-band (or inter-subband) transition between Landau levels at cryogenic temperatures is exploited for THz lasing, as described below.

We all know that a charged particle will be linearly accelerated in the presence of an external electric field. In a slightly more involved situation where a magnetic field is additionally applied in a direction orthogonal to the existing electric field (called the cross field), the charged particle will become under the influence of Lorentz force and experience an acceleration in the transverse direction, ending up with a circular motion with a cyclotron orbit. Interestingly, the energy level associated with the cyclotron orbit will be quantized to meet the quantum condition, and the generated discrete energy levels are called Landau levels. This process will happen to charged particles inside solid-state crystals as well. Therefore, Landau levels will be generated for hole bands in *p*-type Ge crystal when mutually orthogonal external electric

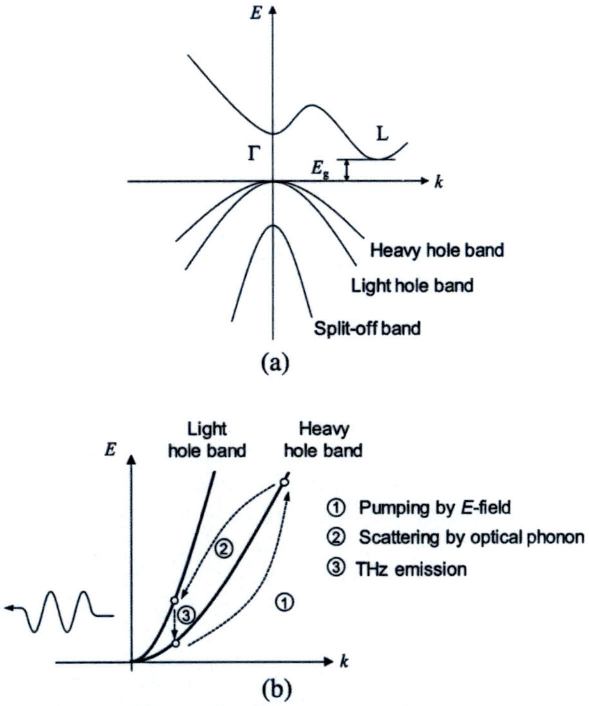

Fig. 5.12 (**a**) Band structure of Ge in the $E - k$ space. (**b**) Process of radiation in Ge laser (polarity of energy is shown in reference to the holes for convenience)

and magnetic fields are applied. Hence, for lasing, you will first need a biased *p*-type Ge with an external magnet in a proper field orientation.

Next, you will need a cryogenic temperature. The readers should be aware that there are two major scattering mechanisms inside semiconductors: phonon scattering and ionized impurity scattering. As for the phonon scattering, there are two modes of phonons, optical phonon and acoustic phonon, the former involving a larger energy than the latter in scattering events. All added up, therefore, there are basically three scattering mechanisms: optical phonon scattering, acoustic phonon scattering, and ionized impurity scattering. For lowly doped *p*-type Ge placed in a cryogenic environment, optical phonon scattering will show a much higher probability than the acoustic phonon scattering and ionized impurity scattering. Interestingly, when a certain combination of the magnitudes for the electric and magnetic fields is attained, the heavy holes will enter a condition where optical phonon scattering will prevail while the light holes will be dominated by acoustic phonon and ionized impurity scatterings. As a result, for such an optimal range of electrical and magnetic fields, the heavy hole states will show a short carrier lifetime due to the high scattering rate of optical phonons, while the lifetime of the light hole states will be relatively long with the low associated scattering rates with the given condition.

With these properties involving the holes in *p*-type Ge in mind, consider a hole accelerated by electric field [31]. It will gradually climb the energy curve of the heavy hole band as described in Fig. 5.12b, which is basically an electric pumping process. Once it gains sufficient energy, it will be scattered soon by an optical phonon and relaxed down to one of the lower states, including a quasi-stable light-hole Landau level with a long lifetime. If this process continues, the amount of holes in the Landau level will grow due to its long lifetime, eventually leading to a population inversion against the lower energy states. Hence, when triggered by photons, stimulated emission will be induced and the holes will transit to the lower energy states in the heavy- or light-hole bands. If the transition energy corresponds to the THz spectrum, which is indeed the case for *p*-type Ge, THz lasing will occur. There are other mechanisms known for the lasing process in *p*-type Ge with slight different details [32], but the overall concept remains the same.

The frequency covered by *p*-type Ge lasers ranges from ~100 GHz up to a few THz, depending on the types of operation mechanisms. A *p*-type Ge laser can also be tuned with the applied magnetic field strength adjustment. Output power can reach up to 10 W, but only with pulsed operation. Full CW radiation is not currently available as the operation would heat up the laser and activate the acoustic phonon scattering, which will prevent the desired process for laser operation.

5.2.3.3 Quantum Cascade Lasers

As mentioned earlier in this section, we need a transition energy that is as tiny as a few meV to achieve a radiation of CW THz by a laser action. One way to create energy levels with such a small difference is to create a quantum well inside the conduction band with an appropriate width that would generate discrete energy levels with the desired inter-level spacings. One effective approach to realize such a quantum well is to alternatively stack two types of semiconductor layers with vastly different energy bandgaps, one with narrow and the other with wide bandgaps. The energy allowed for electrons residing inside the narrow-bandgap semiconductor, which is sandwiched by two wide-bandgap semiconductors, will be quantized. With a proper choice of the well width, an electron transition between a pair of energy levels will generate a photon with an energy in the range of a few meV, thus leading to THz radiation.

The strength of the radiation from a single transition, however, may not be sufficient for practical purposes. One ingenious approach to increase the radiation power is to induce *multiple* transitions with a single electron by employing a superlattice, or an epitaxial structure composed of multiple quantum wells, as shown in Fig. 5.13a. If a bias is applied across the superlattice, there will be a slope established across the structure as depicted in Fig. 5.13b. With a vastly simplified picture, the following description may help: If an electron is injected into the superlattice and completes a transition inside the first quantum well, it will tunnel into the second quantum well through a narrow energy barrier and make another transition, which is followed by yet another tunneling to the next quantum well. This process will repeat until the last quantum well is reached, through which a single electron is "reused" for each transition and corresponding photon generation. Consequently, as many photons as the number of transitions, or the number of

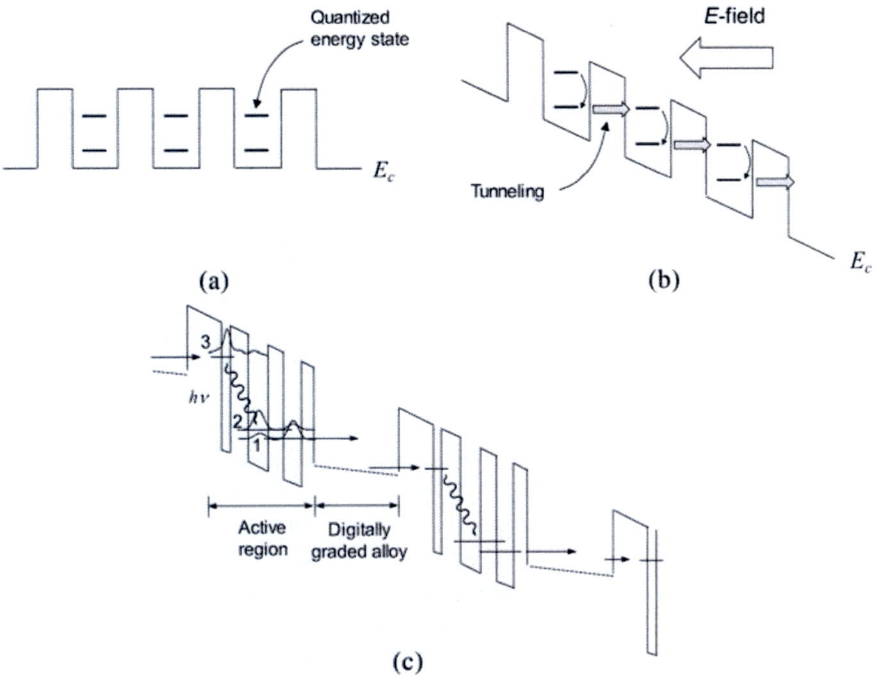

Fig. 5.13 (**a**) Multiple quantum wells in a superlattice, shown with the quantized energy states. (**b**) Multiple quantum wells under bias, shown with electron transitions between energy states and tunneling across energy barriers. (**c**) Actual structure and electron transitions inside a typical QCL (quantum cascade laser). (From [33]. Reprinted with permission from AAAS)

quantum wells, will be eventually generated, leading to a higher output radiation power than the conventional single-well case. As the behavior of the electrons resembles a cascade, a laser based on this process is called the quantum cascade laser (QCL) [33].

A more detailed view can be provided based on the more realistic structure illustrated in Fig. 5.13c. A QCL is composed of a stack of many identical active regions, which may count over a hundred. An active region may be a single quantum well (rare) or a composition of multiple quantum wells (typical), for which various designs have been proposed [34]. An active region typically comprises more than two energy levels, while a three-level system is assumed in Fig. 5.13c, for which the stimulated emission will be associated with an electron transition from the energy level 3 to level 2. After a transition down to level 2 is made inside an active region, the electron is quickly relaxed to level 1 and tunnel into the next active region across the energy barrier(s). Population inversion between level 3 and level 2 is achieved with the help of a quick electron relaxation from level 2 to level 1 after each transition, which helps to lower the electron population at level 2. Additionally, a swift tunneling between active regions is also necessary, which will effectively pump level 3 to maintain a high population. The generated photons travel along the lateral in-plane path (perpendicular to the layer stack direction), which is confined by mirrors on both sides, typically realized with a cleaving of the semiconductor piece containing the QCL structure.

For the operation of THz QCLs, cryogenic temperature is again needed. For THz generation, the involved energy level ranges in the order of a few meV, which can be easily exceeded by the thermal energy of electrons, roughly around kT. For example, the transition energy corresponding to 1 THz is 4.1 meV, which equals to the thermal energy of an electron at around 50 K. A large thermal energy would activate optical phonon scattering, leading to a high probability for electrons to be relaxed by losing energy to optical phonons instead of photons. This issue is not relevant only to THz QCL but to THz lasers in general, imposing a great challenge to the realization of the room-temperature THz lasers. This explains the absence of room-temperature operation of THz QCLs reported so far. At cryogenic temperatures, however, advances in THz QCLs have continued [34]. A QCL with an output power beyond 1 W at 10 K has been reported with a pulsed mode operation around 3.4 THz, which also exhibited 420 mW at 77 K [35]. As for the efforts to increase the operation temperature, a QCL operating near 200 K has been reported at 3.2 THz [36]. There is a large number of other reports on THz QCLs, whose radiation frequency ranges roughly from 1.5 to 4.5 THz.

5.2.3.4 Gas Lasers

While both of the THz lasers described above are solid-state lasers, THz lasing can be achieved with gas lasers as well. Gas lasers employ gas-phase materials as the gain media. While some gas lasers are based on monatomic gases, as in the case of the popular HeNe (helium-neon) lasers, many of them rely on gas *molecules* to serve as the gain media. For laser operation, we need quantized energy levels. For lasers with gas molecules, it is the vibrational and rotational modes (or coupling of the two) of molecules that are exploited for electron transitions required for stimulated photon emission. While both modes typically involve a much smaller excitation energy compared to the atomic energy level transition, it is the rotational modes that provide smaller energy exchange for inter-mode transition, which is needed for long-wavelength radiation. If the transition energy between the rotational modes is small enough, THz radiation can be expected with gas lasers.

There indeed have been such gas lasers reported for radiation in the THz spectrum [37, 38]. The major efforts have been made on CH_3OH (methanol) lasers, the 118.9 μm (2.52 THz) line being the main target among other spectral lines available with CH_3OH [39]. CH_3OH lasers are optically pumped with another laser, typically with a high-power CO_2 laser radiating at around 9.7 μm (31 THz). As illustrated in Fig. 5.14, the pumping with a CO_2 laser excites electrons to an upper vibrational mode of the CH_3OH gas, which is followed by transitions between rotational modes bunched within the vibrational mode, leading to THz radiation. A CH_3OH laser with a CW output power exceeding 1 W at 118.9 μm has been reported with CO_2 laser

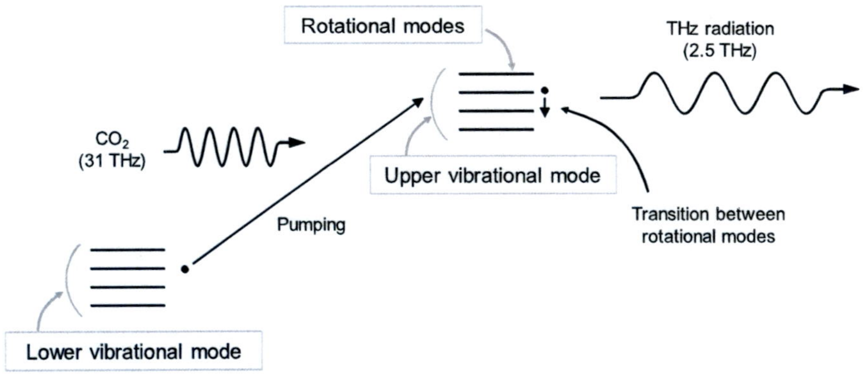

Fig. 5.14 Mechanism of THz radiation with a methanol laser

pumping [40]. Other gases known to be capable of THz lasing with a substantial output power include NH_3 (ammonia), CH_2F_2 (difluoromethane), CH_3Cl (chloromethane), and CH_3I (methyl iodide) [1].

5.3 THz Detection

In Chap. 3, various techniques for THz radiation detection such as thermal detection as well as detection with diodes or transistors were discussed. However, what they detect is the *time-averaged* power of the THz radiation, mostly applicable to CW radiation. In some cases, however, phase-sensitive detection is desired, especially when the THz radiation is in the form of pulses. For example, the recovery of the THz pulse shapes, typically after the transmission through or reflection from a target object, is the core part of THz spectroscopy. Although it is extremely challenging to precisely reconstruct the pulse profiles as short as those of THz pulses, it can be achieved based on principles that are similar to those behind the THz pulse generation, with either photoconductive antenna or electro-optic crystals. The key element employed for the phase-sensitive detection of ultra-short pulses is the sampling (or gating) technique, which is detailed below.

5.3.1 Detection with Photoconductive Antennas

If we recall the generation of a THz pulse with a photoconductive (PC) antenna, as described in Sect. 5.1.1, an optical pulse from a femtosecond laser is incident on the PC antenna with the beam spot landing on a middle gap of the metallic antenna pattern. With the antenna under a DC bias, an electric field is pre-established across the gap and the application of the optical pulse gives rise to a current, the temporal

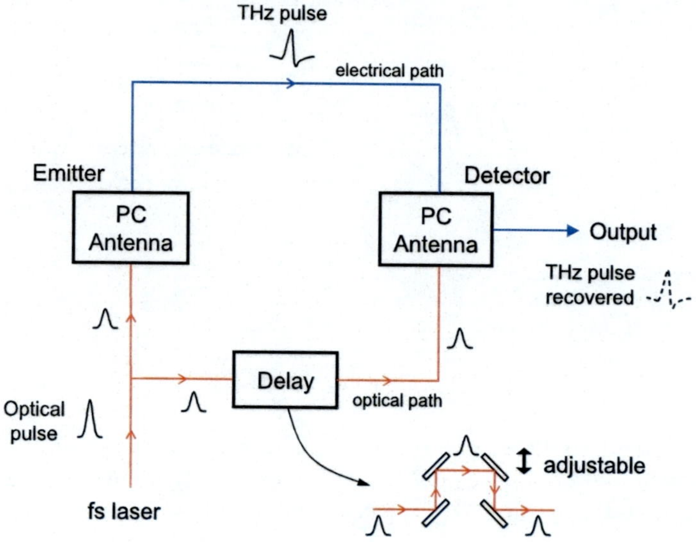

Fig. 5.15 Typical setup for THz waveform detection with photoconductive antennas

variation of which results in THz pulse generation. The detection can be made with the same antenna structure. For the detection, however, the antenna is *not biased*. Instead, the electric field needed to generate the current is provided by the *incoming THz pulse* incident on the antenna that is to be detected. With the incident THz pulse only, there will be no current generation despite the field established across the antenna gap due to the lack of carriers. However, when it is combined with an optical pulse, current will be generated owing to the carrier generation by the photo excitation. Since the current level will be higher with a stronger THz pulse intensity, the measured current level will be a good indicator of the THz pulse strength at the moment of the optical pulse arrival. Now, if such a current generation process is repeated multiple times with progressively shifting arrival time of the optical pulse with reference to the THz pulse, one can reconstruct the THz pulse shape by connecting the current level points over arrival time. This is the basic principle of sampling and the applied optical pulse for sampling is called the sampling or gating pulse.

A typical setup to realize the sampling-based THz detection with PC antennas is depicted in Fig. 5.15. It is basically composed of a pair of PC antennas, one for the generation and the other for the detection of THz pulses, which are called the emitter and detector, respectively. The two antennas are linked by an electrical path on one side and an optical path on the other side. The generated THz pulse at the emitter travels along the electrical path to reach and illuminate the detector, optionally through a target sample. On the other hand, the optical pulse from the femtosecond laser is split into two pulses: one directly arrives at the emitter for THz pulse generation mentioned above, and the other reaches the detector for THz pulse detection after traveling along the optical path. Placed in the middle of the optical path is a time delay component, which is a critical element to precisely control the timing of the optical pulse arrival at the detector. It is typically composed of a set of mirrors, a part of which can be moved with high precision to accurately adjust the total traveling path of the optical pulse (see the sketch in the bottom of Fig. 5.15). The output of the detector antenna is terminated with an ammeter for current detection, which is in contrast with the emitter antenna that is terminated with a voltage source for bias supply. The detected current level is typically amplified to boost the signal strength afterwards.

With the system described above, the generated THz pulse at the emitter can be detected at the detector with the phase information retained as described below. The sampling technique to recover the THz pulse profile, as mentioned earlier, requires a series of identical measurements to trace the shape. In each round of the measurement, a THz pulse is generated and arrives at the detector in synchronization with the arrival of an optical pulse with a controlled delay. The delay is slightly different for each round and sequentially increases in the course of the entire measurement cycle. Hence, if the current level is recorded for each measurement round and the collected current level points are plotted as a function of the delay time for the complete process, it will show the profile of the original THz pulse. This sampling process is illustrated in Fig. 5.16, which shows the sampling of the THz pulse shape with a series of measurements assuming the time delay increases by a step of $\Delta\tau$ each round (in the figure, only even-numbered sampling points are shown due to space limit, although the sampling occurs for the every sampling point). If needed, the obtained time-domain profile can be converted into the frequency-domain profile by Fourier transformation. In practical cases, a target sample is often placed in the middle of the electrical path, in which case the THz pulse is detected after the transmission through the sample. By comparing the profiles of the THz pulses with and without the sample, rich spectral information of the sample can be retrieved. This is in fact the principle of the time-domain spectroscopy (TDS), which will be discussed in Sect. 6.1 in more detail.

While the sampling technique so far discussed has been successfully applied to THz pulse measurements, there is a limitation as well. The extracted temporal variation of the detected photocurrent well represents the THz pulse profile, but it falls short of being truly identical to the original. There exists a finite difference between the original and the recovered profiles, which arises from the fact that the photocurrent is induced by the combined effect of the THz pulse (electric field) and the optical pulse (carrier generation). This is better understood by observing the expression of the photocurrent in the frequency domain as shown below [1]:

$$J_{\text{photo}}(\omega) = \sigma(\omega)E_{\text{THz}}(\omega), \tag{5.8}$$

where σ is the conductivity, which is governed by the carrier concentration and the mobility. If the conductivity is constant over the frequency, the photocurrent profile would exactly match the field profile in the frequency domain, and consequently in the time domain as well. If not, which is the case in reality, the true profile will be filtered by the bandwidth of the conductivity, leading to a finite distortion. One major

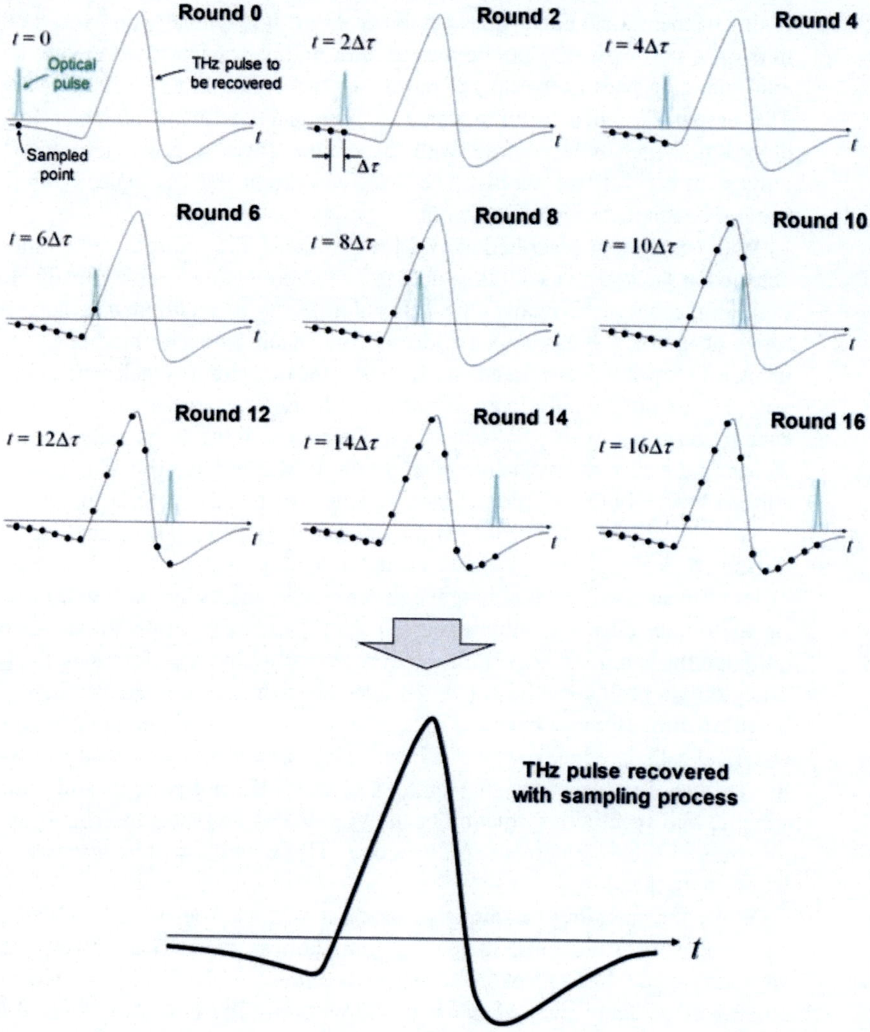

Fig. 5.16 Conceptual diagram for the recovery of a THz waveform with the sampling process. In the figure, only even-numbered sampling points are shown due to the space limit, although the sampling occurs for the entire sampling points in reality

factor that affects the frequency dependence of the conductivity is the bandwidth of the carrier concentration, which is in turn influenced by the pulse width of the incident optical beam. It has been shown that a short optical pulse significantly improves the bandwidth of the carrier concentration [41]. This indicates that the ultrashort pulse generation from the femtosecond laser is critical for both generation and detection of THz pulses.

5.3.2 Detection with Electro-Optic Crystals

Sampling-based phase-sensitive detection of THz pulses can be achieved with electro-optic crystals as well [42]. As the THz pulse generation was enabled with either photoconductive (PC) antennas or electro-optic (EO) crystals, the same pair of methods can be applied to THz pulse *detection*. Both the methods rely on the temporal variation of the current level extracted with the sampling technique to reconstruct the THz pulse profile, the only difference being the actual process to induce the current level that matches the strength of the THz field. When EO crystals were used for THz pulse generation as described Sect. 5.1.2, it was the optical rectification property of the crystals that was exploited. For THz pulse detection, Pockels effect is utilized, which is closely related to but differ from optical rectification. Pockels effect refers to the generation or change of birefringence in accordance with the strength of the electric field present inside electro-optic crystals. This topic requires the readers to understand *birefringence* first. In some crystals, the index of refraction depends on the crystal orientation with respect to the polarization of the incoming light. This property is called birefringence and the materials with birefringence is called birefringent materials. Various optical components are based on birefringence, including those employed in the setup for EO sampling. A couple of them are briefly introduced below (for the details of polarization, see Sect. 4.2.1 and Figs. 4.16 and 4.17).

λ/4 plate is a component that converts a linear polarization into a circular polarization using birefringence. When linearly polarized light enters a $\lambda/4$ plate in 45° angle with respect to the "optic axis", the vertical (parallel to the optic axis) and horizontal (perpendicular to the optic axis) components of the polarization will experience different refractive indices due to the birefringence. The difference will develop a phase difference between the two components as they travel through the material. For $\lambda/4$ plates, the plate thickness is designed, so that the total phase difference matches 90°, or a quarter wavelength, which would enable linear-to-circular polarization conversion. *Wollaston prism* is another component made of birefringent material that splits the physical path of an incoming light depending on the polarization direction. The two orthogonal polarization components of a polarized light that enters a Wollaston prism will experience different refractive indices. Thus, the refraction angle at the crystal surface will differ for the two polarization components, resulting in two separate paths for the light after penetrating through the prism.

The heart of the EO sampling lies in the *field-dependence* of birefringence exhibited by some materials. In EO crystals such as ZnTe, there will be no birefringence in the absence of electric field. However, the application of electric field will induce birefringence, the strength of which will increase with the field intensity. Now assume a linearly polarized optical beam incident on a slab of ZnTe with 45° angle with respect to the optic axis. Without any electric field present, the beam will exit the crystal with linear polarization retained as no birefringence is in effect. However, when the beam is incident together with a THz pulse, the electric field due

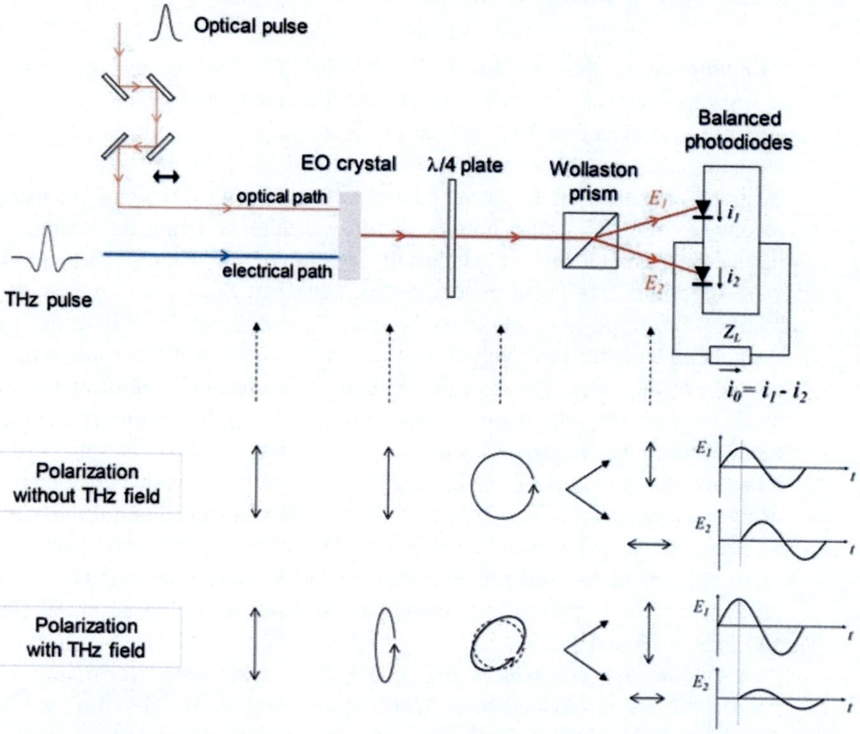

Fig. 5.17 Typical setup for THz waveform detection with EO sampling. Also shown on the bottom are the polarization states at selected locations along the path (the locations are indicated by dotted arrows). (Adapted by permission from Springer Nature: [1] © 2009)

to the pulse will trigger birefringence inside ZnTe, rendering the optical beam under the influence of birefringence. Consequently, one polarization component of the optical beam will experience a phase retardation against the other component, resulting in elliptical polarization for the beam exiting the crystal. The larger the field strength, the larger will be the phase retardation. Hence, if the extent of the phase difference can be converted into measurable quantity, such as current, the field strength of the THz pulse can be estimated, and eventually the pulse shape can be restored with a sequential acquisition of the quantity [13].

A setup for the THz pulse detection with EO-sampling is illustrated in Fig. 5.17 [1]. A linearly polarized probe optical beam is incident on an EO crystal together with a THz pulse with a 45° angle against the optic axis of the crystal. The crystal will be momentarily birefringent due to the field induced by the THz pulse that passes by, and thus the optical beam will emerge from the crystal elliptically polarized. The beam will subsequently travel through a $\lambda/4$ plate and a Wollaston prism. The $\lambda/4$ plate, which would convert linear polarization into circular polarization, will convert the elliptical polarization of the optical beam into nearly circular polarization, with a slight deviation from the perfect. The extent of the deviation, or distortion, from the perfect circular polarization will depend on the strength of the field induced by the THz pulse. With the Wollaston prism, a beam of perfect circular polarization (zero field inside ZnTe or the absence of THz pulse) will be split into two beams with equal intensity. The two beams will have a phase difference of 90°, and the phase information will be lost in the end at the detection with a pair of photodiodes, since the currents induced by the two balanced diodes will be canceled out. For a beam of elliptical polarization (non-zero field inside ZnTe or the presence of THz pulse), whose axis is aligned to the optic axis of the Wollaston prism, the two split beams will have a different intensity (see Fig. 4.17e for the relative intensities of the field in elliptical polarization). The difference increases with increasing deviation from the perfect circular polarization and thus with increasing electric field inside ZnTe. The two beams coming out of the Wollaston prism will be separately incident on a pair of balanced photodiodes and the current difference between the two photodiodes ($i_0 = i_1 - i_2$, where i_1 and i_2 are the currents generated by the two photodiodes) will be extracted at the output. Hence, a larger deviation from the circular polarization, which corresponds to a larger THs pulse intensity, will result in a large out current level. In this way, the output current will serve as the indicator of the THz field at the moment of the optical beam arrival at the EO crystal. Repeated measurements of the output current with progressively increasing time delay imposed on the optical beam will lead to the recovery of the profile of the THz pulse being detected.

References

1. Y.-S. Lee, *Principles of Terahertz Science and Technology* (Springer, Berlin, 2009)
2. K. Sakai (ed.), *Terahertz Optoelectronics* (Springer, Berlin, 2004)
3. E. Bründermann, H.-W. Hübers, M.F. Kimmitt, *Terahertz Techniques* (Springer, Berlin, 2012)
4. D.E. Spence, P.N. Kean, W. Sibbett, 60-fsec pulse generation from a self-mode-locked Ti: sapphire laser. Opt. Lett. **16**(1), 42–44 (1991). https://doi.org/10.1364/OL.16.000042
5. S. Gupta et al., Subpicosecond carrier lifetime in GaAs grown by molecular beam epitaxy at low temperatures. Appl. Phys. Lett. **59**(25), 3276–3278 (1991)
6. D.C. Look, D.C. Walters, G.D. Robinson, J.R. Sizelove, M.G. Mier, C.E. Stutz, Annealing dynamics of molecular-beam epitaxial GaAs grown at 200 °C. J. Appl. Phys. **74**(1), 306–310 (1993). https://doi.org/10.1063/1.354108
7. K.A. McIntosh, K.B. Nichols, S. Verghese, E.R. Brown, Investigation of ultrashort photocarrier relaxation times in low-temperature-grown GaAs. Appl. Phys. Lett. **70**(3), 354–356 (1997). https://doi.org/10.1063/1.118412
8. F.E. Doany, D. Grischkowsky, C.C. Chi, Carrier lifetime versus ion-implantation dose in silicon on sapphire. Appl. Phys. Lett. **50**(8), 460–462 (1987). https://doi.org/10.1063/1.98173
9. Z.-S. Piao, M. Tani, K. Sakai, Carrier dynamics and THz radiation in biased semiconductor structures. Terahertz Spectrosc. Appl. **3617**, 49–56 (1999)
10. L. Duvillaret, F. Garet, J.-F. Roux, J.-L. Coutaz, Analytical modeling and optimization of terahertz time-domain spectroscopy experiments, using photoswitches as antennas. IEEE J. Sel. Top. Quantum Electron. **7**(4), 615–623 (2001)

11. M. Tani, S. Matsuura, K. Sakai, S.-i. Nakashima, Emission characteristics of photoconductive antennas based on low-temperature-grown GaAs and semi-insulating GaAs. Appl. Opt. **36**(30), 7853–7859 (1997). https://doi.org/10.1364/AO.36.007853
12. C.W. Berry, N. Wang, M.R. Hashemi, M. Unlu, M. Jarrahi, Significant performance enhancement in photoconductive terahertz optoelectronics by incorporating plasmonic contact electrodes. Nat. Commun. **4**(1), 1–10 (2013)
13. Q. Wu, M. Litz, X.C. Zhang, Broadband detection capability of ZnTe electro-optic field detectors. Appl. Phys. Lett. **68**(21), 2924–2926 (1996)
14. Q. Chen, M. Tani, Z. Jiang, X.-C. Zhang, Electro-optic transceivers for terahertz-wave applications. J. Opt. Soc. Am. B. **18**(6), 823–831 (2001)
15. R.W. Boyd, *Nonlinear Optics* (Elsevier, Amsterdam, 2003)
16. E. Brown, F. Smith, K. McIntosh, Coherent millimeter-wave generation by heterodyne conversion in low-temperature-grown GaAs photoconductors. J. Appl. Phys. **73**(3), 1480–1484 (1993)
17. S. Preu, G. Döhler, S. Malzer, L. Wang, A. Gossard, Tunable, continuous-wave terahertz photomixer sources and applications. J. Appl. Phys. **109**(6), 4 (2011)
18. S. Verghese, K.A. McIntosh, E.R. Brown, Optical and terahertz power limits in the low-temperature-grown GaAs photomixers. Appl. Phys. Lett. **71**(19), 2743–2745 (1997). https://doi.org/10.1063/1.120445
19. E.R. Brown, K.A. McIntosh, K.B. Nichols, C.L. Dennis, Photomixing up to 3.8 THz in low-temperature-grown GaAs. Appl. Phys. Lett. **66**(3), 285–287 (1995). https://doi.org/10.1063/1.113519
20. S. Verghese et al., A photomixer local oscillator for a 630-GHz heterodyne receiver. IEEE Microw. Guided Wave Lett. **9**(6), 245–247 (1999). https://doi.org/10.1109/75.769535
21. S.M. Duffy, S. Verghese, A. McIntosh, A. Jackson, A.C. Gossard, S. Matsuura, Accurate modeling of dual dipole and slot elements used with photomixers for coherent terahertz output power. IEEE Trans. Microw. Theory Tech. **49**(6), 1032–1038 (2001). https://doi.org/10.1109/22.925487
22. E. Peytavit et al., Milliwatt-level output power in the sub-terahertz range generated by photomixing in a GaAs photoconductor. Appl. Phys. Lett. **99**(22), 223508 (2011). https://doi.org/10.1063/1.3664635
23. T. Ishibashi, N. Shimizu, S. Kodama, H. Ito, T. Nagatsuma, T. Furuta, Uni-traveling-carrier photodiodes. Ultrafast Electron. Optoelectron. **13**, 83–87 (1997)
24. T. Nagatsuma, H. Ito, High-power RF uni-traveling-carrier photodiodes (UTC-PDs) and their applications, in *Advances in photodiodes*, ed. by G. F. D. Betta, (IntechOpen, Rijeka, 2011)
25. H. Ito et al., High-power photonic millimetre wave generation at 100 GHz using matching-circuit-integrated uni-travelling-carrier photodiodes. IEE Proc. Optoelectron. **150**(2), 138–142 (2003)
26. H. Ito, T. Furuta, Y. Muramoto, T. Ito, T. Ishibashi, Photonic millimetre-and sub-millimetre-wave generation using J-band rectangular-waveguide-output uni-travelling-carrier photodiode module. Electron. Lett. **42**(24), 1424–1425 (2006)
27. H. Ito, F. Nakajima, T. Furuta, T. Ishibashi, Continuous THz-wave generation using antenna-integrated uni-travelling-carrier photodiodes. Semicond. Sci. Technol. **20**(7), S191 (2005)
28. W. Shi, Y.J. Ding, N. Fernelius, K. Vodopyanov, Efficient, tunable, and coherent 0.18–5.27-THz source based on GaSe crystal. Opt. Lett. **27**(16), 1454–1456 (2002). https://doi.org/10.1364/OL.27.001454
29. S. Komiyama, Far-infrared emission from population-inverted hot-carrier system in p-Ge. Phys. Rev. Lett. **48**(4), 271–274 (1982). https://doi.org/10.1103/PhysRevLett.48.271
30. S. Komiyama, N. Iizuka, Y. Akasaka, Evidence for induced far-infrared emission from p-Ge in crossed electric and magnetic fields. Appl. Phys. Lett. **47**(9), 958–960 (1985)
31. J. Hovenier, A. Muravjov, S. Pavlov, V. Shastin, R. Strijbos, W.T. Wenckebach, Active mode locking of a *p*-Ge hot hole laser. Appl. Phys. Lett. **71**(4), 443–445 (1997)

32. I. Hosako, N. Hiromoto, Research and developments on p-type germanium lasers in the frequency range from 0. 1 terahertz to few terahertz. J. Commun. Res. Lab. **49**(1), 81–94 (2002)
33. J. Faist, F. Capasso, D.L. Sivco, C. Sirtori, A.L. Hutchinson, A.Y. Cho, Quantum cascade laser. Science **264**(5158), 553–556 (1994). https://doi.org/10.1126/science.264.5158.553
34. B.S. Williams, Terahertz quantum-cascade lasers. Nat. Photonics **1**(9), 517 (2007)
35. L. Li et al., Terahertz quantum cascade lasers with >1 W output powers. Electron. Lett. **50**(4), 309–311 (2014). https://doi.org/10.1049/el.2013.4035
36. S. Fathololoumi et al., Terahertz quantum cascade lasers operating up to ~200 K with optimized oscillator strength and improved injection tunneling. Opt. Express **20**(4), 3866–3876 (2012). https://doi.org/10.1364/OE.20.003866
37. G. Dodel, On the history of far-infrared (FIR) gas lasers: thirty-five years of research and application. Infrared Phys. Technol. **40**(3), 127–139 (1999). https://doi.org/10.1016/S1350-4495(99)00005-5
38. N.G. Douglas, *Millimetre and Submillimetre Wavelength Lasers: A Handbook of Cw Measurements* (Springer, Berlin, 2013)
39. G. Moruzzi, J.C. Silos Moraes, F. Strumia, Far infrared laser lines and assignments of CH3OH: a review. Int. J. Infrared Millim. Waves **13**(9), 1269–1312 (1992). https://doi.org/10.1007/bf01009988
40. J. Farhoomand, H.M. Pickett, Stable 1.25 watts CW far infrared laser radiation at the 119 μm methanol line. Int. J. Infrared Millim. Waves **8**(5), 441–447 (1987)
41. S. Kono, M. Tani, K. Sakai, Ultrabroadband photoconductive detection: comparison with free-space electro-optic sampling. Appl. Phys. Lett. **79**(7), 898–900 (2001). https://doi.org/10.1063/1.1394719
42. A. Yariv, P. Yeh, *Optical Waves in Crystals* (Wiley, New York, 1984)

Chapter 6
THz Applications

In the previous chapters, basic principles of the topics related to the THz frequency band have been overviewed with a main focus on the electronics-based approaches. While the efforts on exploiting the THz band have been so far rather skewed toward the academic side, there are recent indications that the theories and knowledges accumulated by THz research are now quite matured, so that they can be applied to the real world with more practical applications. In this chapter, four most representative THz applications will be reviewed: spectroscopy, imaging, communication, and radars. The first two topics are rather traditional application fields of the THz band, and already a great amount of accomplishments have been made in the labs and are moving toward the application-oriented front. The following two topics are relatively new and emerging fields of THz application and expected to have a more practical impact on our daily lives. For each application topic, basic principles will be first provided, followed by notable example cases.

6.1 THz Spectroscopy

6.1.1 Overview

Spectroscopy refers to the analysis of the interaction between matters and electromagnetic waves presented as a function of frequency (or wavelength). The emission or the absorption of electromagnetic waves by matters are characteristic consequences of such interaction. Spectroscopy has long played a great role in various scientific areas including physics, chemistry, astronomy, as well as in biology, owing to the rich information it provides related to the composition and physical structure of a wide range of materials in various scales. We know that, for example, the atomic structure became understood to us through the analysis of the electromagnetic wave emission lines from transitions between atomic energy states. THz spectroscopy concerns spectroscopy in the range of the THz spectrum. It has been

© Springer Nature Switzerland AG 2021

J.-S. Rieh, *Introduction to Terahertz Electronics*,
https://doi.org/10.1007/978-3-030-51842-4_6

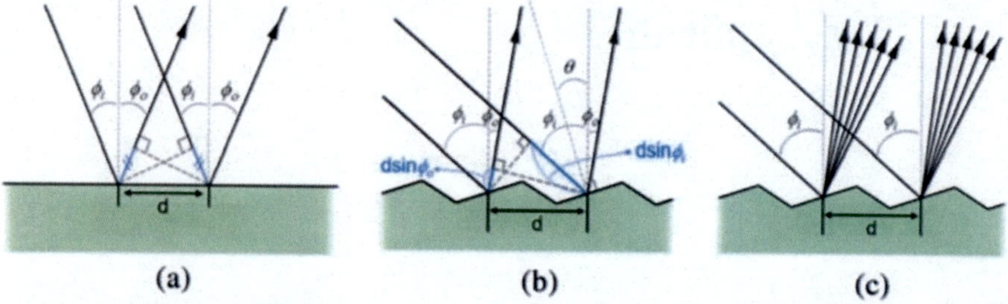

Fig. 6.1 Reflection from various surfaces: (**a**) Flat surface, (**b**) Grating with specular reflection, (**c**) Grating with diffuse reflection

proven useful in investigating the properties of a variety of materials in gas, liquid, or solid phase on the Earth or in the outer space.

In most spectroscopy, the main task is to obtain the frequency dependent properties of an object under study. The most straightforward method to achieve the goal is to physically isolate a certain frequency component of interest from a broadband spectrum source and apply the selected component to the target object to observe the response. The most traditional approach is to induce a physical dispersion of light with a dispersive element and locate a slit in the position where the desired frequency (wavelength) is illuminating, so that only the selected portion of the spectrum passes through the slit and works on the target. In the early stage of spectroscopy developed in the late nineteenth century, *prisms* were a natural selection for this purpose, which makes use of the frequency dependent refractive index of the material that composes the prisms. Most of the transparent materials exhibit increasing refractive index with increasing frequency (called "normal dispersion," as opposed to "abnormal dispersion" with the opposite trend), which results in a dispersion that enforces a higher frequency component to undergo a stronger refraction at the material boundaries. Since the readers should be familiar with how the prism works, no additional description will be provided here.

Another dispersive element that came into use in the early twentieth century is the *diffraction grating*, which had been a major tool for spectroscopy until the mid-twentieth century [1, 2]. Since the operation of diffraction gratings is not as straightforward as prisms and they are still in a wide use, it would merit a brief description here. A diffraction grating is a plate with periodic gratings or grooves formed on the surface. While two types of gratings are available, reflective and transmissive, let us focus on the reflective type and understand how it works. First consider a parallel beam of light incident on a flat plain surface without gratings as shown in Fig. 6.1a. There will be no wavelength-dependent interference between adjacent light components after the reflection as they will maintain the same phase with each other. This is clear from the fact that the traveled distance is the same for each light component as the incidence and reflection angles are identical, leading to $d \sin \phi_i = d \sin \phi_o$ in Fig. 6.1a. Now assume the beam is incident on a surface with periodic gratings, which is depicted in Fig. 6.1b. As the beam components on each grating will go through different paths, interference is likely to happen between adjacent components. To be more specific, the constructive interference will occur when the path difference between two adjacent components, Δl, satisfies the following relation, which is called the grating equation:

$$\Delta l = d(\sin \phi_i - \sin \phi_o) = n\lambda, \tag{6.1}$$

where n is an integer and λ is the wavelength.

One may notice, however, that the configuration given in Fig. 6.1b may not explain the actual dispersion observed for a diffraction grating, which would show a rainbow-like physical separation of different wavelength components on the screen illuminated by the reflected beam. This is because, with the picture shown in Fig. 6.1b, the reflected beam will travel toward a single fixed direction, in which case the spatial separation of different wavelengths cannot be obtained unless the grating is rotated and the incident angle changes. With a fixed grating without rotation, simply a certain wavelength that is lucky to satisfy Eq. (6.1) with the given angles and grating spacing will be highlighted in that fixed direction. For spectroscopy, on the other hand, we wish the dispersion, or the physical separation of different wavelengths, can be obtained with a fixed grating, which in fact happens with simple gratings. So what is wrong about the picture described in Fig. 6.1b? This apparent inconsistency is resolved when we acknowledge that the type of the reflection on the grating surface is *diffuse reflection* (beam reflected toward multiple directions with surface scattering), not *specular reflection* (beam following the law of reflection, or $\phi_i - \theta = \phi_o + \theta$ as in the picture in Fig. 6.1b). This more realistic configuration is illustrated in Fig. 6.1c. Since different reflection angles will select different wavelengths for the constructive interference, for which the incidence angle is still fixed, the desired dispersion can be attained without any rotating part. One additional note is that there is a chance for a single wavelength to satisfy Eq. (6.1) with more than one integer value for m, resulting in a set of identical dispersion patterns repeated along the angle variation, in which $n = 0$ case is called the zero-order mode. The observed interference-based phenomenon responsible for the dispersion obtained with gratings is in fact the diffraction, and this is why the dispersive element is called the diffraction grating.

The configuration for a typical spectrometer based on a diffraction grating is shown in Fig. 6.2. The exit slit passes only the desired frequency (wavelength) component (or a narrow band around it), which can be selected by shifting the slit or tilting the diffraction grating. Alternatively, an array of detectors can be used in place of the slit, which will allow the simultaneous detection of a range of spectrum without any moving part in the configuration. It is noteworthy that diffraction gratings were particularly welcomed for spectroscopy with longer wavelengths (compared to optical bands) such as far-infrared and THz, which were not covered by prisms as the materials employed for early prisms such as NaCl (sodium chloride) or KBr (potassium bromide) become opaque below a certain frequency boundary.

A more sophisticated device developed afterwards for spectroscopy was the *Fourier-transform spectrometer* [3]. It is based on the interferometry and requires demanding calculation to handle multiple spectrum peaks. Hence, the adoption of

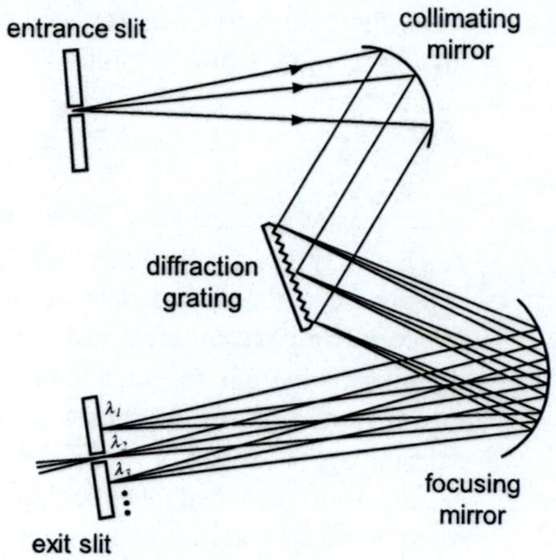

Fig. 6.2 Spectrometer based on a diffraction grating

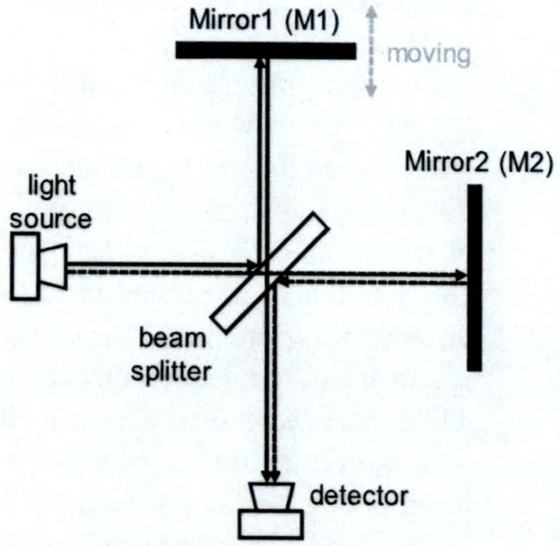

Fig. 6.3 Structure of Michelson interferometer

Fourier-transform spectrometer for practical use was enabled only when digital computation techniques became readily available in the mid-twentieth century. The core structure of Fourier-transform spectrometers is the Michelson interferometer as illustrated in Fig. 6.3. It is composed of a beam splitter, a light source, a detector, and a pair of mirrors, one of which is movable (M1). A beam from the light source is split into two orthogonal branches by the beam splitter, which are combined again after reflections by the pair of mirrors. The combined beam is eventually detected by the detector, optionally passing through a target object. As the paths traveled by the two beams are different, there will be interference between the two beams upon arrival at the detector. As a result of the interference, the maxima will occur when the following relation is satisfied:

$$\Delta l = l_1 - l_2 = n\lambda \tag{6.2}$$

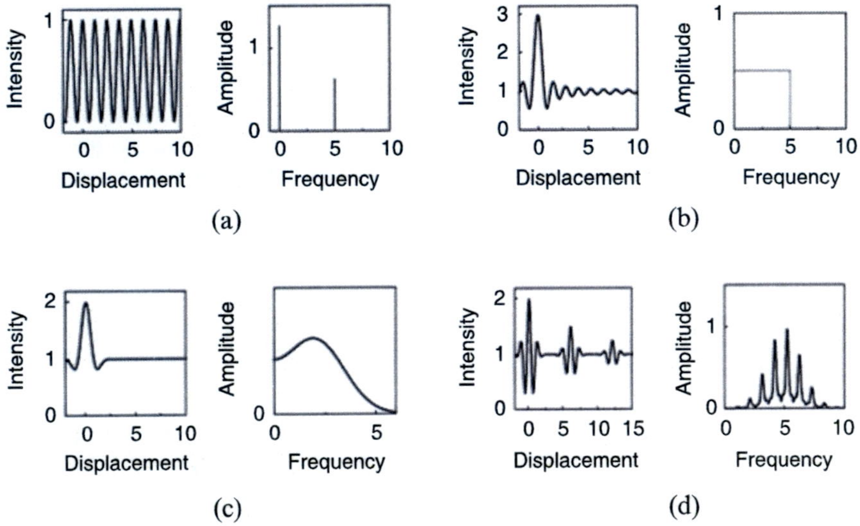

Fig. 6.4 Interferograms and corresponding frequency domain profiles for various cases (not to scale): **(a)** Sine wave, **(b)** Sinc function, **(c)** Pulse, **(d)** Multiple pulses after multiple reflections. (Reprinted by permission from Springer Nature: [1] © 2012)

where Δl is the path difference, l_1 and l_2 are the total paths of the beams reflected by M1 and M2, respectively (each designated by a solid- and dotted-lines in Fig. 6.3), n is an integer, and λ is the wavelength.

If the light source is monochromatic, maxima and minima will be alternately observed at the detector as one keeps moving M1 toward the detector or away from it. If the light source is broadband, which is mostly the case, maxima and minima will be alternately observed for *every* wavelength component with M1 movement, but with a different period, a longer period for a longer wavelength. This implies that, for a given position of M1, a set of various wavelengths will be simultaneously detected, and, as M1 position is varied, the combination of the detected wavelengths will change accordingly. Now, if a target object is placed in the middle of the combined beam path, we can record the absorption level by the object as a function of the *mirror position*, which is called the interferogram. However, our ultimate goal is to obtain the absorption given as a function of the *wavelength*. The required conversion from the mirror-position domain (in cm) to the wavelength domain (in cm^{-1}) (or, equivalently, the frequency domain in Hz) turns out to be the Fourier transform, as the name of the spectrometer indicates. Figure 6.4 shows interferograms and the corresponding frequency domain profiles for various cases [1].

There are various versions of Fourier-transform spectrometers, and the one for the infrared spectrum is called FTIR (Fourier-Transform InfraRed). Figure 6.5 shows the optical configuration of a typical FTIR setup. For the light source, a broadband IR source such as a Globar is typically used. For the detector, cryogenic bolometers are a popular choice, while a room-temperature Golay cell can also be used, partially sacrificing the sensitivity in favor of the measurement convenience. After dozens of

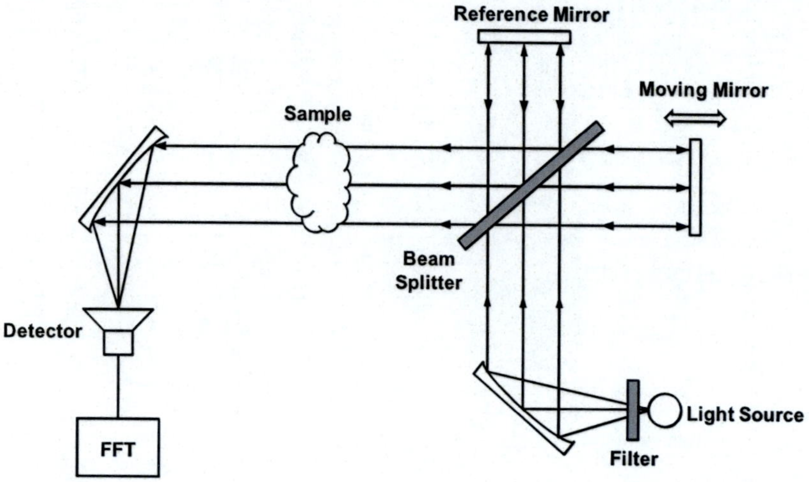

Fig. 6.5 Optical configuration of a typical FTIR setup

years since its first appearance, FTIR still serves as a useful tool for infrared and THz spectroscopy.

The landscape of THz spectroscopy went through another major transformation when *THz time-domain spectroscopy* (TDS) was invented in 1980s [4, 5]. Contrary to the previous spectroscopic methods that were based on the frequency domain, THz TDS is a time-domain technique making use of ultrashort time pulses. Owing to a set of advantages over the Fourier-transform spectroscopy which will be discussed later, THz TDS rapidly arose as a highly favored choice for a wide range of THz spectroscopic applications [6]. Being a major topic, THz TDS will be treated as a separate subject in Sect. 6.1.2. Yet another technique is the *CW-based THz spectroscopy* that is performed in the frequency domain, called CW THz frequency-domain spectroscopy (FDS). The CW signals required for this technique may be generated by optical methods but also with electronic components, which provides an opportunity for compact and affordable solutions with a low power consumption. This will also be described separately in Sect. 6.1.3.

6.1.2 Time-Domain Spectroscopy

Understanding time-domain spectroscopy should start with recalling the relation between the functions presented in the time domain and the frequency domain, which can be converted into each other by the Fourier transform. As most readers should be aware, the delta function in the time domain corresponds to a constant value, or an infinite bandwidth, in the frequency domain (and vice versa, of course, due to reciprocity). Unfortunately, there is no practical way to realize the ideal delta function in the time domain in the real world. We understand, however, that a short

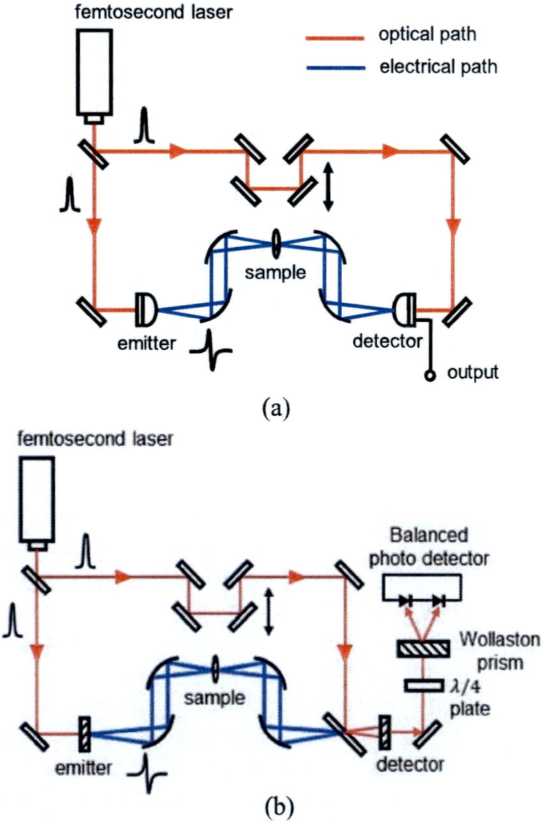

Fig. 6.6 Setup for TDS: (**a**) Based on photoconductive (PC) antennas, (**b**) Based on electro-optic (EO) crystals

time pulse resembles the delta function, and the shorter the pulse duration, the wider the corresponding bandwidth in the frequency domain. Hence, if we know the response of a material to a short time pulse, we can extract the corresponding frequency response by the Fourier transform. The obtained frequency response will be over a wider bandwidth if the time pulse is shorter. This is the principal idea behind THz TDS, and the availability of sub-picosecond THz pulses with the invention of femto-second lasers naturally led to a very useful spectroscopy technique with a wide frequency coverage.

From a technical point of view, the generation and detection of ultrashort THz pulses are the key elements of THz TDS. The related details can be found in Chap. 5, while the main concept is recaptured here along with the actual setup diagram for THz TDS. There are two approaches for the generation and detection of THz pulses, one with photoconductive (PC) antennas and the other with electro-optic (EO) crystals, both driven by a femtosecond laser. The TDS setup for the first approach is depicted in Fig. 6.6a. A THz pulse is radiated from the emitter when an optical pulse from a femtosecond laser is applied on a photoconductive antenna with the help of the very short carrier lifetime of the photoconductive material in the photoconductive antenna, for which low-temperature (LT)-grown GaAs is typically employed. The generated THz pulse passes through (transmission mode) or is reflected by (reflection mode) the target object under test, eventually reaching the detector, with its path guided by optical components such as mirrors or lenses as needed. At the detector, which is also based on a photoconductive antenna similar to the one employed for the emitter, the THz pulse shape is recovered with a sampling process, which makes use of the field-dependent photo current generated (see Sect. 5.3.1 for the details of the sampling process). The interaction between the THz pulse and the target sample placed in the middle of the beam path alters the pulse profile. The comparison of the altered profile with the reference pulse profile, the one obtained with the identical TDS setup without the sample, provides the desired spectroscopic information of the sample in the time domain (in seconds), as well as in the frequency domain (in Hz) after the Fourier transform.

The second approach for TDS, which involves electro-optic crystals, is implemented with a setup described in Fig. 6.6b. THz pulses are generated when optical pulses from the femtosecond laser are applied on a piece of an electro-optic crystal, such as ZnTe, due to the optical rectification induced by the second-order nonlinearity of the crystal. The configuration between the emitter and the detector is similar to the one for the photoconductive antenna setup described above. The generated THz pulses interact with the target sample by either transmission or reflection, and then travel toward the detector. At the detector, the THz pulse shape is recovered with an electro-optic sampling process, which exploits the field-dependent birefringence of the electro-optic crystal with the help of a $\lambda/4$ plate, Wollaston prism, and a balanced photo detector (see Sect. 5.3.2 for the details). As was the case with the photoconductive antenna, the extraction of the spectroscopic information from the measurement is made by comparing the profiles of the measured THz pulse and the reference pulse.

THz TDS and Fourier-transform spectroscopy are two different techniques, and it may be tricky to make a direct comparison between the two. However, there have been attempts, although limited, to compare the two approaches [7]. According to those comparisons, there are a few aspects that favor THz TDS over Fourier-transform spectroscopy. TDS provides a much larger source power, which can be a few orders higher than what can be obtained with Fourier-transform spectroscopy in terms of the peak power. On the detector side, the room-temperature coherent time-gated techniques employed for TDS are less sensitive to thermal backgrounds than the thermal detectors typically used for Fourier-transform spectroscopy. This leads to a higher signal-to-noise ratio (SNR) for TDS systems when combined with the advantage of higher source power at the emitter side. Additionally, the THz source for TDS shows a better stability than those typically employed for Fourier-transform spectrometers. These factors explain the favors toward THz TDS systems since its first appearance. One relative shortcoming that can be mentioned for THz TDS is the range of the spectrum coverage that is limited to far infrared and THz, while Fourier-transform spectroscopy may reach up to visible light regime.

Many useful applications of THz TDS can be found in various fields, some examples being introduced below. In one of the classic works in THz TDS carried

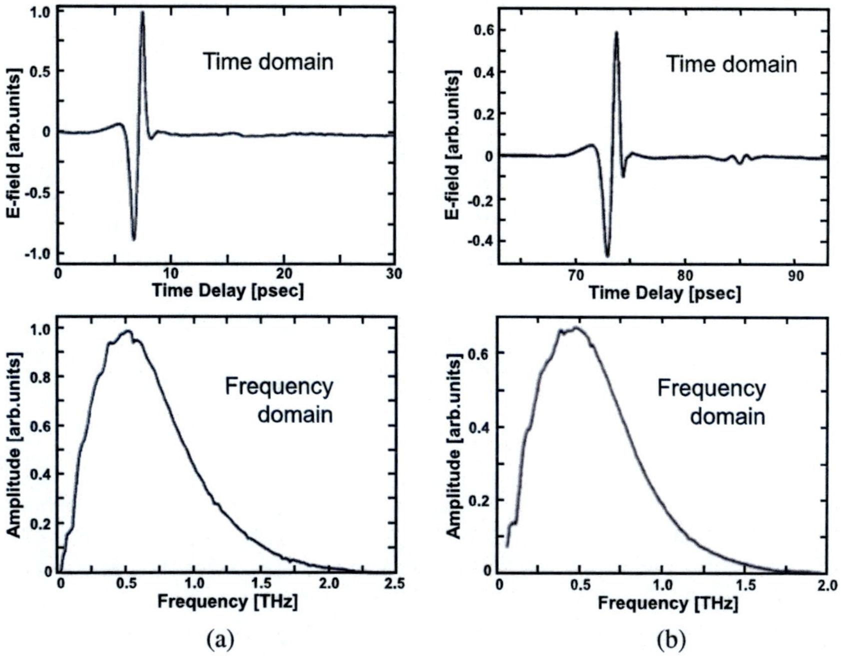

Fig. 6.7 THz waveforms in time domain and frequency domain: (**a**) Through free space, (**b**) Through sapphire sample. (Reprinted with permission from [8] © The Optical Society)

out in IBM [8], the properties of various dielectric crystals and semiconductors have been investigated for a frequency range of 0.2–2 THz. Based on 70-fs optical pulses with a repetition rate of 100 MHz emitted from a colliding-pulse mode-locked dye laser, THz pulses were generated with a GaAs-based photo-conductive antenna illuminated by the optical pulses. THz pulse detection was made with a similar antenna structure built on an ion-implanted silicon-on-sapphire substrate. The time-domain waveform and its frequency profile obtained from the Fourier transform are shown in Fig. 6.7a for a THz pulse that has propagated through a free space. In comparison, Fig. 6.7b presents the same set of data obtained for a THz pulse that has traveled through a 9.6-mm-thick sapphire sample, one of the materials examined in the study. By comparing the two data set obtained with and without the sample, various properties of the sample can be obtained. For example, the refractive index of the sapphire sample can be extracted as shown in Fig. 6.8 [8]. As indicated by the plot, a reasonable agreement with previously reported data points is obtained by THz TDS. Additional information such as conductivity and carrier concentration of semiconductors can also be provided by THz TDS with a similar setup [9, 10].

In another experiment reported in [11], various explosives and related compounds were studied with THz spectroscopy. Based on a 800-nm Ti:Sapphire femtosecond laser with a pulse duration of 80 fs and a repetition rate of 80 MHz, THz pulses were generated using the surface effect (see Sect. 5.1.3) by focusing the optical pulses on a

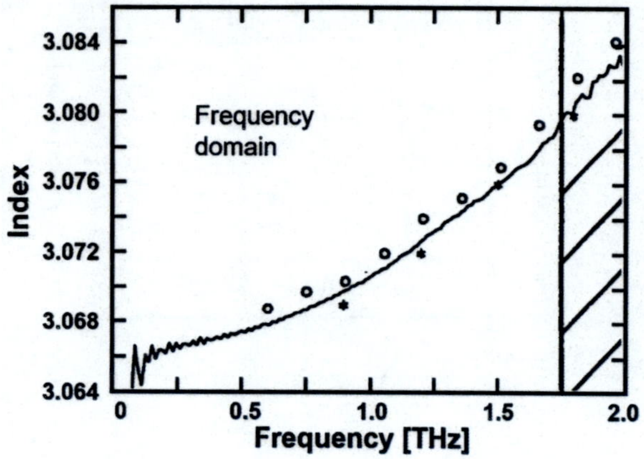

Fig. 6.8 Refractive index of sapphire obtained with THz TDS. Also included for comparison are data points reported in other literatures. (Reprinted with permission from [8] © The Optical Society)

p-type InAs crystal. With the generated THz pulses and electro-optic sampling process based on ZnTe crystal for pulse detection, THz TDS was performed. A frequency range of 0.1–2.8 THz was achieved after a fast Fourier transform with a frequency resolution of 50 GHz. Seventeen compounds have been examined for absorption properties with the setup in the study. Three of them are introduced in Fig. 6.9 as an example to provide a basic idea [11]. It shows the absorption spectra of trinitrotoluene (TNT), cyclotrimethylene trinitramine (RDX), and pentaerythritol tetranitrate (PETN) along with their chemical structures. Multiple characteristic absorption peaks were obtained for each compound as indicated by arrows, which align well with the peak positions from the existing literature.

While THz TDS has been dominated by optical techniques, there have been efforts to obtain THz TDS based on electronic approaches as well. A work in UCLA demonstrated a 95–105 GHz spectroscopy system based on a CMOS transmitter and receiver [12]. It detects the re-emission from gas molecules after excitation by the pulse-modulated CW waves, instead of the absorption by the molecules, with a mirror-ended cavity gas cell that spans only several centimeters. With the system, the spectral line of NO_2 was obtained. Although this approach involves the generation of CW waves, the spectroscopic data were acquired in the time domain with the pulse-modulated CW waves, indicating it is a TDS technique.

6.1.3 *Frequency-Domain Spectroscopy*

While the time-domain technique is a dominating approach for THz spectroscopy, CW-based frequency-domain technique can also be adopted. In this technique, the frequency of a CW signal source is swept for a certain range and the response is detected as a function of frequency. As the measurement is directly performed in the frequency domain, no additional domain transformation is needed. Compared to TDS, frequency-domain spectroscopy (FDS) based on CW signal benefits from a

ERCs and structures	Absorption spectra[a]	Features and absorption coefficients
2,4,6-Trinitrotoluene (TNT) 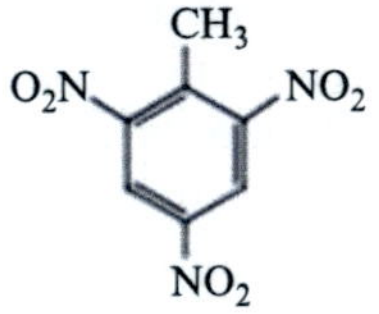	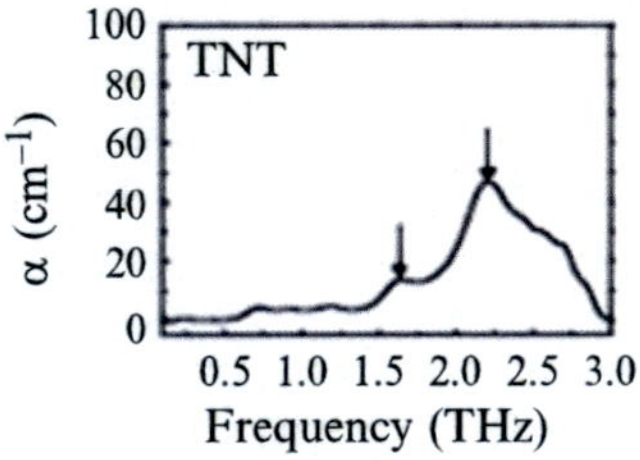	1.62 THz 13.5 cm^{-1} 2.20 THz 46.7 cm^{-1}
Cyclotrimethylene-trinitramine (RDX) 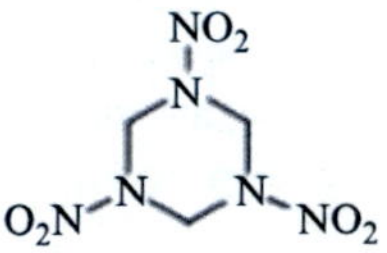	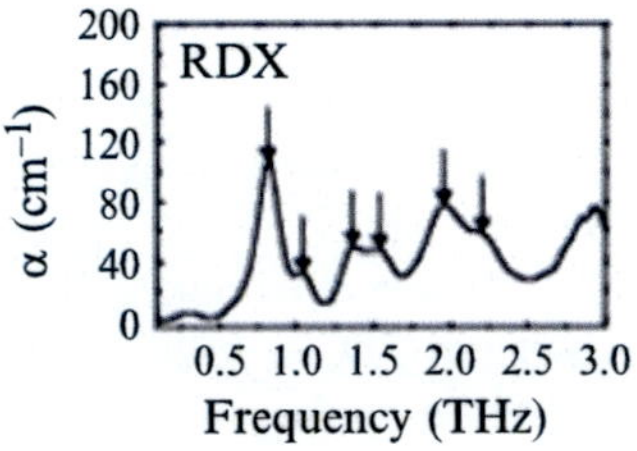	0.82 THz 108.7 cm^{-1} 1.05 THz 31.7cm^{-1} 1.36 THz 49.4 cm^{-1} 1.54 THz 47.0 cm^{-1} 1.95 THz 77.7 cm^{-1} 2.19 THz 59.0 cm^{-1}
Pentacrythritol tetranitrate (PETN) 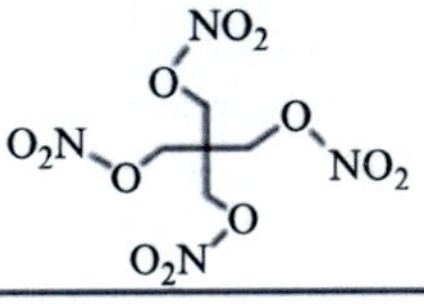	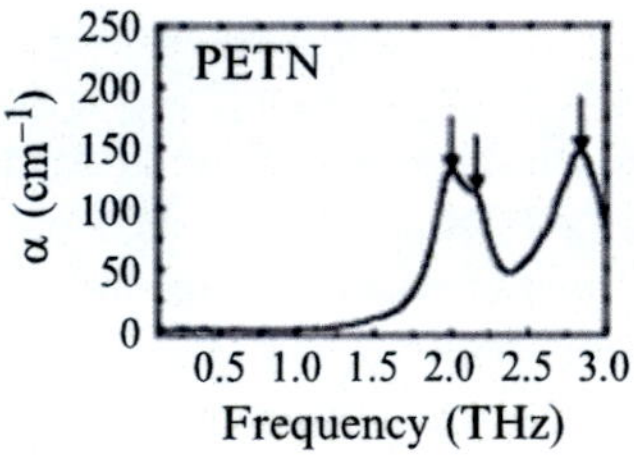	2.00 THz 131.8 cm^{-1} 2.16 THz 113.4 cm^{-1} 2.84 THz[b]

Fig. 6.9 Absorption spectra of various explosive materials obtained with THz TDS. (Adapted with permission from [11] © The Optical Society)

few advantages, including a better frequency selectivity and frequency resolution as well as intensity stability [13]. Also, CW FDS opens a possibility of obtaining phase information in addition to intensity information, which is useful for some spectroscopic applications [14]. You may be concerned that CW spectroscopy will take a much longer time to obtain spectroscopic data than TDS, since the former needs a scan over a large number of frequency points, while the latter provides data over a wide frequency range with a single pulse. It should be remembered, however, that a time-delay scan is still needed in TDS to recover a pulse shape with a sampling process (see Sect. 5.3.1), which may take a finite time ranging from tens of millisecond up to tens of second per waveform [15]. The frequency scan time for CW FDS will vary widely, which is determined by several factors including the frequency range, desired frequency resolution, and the integration time that is affected by the source power and detector sensitivity. It may take a few seconds with a special technique such as cavity ringdown [16], while it can be as long as hours if an extremely fine frequency resolution is desired.

There can be two approaches for CW FDS operating in the THz band, one based on optics and the other on electronics, each of which will be described below. *Optics-based CW FDS* usually employ a photomixing technique to attain the required CW generation (see Sect. 5.2.1 for the details on photomixing). The beating signal from two laser sources can provide a bandwidth exceeding 1 THz with a wavelength tuning less than 1%, with typical near-infrared lasers employed for spectroscopy. A frequency resolution in the order of MHz may be achieved, although the actual resolution obtained in the practical setup may be much larger than this [13]. It is noted that the typical frequency resolution of THz TDS is limited around 1 GHz, which is dictated by the span of the delay stage (the resolution is given by $c/(2l)$, where l is the moving stage span and c is the speed of light [17]). Recently emerging as an alternative is *electronics-based FDS* that employs an electronics-based CW THz signal source, which is under rapid development as described in Chap. 2. Compared to optics-based FDS, the available frequency range with electronics-based FDS is still lower, limited to a few hundred GHz range. However, it offers a much better frequency resolution, which can reach as small as sub-kHz range [18]. Additionally, a low-cost compact spectroscopy system with a tiny power consumption can be realized based on electronics-based FDS, which is favored for various practical applications.

One application that attracts recent interests is the detection of the rotational states of gas molecules. The interest was partially driven by its potential application to the detection and analysis of the human exhaled breath. A breath of a human may contain hundreds of volatile organic compounds (VOCs) [19], some of which can serve as diagnostic biomarkers of various diseases [20]. For the spectroscopy to detect the rotational states of gas molecules, which is called the rotational spectroscopy, electronics-based CW FDS is very well suited. While the vibrational state transition of gas molecules falls in the range of mid-infrared spectrum, the rotational state transition corresponds to mm-wave and THz region due to its smaller energy transfer involved, which can now be covered by electronics-based FDS. Besides, the linewidth for rotational spectral lines is only in the order of 1 MHz, which requires a frequency resolution well below this linewidth. This demanding target is also satisfied by electronics-based FDS. The bandwidth that can be obtained with the currently available electronic THz systems for spectroscopy, up to around 100 GHz [18, 21], may limit the applications that require a wide frequency range. However, most of chemical species of interest show repeated response pattern in the frequency domain. Thus, most of the key spectroscopic information of those species is likely to be included in the frequency window offered by the current electronics-based systems [18].

There have been recent reports that demonstrated rotational spectroscopy based on semiconductor-based THz sources and detectors [18, 22]. In a work in IHP [22], transmitters and receivers based on SiGe BiCMOS technology operating at 238–252 GHz and 494–500 GHz were developed for gas spectroscopy of exhaled breath. For the detection experiment, the gas sample was contained in a folded

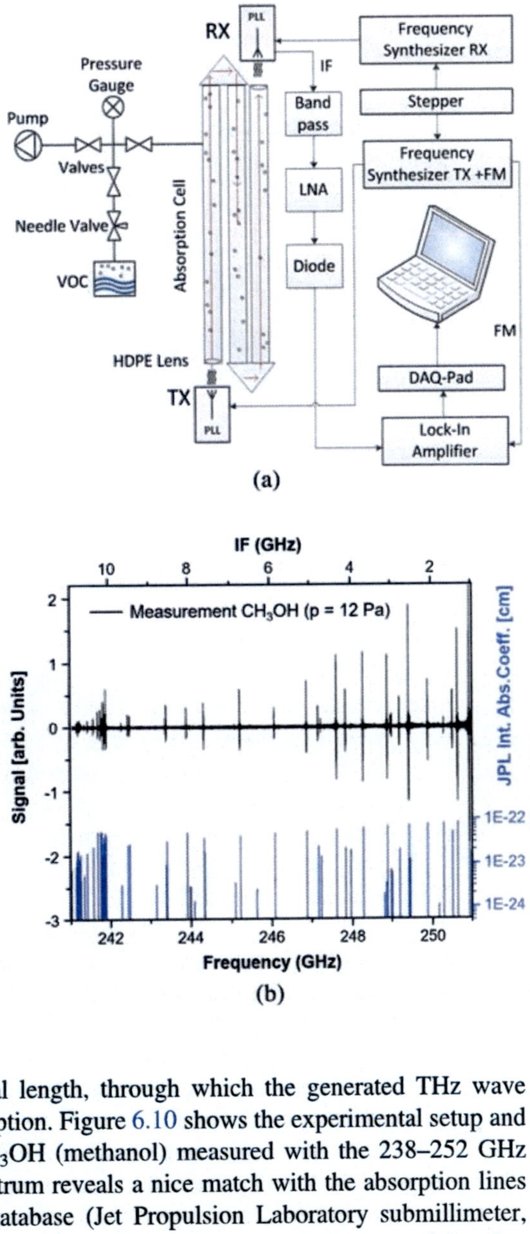

Fig. 6.10 (**a**) Experimental setup for spectrometer based on SiGe BiCMOS TX/RX, (**b**) Measured spectrum of CH_3OH (methanol). (© 2017 IEEE [22])

absorption cell of a 1.9-m total length, through which the generated THz wave traveled to experience the absorption. Figure 6.10 shows the experimental setup and the absorption spectrum of CH_3OH (methanol) measured with the 238–252 GHz system [22]. The obtained spectrum reveals a nice match with the absorption lines calculated based on the JPL database (Jet Propulsion Laboratory submillimeter, millimeter, and microwave spectral line catalog) shown on the bottom of the plot. Also, an MIT group developed a spectrometer operating in the range of 220–320 GHz based on a transceiver in CMOS technology [18]. The transceiver, which serves as both transmitter and receiver, generates a signal with 10 comb spectral lines over two frequency bands, reducing the scanning time with a factor of 20. With a 70-cm-long gas cell whose optical length depends on the propagation loss in the cell, the absorption lines of CH_3CN (acetonitrile) and OCS (carbonyl sulfide) were obtained, which were consistent with the JPL database.

6.2 THz Imaging

6.2.1 Overview

Imaging can be defined as a process of mapping the locational information of an object on to 2D (sometimes 3D) space, typically based on the radiation from the object through reflection, transmission, or self-generation. The mapping is typically made upon a 2D space, but 3D space mapping is also possible as will be described later. We are currently living in a world flooded with images owing to the rapidly developing imaging technologies over affordable devices, partly fueled by easy sharing with broadband communication networks. Such general-purpose imaging, however, is based on the visible light band of spectrum and provides no more information than can be seen with human eyes. It simply serves to overcome the time and space limits in vision. On the other hand, special-purpose imaging modalities, typically utilizing the spectra outside the visible light band, provides extra information beyond the detection capability of the human electromagnetic sensor, the eyes. Well known examples include X-ray and infrared imaging, widely employed for medical and military purposes.

Recently arising as an alternative modality is THz imaging, sometimes called T-ray imaging. It benefits from the unique properties of THz waves as briefly discussed in Chap. 1. THz waves are known for high absorption by water, which can be exploited for imaging that requires strong contrast for the regions with high water content. Also, when incident on the material surface, THz waves show higher reflection on the region with higher water content, making it useful for reflection imaging as well. Another attractive property is the transparency of various dielectric materials, such as paper, clothe, leather, wood, and most of plastics and ceramics, against THz waves. It is a favored property for security imaging as it can help reveal concealed items under package or clothes that are potentially dangerous. In this regard, THz imaging shares the main advantage of X-ray imaging, as both can be used for detecting concealed objects. However, when it comes to human applications, THz imaging is clearly favored over X-ray imaging, since it does not impose harmful radiation effects on humans owing to its non-ionizing property (involved photon energy in the order of meV for THz, in contrast to ~10 eV for X-rays). The lower resolution due to the longer wavelength is one downside, though. When compared to microwave imaging, THz imaging benefits from the straight propagation path through target objects, a result of the less pronounced diffraction effect for a given aperture dimension. Also, the availability of a higher resolution will be a big advantage of the shorter wavelength, which also enables a smaller component sizes, including that of antennas.

There are various types of THz imaging, and it can be categorized from various aspects. The types of THz imaging will be described below with relative strong and weak points for each. Note that a large part of the categorization listed below can be applied not only to THz but also to other imaging modalities as well. First, THz imaging can be grouped into *active imaging* and *passive imaging*. In active imaging, the target object is illuminated by THz waves generated from a signal source, and the 2D (sometimes 3D) spatial distribution of the response by the object is taken as an image. It can be further sub-grouped into *transmission imaging* and *reflection imaging*. In the former case, the image is obtained based on the THz waves transmitted through the object, while the latter case makes use of the waves reflected from the object. Active imaging benefits from a better control of the wave properties such as frequency and intensity, as well as the availability of waveform modulation if needed. However, it requires a dedicated signal source, preferably with a large output power to attain a high SNR and a longer imaging range. Passive imaging is based on the detection of the THz radiation from the target object, which obviates the need for the signal source. The radiation can be either thermal radiation from the object or the reflection of ambient THz waves by the object. While passive imaging benefits from the simple system scheme without the source, the detector should exhibit an extremely high sensitivity along with a wide bandwidth to catch the weak signal from the object, a major challenge for THz passive imaging.

THz active imaging can also be grouped based on the type of the source signal: *pulse imaging* and *CW imaging*. The basic properties of the pulse and CW signals reviewed in the previous section for spectroscopy apply to imaging as well. Ultra-short pulses inherently contain information over a broad frequency range. Hence, imaging with such short pulses naturally provides images over a wide range frequency, as is called spectral imaging. This implies that images can be reconstructed at any desired frequency point included in the wide bandwidth offered by the pulses, once the imaging is performed with the pulses. On the other hand, CW imaging can be carried out with a simpler setup than pulse imaging owing to the absence of the delay stage needed for the sampling process, possibly with both source and detector realized with electronic components. Spectral imaging is in principle possible with CW imaging as well, but the image capture needs to be repeated for every frequency point desired. In both pulse imaging and CW imaging, images can be constructed based on either attenuation or phase delay of the signal after passing through the target object for the transmission-mode imaging.

Yet another categorization may divide THz imaging into *2D imaging* and *3D imaging*. Conventional images are based on 2D imaging, which is basically a process of mapping the target image seen from a single direction into a 2D space. For some applications, however, 3D images of the target object are needed, in which 3D mapping of the object is carried out, as is the case for tomography. One well-known modality of tomography is the computed tomography or CT, which is ubiquitous in hospitals for medical inspection purposes. While CTs in medical fields are based on X-ray to view the human body interior, THz CT can be realized with

THz waves based on the same imaging principles. There are other modalities for THz tomography as well, based on both transmission and reflections imaging. They will be detailed in Sect. 6.2.5.

The last categorization of THz imaging to be discussed here is *single-pixel imaging* vs *array imaging*. A stand-alone detector, the simplest detector scheme, can detect THz signal incident on a single point at a time. Hence, it can provide information corresponding only to a single pixel of an image. For this reason, it is called a single-pixel detector. To construct 2D (or higher-order) images with such single-pixel detectors, scanning process is indispensable, either of the target object or the detector. For some applications, real-time imaging is desired, which cannot be supported by single-pixel detectors. To meet such a demand, array detectors are necessary, with which a simultaneous acquisition of multiple-pixel data is possible. 1D detector arrays may significantly reduce the scanning time, but fall short of providing real-time imaging with a reasonable video frame rate. From a practical viewpoint, 2D arrays, preferably with sufficient pixel counts, are needed to realize real-time imaging. As the pixel dimension increases with the wavelength, realizing THz 2D arrays with a large pixel count is more challenging than those for visible light, especially when size is considered as a constraint. Nevertheless, there have been efforts to make large THz 2D detector arrays, which resulted in some successful real-time imaging, as will be discussed in this section.

In the following subsections, various topics for THz imaging will be discussed in further detail. The description will be first made for active imaging, including both transmission and reflection modes, and passive imaging, followed by real-time imaging and tomography. The focus will be made largely on CW imaging with electronic sources and detectors.

6.2.2 Active Imaging

As briefly mentioned above, active imaging provides a high SNR for THz imaging by illuminating the target object with a THz beam of a desired frequency from a signal source. With the apparent benefits, most of the reports on THz imaging are currently based on active-mode imaging. Both transmission and reflection modes of active imaging will be treated in this section.

Before moving on to the actual imaging setup and example results, it would be helpful to review the definition of the *resolution* in imaging. The resolution is not a term strictly reserved for technical communities as it is widely spread in our daily lives. When you want sharp images on your phone or monitor/TV screens, you say you need a higher resolution. In such cases, where the resolution concerns the display performance, it is given as pixel counts, such as 3840 × 2160 pixels or ~8.3 mega pixel for UHD (ultra high definition, also called 4k display). There may be various definitions of the resolution depending on specific applications. In optics, the resolution is represented by different measures as described below.

One convenient and widely used parameter is the angular resolution, which indicates the minimum angle required for two point objects to be resolved or

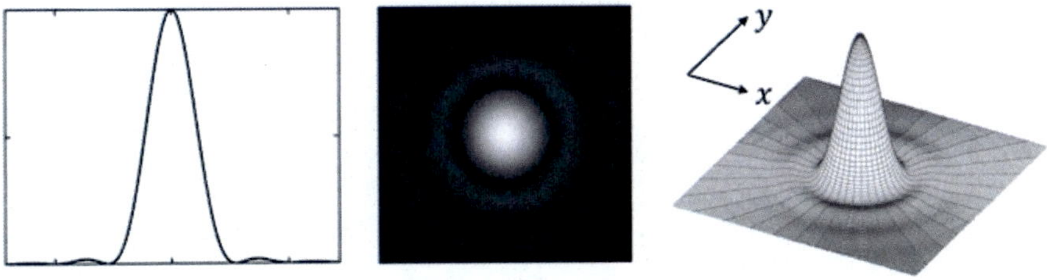

(a)

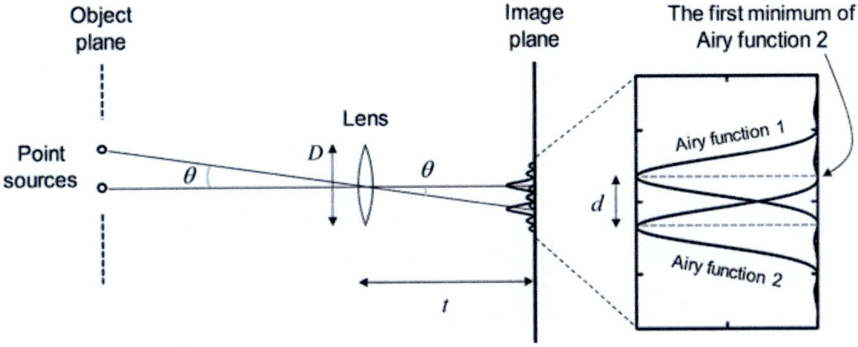

(b)

Fig. 6.11 (**a**) A typical Airy function, shown together with the 2D and 3D representation of the intensity. (**b**) The definition of the angular resolution θ, which corresponds to the case where the center of Airy function 1 and the first minimum of the Airy function 2 coincides

discernable. In many cases, the resolution is limited by the diffraction in optical systems, typically that of lenses. Remember that light passing through a small aperture experiences diffraction, ending up with interference-like patterns on the image plane. As this happens even for a perfect lens if it has a finite dimension (diffraction-limited case), the image of a point source through a lens shows a pattern that corresponds to the point spread function (PSF) of the lens. PSF of a circular lens is known as the Airy function (or Airy pattern or Airy disk if represented on 2D space), which is shown in Fig. 6.11a. As a result, when two point sources are located close to each other, the corresponding two images obtained with a diffraction-limited lens cannot be resolved below a certain view angle due to the partial overlap of the two Airy functions formed on the image plane. The diffraction-limited angular resolution is defined as the angle formed between the centers of the two Airy functions when the center of one Airy function coincides with the first minimum (first dark circle) of the other Airy function (see Fig. 6.11b), which is given as:

$$\theta \approx 1.22\frac{\lambda}{D} \tag{6.3}$$

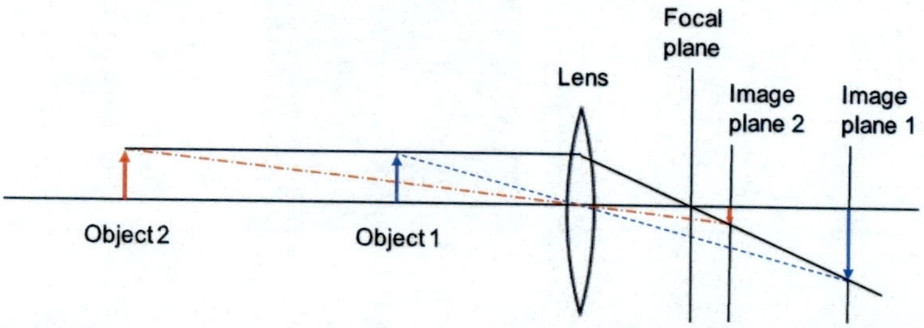

Fig. 6.12 The location of the focal plane and the image planes. As an object moves away from the lens, its corresponding image plane approaches the focal plane. Note that, in geometric (ray) optics, the incident light parallel to the lens axis passes through the focal point after the lens, and the light passing through the center of the lens maintains a straight path. The image shows sharp patterns, or "focused," on the image plane where the multiple lights from a point source merge after different paths

where λ is the wavelength and D is the lens diameter. This is called the Rayleigh criterion (the factor of 1.22 is for a circular aperture, which becomes 1 for a single slit). For instance, an image obtained with a THz wave of $\lambda = 1$ mm (300 GHz) and a lens with $D = 50$ mm, will show a resolution of $\theta = 0.024$ rad or 1.4°. The resolution will be improved with a small λ/D ratio, which can be easily understood since the diffraction will be less pronounced for a small λ/D ratio. It is noted, though, that as λ/D becomes smaller and the diffraction becomes less dominating, other factors, such as aberration, begins to affect the resolution, resulting in an angular resolution worse than Eq. (6.3).

If needed, the resolution may also be represented by the minimum *distance* between the images of two point sources projected on the image plane for them to be resolved. With the help of Fig. 6.11b, this spatial resolution can be simply obtained by multiplying the angular resolution by the distance from the lens to the image plane, t, as follows:

$$d \approx \theta t \approx 1.22 \frac{\lambda t}{D} \tag{6.4}$$

For example, d of 1.22 mm is obtained with $t = D = 50$ mm at $f = 300$ GHz. At this point, it would be useful to discuss the difference between the "image plane" and the "focal plane". The former is the plane where the image appears sharp and "focused", while the latter is the plane that contains the focal point. Hence, in general, the two planes are different, which is clearly shown in Fig. 6.12. However, as the object plane (on which objects are located) moves away from the lens, the image plane approaches the focal plane, which is also obvious with Fig. 6.12. As an extreme case, when the object is located at the infinity, two planes eventually overlap with each other. In this special case, which often occurs in astronomy as stars can be assumed to be located at an infinite distance, t becomes identical to the focal length f_c of the lens. In this situation, Eq. (6.4) is reduced to $d = 1.22\ \lambda N$, where $N = f_c/D$ is the *f*-number of the lens. There are other definitions for spatial resolution, such as the microscope resolution ($d = 0.61\lambda/\text{NA}$, where NA is the numerical aperture). Based on different assumptions, the various spatial resolution definitions may not show the exactly same values, but they fall within an acceptable range.

6.2.2.1 Transmission Imaging

THz transmission imaging provides 2D or higher order images of the absorption or other dielectric properties of the object based on the attenuation level or the phase delay of the transmitted signal. In a sense, the process is similar to the transmission-mode THz spectroscopy described above. The key difference is that the spatial distribution of such properties is obtained for imaging, whereas only the information averaged over the illuminated region is acquired for spectroscopy. The frequency distribution (or profile) as offered by spectroscopy is not required for imaging, but it would be helpful to have such information as well for imaging, which is in fact called spectral imaging. THz transmission imaging is widely adopted and a prevailing modality of THz imaging as it carries the information internal to the object. However, its application is limited with objects of highly attenuating material, such as human body with high water content. Also, the system configuration requires the source and detector to be located on the opposite sides of the target object. For these cases, reflection imaging is better suited, which will be the next topic to be discussed.

2D imaging with an object scan is the most standard type of THz transmission imaging of today. Such images are acquired with a single-pixel (or low-pixel-numbered) detector by scanning the target object. A typical setup for the THz transmission imaging is illustrated in Fig. 6.13. A target object is placed in the middle and a source and a detector are placed on either side of the object. The THz CW beam from the source passes through a pair of lenses and focused on the surface of the object. The THz beam, emerging from the other side of the object after penetration, travels through another pair of lenses and arrives at the detector located at the focal point. The THz beam is typically AM-modulated for the suppression of $1/f$ noise and DC offset at the detector, as facilitated by an electrical modulator or a

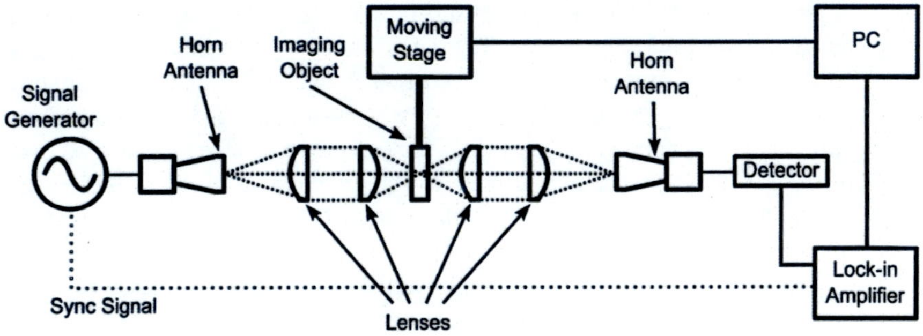

Fig. 6.13 Typical setup for THz transmission imaging

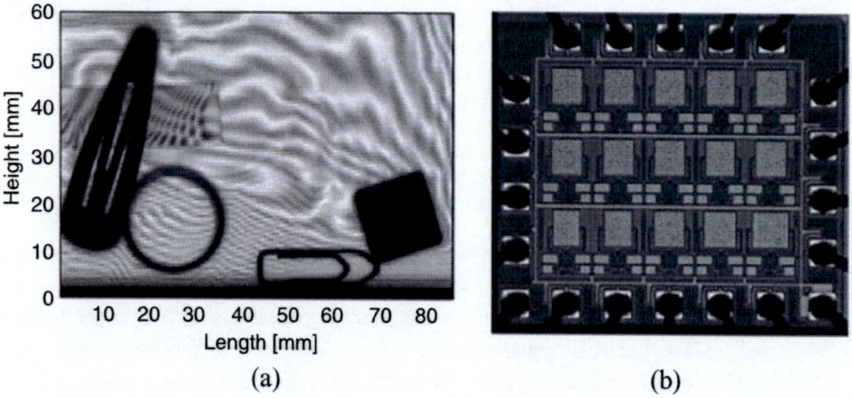

Fig. 6.14 **(a)** Image of various object hidden in a paper envelope obtained from electronics-based CW THz transmission imaging. **(b)** Chip photo of the detector array used for the image. (© 2009 IEEE [24])

mechanical chopper. This configuration for transmission imaging so far described is largely similar to the spectroscopy setup, but the main difference comes with the object scan, which is enabled with the computer-controlled moving stage that holds the object. The object is scanned line-by-line with a zigzag pattern, which is called the raster scan. The data acquired by the detector with the raster scan, which contain the information on the entire pixels, are transferred to a computer after digitalized by an ADC (analog-to-digital converter). With the help of the computer, a 2D image is reconstructed. Although the described setup is based on CW imaging, optics-based pulse imaging setup can be readily obtained by modifying the pulse-based spectroscopy setup (see Fig. 6.6) with addition of the scanning scheme.

Figure 6.14a shows one of the first images obtained with electronic-based CW THz transmission imaging reported by University of Wuppertal [23, 24]. It is an image of various objects hidden in a paper envelope. For the THz source, a 650-GHz diode frequency multiplier with 0.5-mW output was employed to upconvert the frequency of the signal supplied from an external source. To detect the signal, a CMOS array detector of an NEP of 300 $pW/Hz^{1/2}$ integrated with on-chip patch antennas was used (Fig. 6.14b). The data from four pixels in the detector array were simultaneously processed for reduced scan time. In the image, the shapes of the enclosed objects such as a hair clip, key ring, paper clip, and a lump of sugar, which are made of materials blocking the THz beam, are clearly visible. The image also manifests the transparency of the paper envelope against the THz beam. The apparent interference pattern in the background is ascribed to standing waves formed in the imaging setup.

Another set of example images obtained with electronics-based CW 2D transmission imaging are presented in Fig. 6.15a [25]. The images were taken for a half of a plastic floppy disk with various imaging conditions for comparison, revealing the internal structure of the disk through the plastic cover. For the THz source, an external 300-GHz signal source was employed. For the detector, two versions of

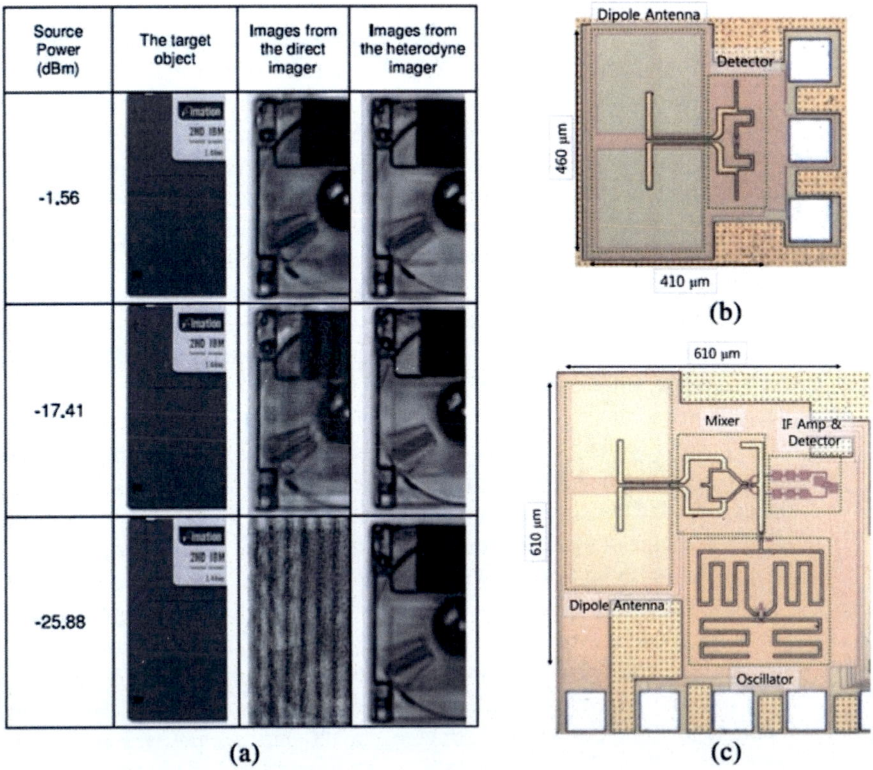

Fig. 6.15 (**a**) Images of a half-sided floppy disk with various input power level obtained with two different types of detectors. (**b**) Chip photo of the 300-GHz SiGe HBT direct detector. (**c**) Chip photo of the 300-GHz SiGe HBT heterodyne detector. (© 2017 IEEE [25])

detectors, direct and heterodyne, both designed for 300-GHz operation based on SiGe HBT, were adopted and compared. The direct detector consists of an on-chip dipole antenna and a differential-pair-based HBT detector circuit (Fig. 6.15b). The heterodyne detector is composed of an on-chip dipole antenna integrated with a mixer, an on-chip oscillator as LO, an IF amplifier, and an IF detector (Fig. 6.15c). Both detectors were mounted on a Si lens for efficient backside radiation from the chip. The imaging results show that clear images can be maintained down to much lower input power levels for the heterodyne case, which is an obvious consequence of the lower NEP of the heterodyne detector than that of the direct detector (3.9 pW/Hz$^{1/2}$ vs 21.2 pW/Hz$^{1/2}$).

6.2.2.2 Reflection Imaging

THz reflection imaging provides images of the reflectance on the object surface based on the signal strength reflected back from the various regions over the object after THz beam illumination. For the reflection imaging, the acquired information is limited to the shallow layer close to the surface of the object. Also, it is highly sensitive to the reflection angle on surface and affected by the surface scattering as well. On the positive side, however, as opposed to the case of the transmission mode, the reflection-mode imaging can be applied to objects with high THz attenuation and/or large thickness. For example, biomedical samples, which contain high level of water content, should be sliced thin for THz transmission imaging, allowing only ex vivo analysis. With reflection imaging, on the other hand, in vivo imaging is possible. Also, THz reflection imaging benefits from the fact that both THz source and detector can be located at the same side of the object, which is an attractive feature for many practical applications especially for the stand-off imaging. In fact, radar systems require the source and detector to be located on the same side, and the imaging radar belongs to the domain of reflection imaging. These observations clearly indicate that there do exist THz applications that prefer the reflection imaging, although the transmission imaging is still the dominant modality. As is the case for the transmission mode, 2D imaging with object scan is the most widely adopted scheme of the reflection mode, while real-time imaging and 3D tomography imaging can also be realized with the reflection mode.

There have been quite a lot of reports on THz reflection imaging based on THz pulses generated with optical means [26, 27], whereas CW-based results with electronic sources and detectors have been relatively sparse. A reflection-mode THz imaging based on an electronics-based CW source and detector has been reported in [28]. For the THz source, an on-chip integrated 300-GHz oscillator built with InP HBT was used, which exhibited a peak output power of 3.4 mW [29]. The InP detector based on a differential pair exhibited an NEP lower than 35 pW/Hz$^{1/2}$. The imaging setup is illustrated in Fig. 6.16a. The radiation and the detection were made through a pair of horn antennas with a gain of 26 dBi via on-chip probing of the source and detector circuits, which were placed on mobile probe stations that enable beam alignment. The radiated beam was incident on the sample fixed on a moving stage for raster scan and then consequently reflected and collected by the detector. The beam path was controlled by lenses and off-axis parabolic mirrors, with a mechanical chopper inserted for beam intensity modulation. Images for various bio samples were obtained including the one shown in Fig. 6.16b. It provides the surface images of a rat brain with a tumor bisected for the experiment, obtained with both regular and THz reflection imaging. A clear image with a high contrast for the tumor region was obtained with the THz imaging, which is due to the slightly higher water content in tumor region. The measured reflectivity was 45–46% for tumor and 40–41% for healthy tissue, the difference of which was converted into the clear image contrast. The results indicate the possibility of applying THz reflection imaging to medical applications to precisely identify the cancerous regions.

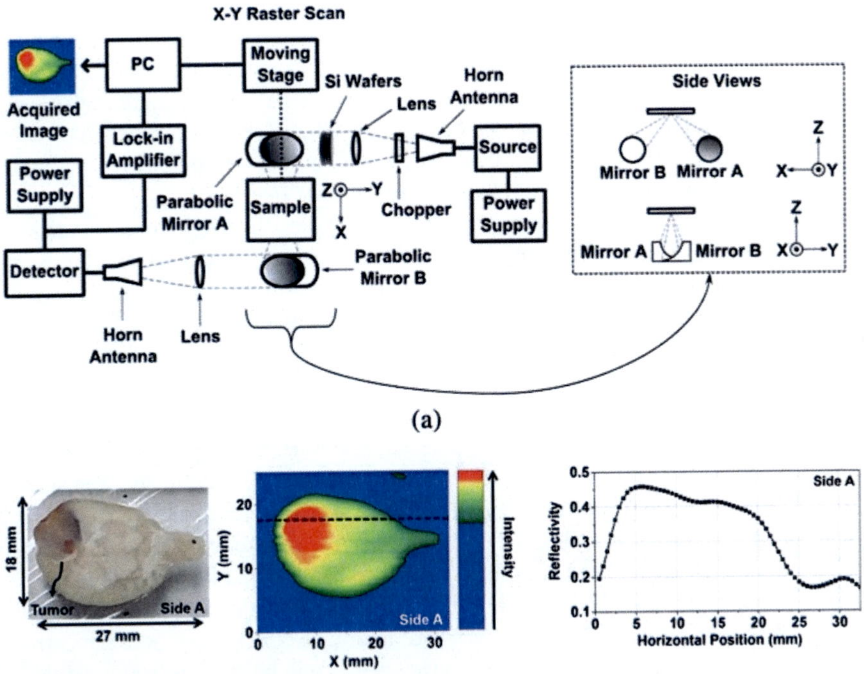

Fig. 6.16 (**a**) Setup for THz reflection imaging. (**b**) THz reflection image of a rat brain with tumor cut by half. Also shown are the regular image and the reflectivity profile along a cutline. (© 2017 IEEE [28])

6.2.3 *Passive Imaging*

As briefly mentioned earlier, passive imaging relies upon the natural radiation from the object. The conventional photography, which has become a core part of our daily lives, is a good example of passive imaging (images with flashlight technically can be considered as an active imaging, though). The imaging system is different from that of active imaging. It does not require a signal source and consists of only a detector. Also, because of this, the optics for signal path control is different, typically with a simpler configuration. The performance requirement for the detectors is also different from those used for active imaging. In addition to extremely high sensitivity to pick up the tiny signal, a large bandwidth is crucial to collect the incoming signal spread over a wide spectrum range. Besides, a feature to suppress the long-term fluctuation of the received signal level is desired. An overview of THz passive imaging systems will be presented below, followed by examples obtained based on electronic approaches.

Passive imagers can be considered as an imager based on a *radiometer*. This justifies us to start the topic with a brief review of the radiometer theory. A radiometer is defined as a device that detects the electromagnetic radiation. Hence, it may refer to a detection device working on any spectrum range. The description below, however, will mostly concern microwave radiometers [30], which well applies to mm-wave and THz detections we are interested in.

In active imaging, where the radiation from the target object, transmitted or reflected, is monochromatic in most cases and the signal power is confined in a very narrow band, if not a singular frequency point. In contrast, in passive imaging, the radiation from the object is typically spread over a wide band and can be considered as a noise by nature. In this case, the *signal* to be detected for passive imaging will be the *spatial variation of the noise*, which is often represented by the change in the radiation temperature of the object, or ΔT. From this viewpoint, the signal power to be detected by the radiometer is given as $P_S = k\Delta TB$, where k is Boltzmann constant and B is the RF bandwidth of the radiometer. The sensitivity of radiometers and passive imagers in this case can be defined by the smallest ΔT that can be detected, which is called the noise equivalent temperature difference (NETD, or sometimes referred to as NEΔT). This parameter can be compared to NEP, or the noise equivalent power, which is defined as the smallest *input power* that can be detected (see Sect. 3.1 for a formal definition of NEP). They both indicate the noise level of detectors and are highly correlated as will be shown shortly, while NETD is rather a specific parameter reserved for radiometers and passive imagers.

NETD depends on the system noise temperature (T_{SYS}) as well as its RF bandwidth (B) and integration time (τ), as is given below [31]:

$$\mathrm{NETD} = \frac{T_{SYS}}{\sqrt{B\tau}}. \tag{6.5}$$

The system noise temperature is the sum of the effective antenna noise temperature (T_A) and the radiometer noise temperature (T_R), so that $T_{SYS} = T_A + T_R$. A rule of thumb for passive imaging is that NETD needs to be below 1 K, while even much smaller values may be required for enhanced image quality. A simple relation also holds between NETD and NEP [32]:

$$\mathrm{NETD} = \frac{\mathrm{NEP}}{kB\sqrt{2\tau}}. \tag{6.6}$$

According to Eq. (6.5), the integration time directly affects NETD, along with the system noise temperature and the bandwidth of the radiometer. A longer integration time would reduce NETD, due to the noise averaged out during the integration. In reality, however, the sensitivity improves with the integration time only to a certain point (called Allan time), beyond which the sensitivity actually begins to degrade, a typical pattern observed for Allan variance [33, 34]. This is due to the gain fluctuation of the amplifier in the radiometer, which hinders the detection of ΔT as the gain fluctuation can hardly be distinguished from object temperature fluctuation. Such a

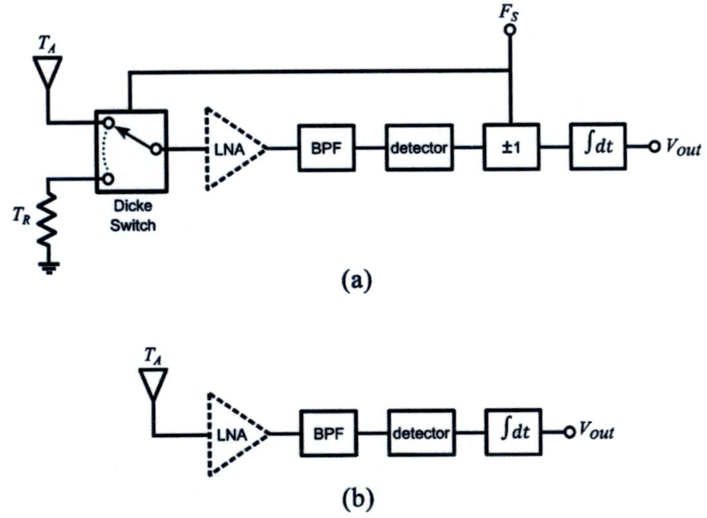

Fig. 6.17 Schematic diagram of a typical (**a**) Dicke radiometer, (**b**) Total power radiometer

gain instability is independent of the ratiometer noise, and Eq. (6.5) can be modified to include the effect of the gain fluctuation as follows [31, 32]:

$$\mathrm{NETD} = T_{\mathrm{SYS}}\sqrt{\frac{1}{B\tau} + \left(\frac{\Delta G}{G_0}\right)^2}, \tag{6.7}$$

where ΔG is the effective gain variation and G_0 is the averaged gain.

If the gain fluctuation is significant, NETD becomes dominated by this gain instability rather than the noise level of the system itself. The gain fluctuation exhibits a 1/*f* behavior, with its strength increasing with decreasing frequency. This offers an opportunity to suppress the effect of the fluctuation by modulating the received signal at a frequency higher than the corner frequency of the fluctuation. This modulation technique was first proposed by Dicke [35], and still is very popular for various types of radiometers. The radiometer employing this technique is referred to as a Dicke radiometer. The schematic diagram of a typical Dicke radiometer is shown in Fig. 6.17a. In contrast, a radiometer without the modulation is called the total power radiometer, as shown in Fig. 6.17b for comparison. Both systems share the core part of the radiometer, composed of an antenna, amplifier, bandpass filter, detector, and integrator (some components are optional). The key difference is the switch (called "Dicke switch") featured only in Dicke radiometers, which toggles between the antenna and the reference load, working as a modulator.

For Dicke radiometers, the switched modulation will result in the output taken as the difference between the signal detected during the on-state (switch connected to antenna) and the off-state (connected to the reference). Hence, the common noise component will be eliminated if the switching is faster than the fluctuation rate.

Hence, the effect of the gain fluctuation as well as the low frequency component of the system noise will be suppressed. In fact, this function is similar to the chopping of the signal source often employed for active imaging. Only difference is that, for passive imaging, the modulation ought to be made at the detector side, requiring a switch integrated in the receiver. While the modulation helps to suppress the low frequency fluctuation, the received power is reduced by half as the detector is connected to the antenna only for a half the cycle. In this case, the sensitivity is degraded by a factor of two as follows:

$$\mathrm{NETD_{Dicke}} = \frac{2T_{\mathrm{SYS}}}{\sqrt{B\tau}}. \tag{6.8}$$

This can be compared with the case for total power radiometers, in which the object signal is received for the entire period of the integration time, rendering the "total power" of the object to be integrated and NETD given as Eq. (6.5). However, this apparent sensitivity degradation by a factor of 2 for Dicke radiometers applies only to the ideal case of no gain fluctuation. At the presence of the gain fluctuation, NETD is likely smaller for Dicke radiometers than the total power radiometers as the gain fluctuation term in Eq. (6.7) will be negligible with the modulation, which is the whole purpose of employing a Dicke switch.

There has been a recent surge of efforts to realize passive imager circuits based on Si technologies at the frequency range around 100 GHz [32, 36–43]. There was also a hybrid approach in which an InP-based LNA is co-packaged with a Si CMOS receiver [44]. While most of them exhibited NETD of around 1 K or even below 0.5 K [38, 40–44], reports with real images taken have been rare so far [40, 44]. This indicate the challenges in passive imaging at high-frequency bands. One of the results is briefly introduced here [44]. The total power radiometer consists of two chips: an InP chip with a low-noise pre-amplifier and a CMOS chip with a receiver, which is driven by an on-chip LO composed of a frequency synthesizer followed by a frequency doubler and a power amplifier. The photo of the CMOS chip is shown in Fig. 6.18a. For a bandwidth of 90–100 GHz, the stand-alone CMOS chip showed a noise temperature of 3000–25,000 K. However, when combined with the InP pre-amplifier in a separate waveguide package, the entire system exhibited a noise temperature of 300–500 K, reduced by a factor larger than 10. The NETD was 0.5 K with a 50-ms integration time. The DC power consumed by the entire radiometer system is 257 mW, while its weight is only 334 g, one of the main advantages of chip-based radiometer system. With the radiometer system, a passive image of a person was successfully taken with a scan, as is shown in Fig. 6.18b. While the resolution is limited, it shows a possibility of passive imaging around 100 GHz based on a chip-based radiometer.

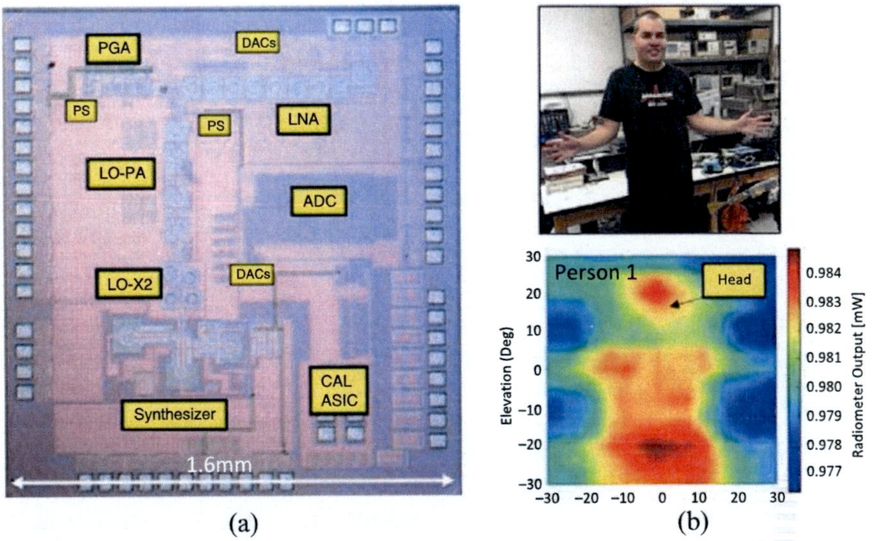

Fig. 6.18 (**a**) Chip photo of the CMOS receiver for passive imaging. (**b**) Passive image of a person obtained with the receiver. (© 2016 IEEE [44])

6.2.4 *Real-Time Imaging*

Scanning for imaging is a time-consuming process, and it may take several minutes or even more than an hour for THz imaging with reasonable resolution depending on the imaging size. The most effective way for scan time reduction is to employ an array detector composed of multiple detector pixels, as was discussed in Sect. 3.4. A simple math indicates that the scan time will be reduced by a factor corresponding to the number of array pixels processed in parallel. That is, for an array detector with a pixel count of n, the scan time will be $1/n$ of the single-pixel detector case. A popular practical approach for scan time reduction is to employ a 1D array. With a sufficient number of elements along the 1D array, only a scan over the direction normal to the array axis will be required, which will significantly reduce the scan time. However, for a full real-time imaging, 2D array will be eventually needed, which will be described below.

The key requirement for real-time imaging is for the entire target scan to be processed within a reasonable time span, or the inverse of the reasonable video frame rate. The typical frame rate for practical imaging systems is a few tens of frame per second (fps). For movies, the standard frame rate is 24 fps. For TVs, it ranges between 24 and 60 fps depending on the standard. If we take 24 fps as a norm for the frame rate, the allowed time for a full 2D scan of the object would be roughly around 40 ms. One may argue that real-time imaging is theoretically possible with a single-pixel detector (or a detector with a small number of pixels) if the integration time spent on each pixel is very small, or, equivalently, if the video bandwidth is very high. However, from a practical viewpoint, a certain duration of integration time

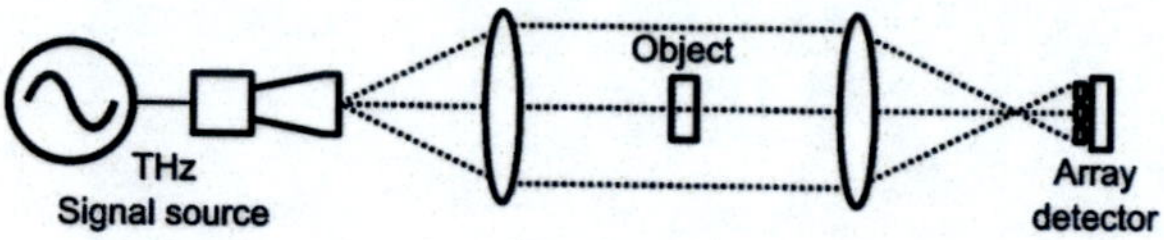

Fig. 6.19 Typical imaging setup for real-time imaging with a 2D array detector

should be spent to obtain a sufficient SNR for each pixel. Furthermore, extra time is needed to mechanically move the moving stage for a raster scan, which would steal time from the signal integration process at the detector. For example, the moving stage adopted for the setup shown in Fig. 6.13 takes ~20 ms to move a 1-mm step at the maximum rate, rendering the single-pixel option virtually impossible. Therefore, the most desired and practical option for real-time imaging is to employ a 2D array detector with a sufficient pixel counts, which can capture the full object image in a "single shot" without mechanical scan (electronic scan over pixels internal to the array is still needed). Figure 6.19 shows a typical imaging setup for real-time imaging with a 2D array detector.

There are challenges to realize a large-sized THz 2D array detector, or THz camera as is often called. First, the constraint on the array size would limit the maximum number of pixels integrated in a single array. This limitation will become more stringent as we move toward the lower frequency side of the THz band with a longer wavelength. For on-chip detectors we are interested in, a single pixel dimension will be dictated by the antenna size rather than the circuit area, assuming resonant antennas are employed. The adoption of electrically small antennas will degrade the sensitivity (see Sect. 4.2.2). Besides, for diffraction-limited imaging systems, the resolution will be roughly the dimension of resonant antennas, making highly integrated small antenna options less effective due to redundancy. A drastically improved resolution can be obtained with near-field imaging [45], but we will stay with the far-field case for the current discussion. Second, the THz signal power delivered per pixel will be reduced as the incoming beam power will be spread over the entire pixels, as was discussed in Sect. 3.4. In contrast, for a single-pixel imaging described earlier, roughly the entire source power would be delivered to the single pixel if we assume near-perfect focusing and negligible path loss through the imaging system. If detectors with a similar sensitivity is used for arrays, this would require a source with a much larger power than the single-pixel case. Third, DC power consumption of the array detector will be larger. With direct detectors in passive-mode operation, the power dissipation per pixel will be tiny, but the accumulated total power may be considerable if the array size is large. A smart biasing scheme to partially turn on detectors at a time in sequence and sharing as many elements (such as amplifiers) as possible between pixels would be desired to reduce the power dissipation. Additional challenges include longer data processing time for entire pixels, extended signal loss and added noise along the path and the readout circuit, nonuniformity over pixels, etc. (see Sect. 3.4.1 for related issues). The cost could be an issue for 2D arrays if they are based on individually fabricated and packaged detectors, leading to needs for imaging techniques with reduced number of detectors, such as the interferometric imaging [46] or compressed sensing [47]. Single-chip 2D arrays will not need techniques to reduce the number of detectors for a

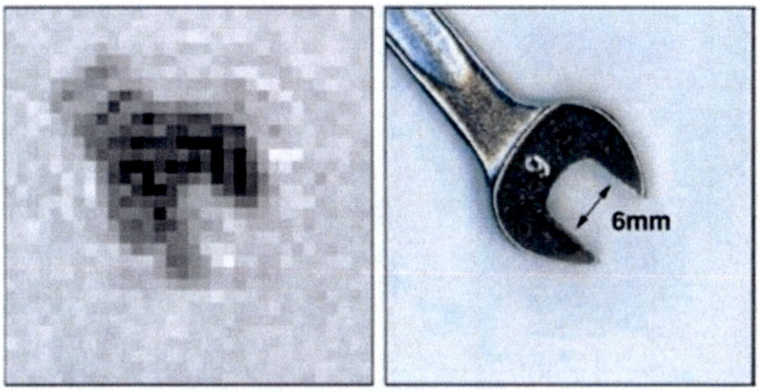

Fig. 6.20 Snapshot of a real-time image taken with a Si CMOS 1 kilo-pixel THz camera, shown together with the regular image of the wrench used as a target object. (© 2012 IEEE [49])

given array area, as partial removal of detectors within the area will not necessarily reduce the cost.

Despite the challenges, there have been promising reports on real-time imaging based on array detectors recently. The array detectors introduced in Sect. 3.4.2 are well suited for real-time imaging. The Si CMOS single-chip 1k-pixel array detector operating at 0.7–1.1 THz has demonstrated its successful function as a camera, which is capable of acquiring still images at a single shot and, more importantly, real-time images [48, 49]. Being composed of 32 × 32 detectors, each with an 80 × 80 μm^2 area, the entire chip occupies an area of 2.9 × 2.9 mm^2. With the THz camera fabricated in a 65-nm CMOS technology, transmission-mode real-time imaging was successfully performed at a frame rate of 25 fps. Figure 6.20 shows a snapshot of a real-time image, which was taken on a metallic wrench as a test object. Another 2D array described in Sect. 3.4.2, a Si CMOS 28 × 28 pixel multi-chip array detector based on a modular approach with 16 sub-array chips each comprising 7 × 7 detectors, has also demonstrated its capability of real-time imaging at 300 GHz [50]. Each sub-array chip has a dimension of 4 × 4 mm^2, leading to a total array size of 16 × 16 mm^2. Figure 6.21 shows a snapshot of the real-time image taken with the array, displaying the transmission-mode image of a metallic blade. The image is based on total 961 pixels including the virtual pixels, which fall on the dead zone along the peripheral area of each sub-array chip that is consumed by bonding pads for inter-chip wiring. The image data on those virtual pixels were created by interpolating the data from the adjacent pixels. The image shows no obvious indication of the virtual pixels, nor the trace of the sub-arrays, demonstrating the seamless integration of the multiple chips based on a modular approach.

6.2.5 Tomographic Imaging

The discussions so far have been made for 2D images, but it is obvious that 3D images would provide richer information on the target object. There can be various ways to obtain 3D images, while the most widely adopted modality is the tomography. Tomographic imaging, or tomography, provides a set of cross-sectional slice images of a target object, or tomograms. Based on the tomograms, the 3D internal

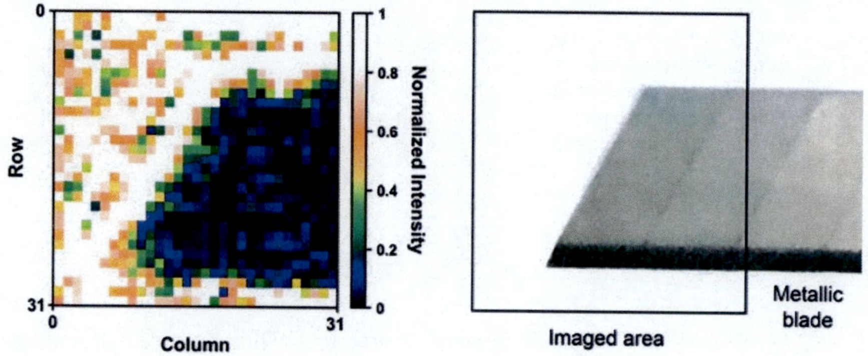

Fig. 6.21 Snapshot of a real-time image taken with a Si CMOS 300-GHz multi-chip 28 × 28 pixel camera (total 31 × 31 = 961 pixels including the virtual pixels). The boxed area of the metallic blade on the right is imaged

structure of the object can also be reconstructed as a 3D image. Tomography based on THz waves is called THz tomography. There are various known methods to facilitate tomographic imaging based on the THz waves. Widely employed are the computed tomography (CT), diffraction tomography (DT), and time-of-flight (TOF) tomography. There are other techniques as well, such as tomosynthesis, holography, Fresnel lens technique, and so forth [51]. Most of THz tomography techniques can be realized based on either THz pulses or THz CW signals, which are based on basically the same principle. In this section, the fundamentals of selected techniques (CT, DT, TOF) will be briefly introduced in general terms, with a review on imaging examples based on CW-based THz tomography.

The *computed tomography* (CT) is the most widely employed tomographic technique, especially combined with X-ray imaging. You may have had a chance to see CT images at a doctor's office. While they are most likely X-ray CT images, CT is a popular technical choice for THz tomography as well. The basic principle of CT imaging is as follows. Assume a 3D target object with an arbitrary distribution of the absorption coefficient inside. The beam emitted from the source transmits through the target and its 2D projection image is obtained at the other side of the target by the detector, either through a raster scan or a single shot with array. This is a conventional 2D imaging. Now, if you rotate the source–detector pair around the target object, or, equivalently, rotate the object with the source-detector fixed, a series of projection images from different angles can be obtained. Based on this set of projection images acquired with various angles, the 3D structure of the target object can be reconstructed by applying a proper recovery algorithm. This is the basic principle behind CT, which will be better described with a little bit of mathematics as follows:

Consider a target object placed in the tomography setting depicted in Fig. 6.22, which explicitly show the distribution of the absorption coefficient inside the object at a fixed height (z-axis). The shown cross-sectional 2D distribution of the absorption coefficient will be given as a function of x and y on the $x - y$ plane, denoted as $f(x, y)$. When $f(x, y)$ is projected on the detector side by illuminating the target with the source on the other side, the intensity profile will be given as a 1D function along l,

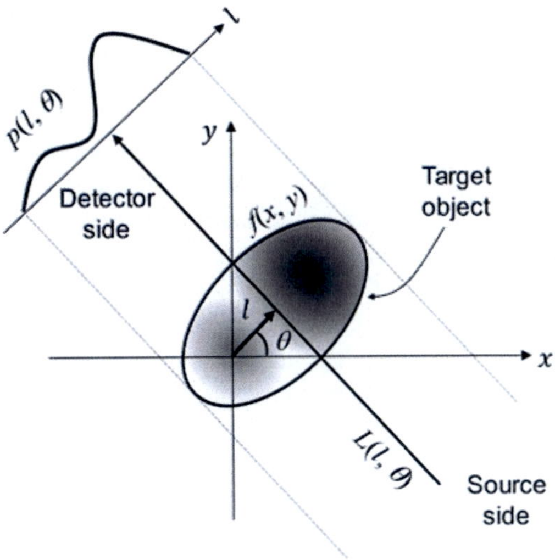

Fig. 6.22 Principle of Radon transform. The arbitrary distribution of the absorption coefficient inside the target object is transformed from *f*(*x*, *y*) to *p* (*l*, *θ*)

or $p(l)$, where l is the shortest distance from the origin to the projection line. If the projection is repeated with multiple projection angles θ, p is eventually given as a function of both l and θ, or $p(l, \theta)$. Hence, $f(x, y)$ defined in the $x - y$ plane can be transformed to $p(l, \theta)$ defined in the $l - \theta$ plane. This process is called the Radon transform and can be simply expressed as a line integral:

$$p(l,\theta) = \int_{L(l,\theta)} f(x,y)\mathrm{d}L, \tag{6.9}$$

where the integration is made along a straight line L connecting the source and the detector. In practice, $p(l, \theta)$ is the raw data and we need to obtain $f(x, y)$ from the raw data. A widely adopted algorithm to reconstruct $f(x, y)$ from $p(l, \theta)$ is the filtered back-projection (FBP), which can be described as follows [52, 53]:

$$f(x,y) = \int_0^{\pi}\int_{-\infty}^{\infty} P(\omega,\theta)|\omega| \exp\left(2\pi j\omega(x\cos\theta + y\sin\theta)\right)\mathrm{d}\omega\,\mathrm{d}\theta, \tag{6.10}$$

where $P(\omega, \theta)$ is the Fourier transform of $p(l, \theta)$ with respect to l, and ω is the spatial frequency in the direction of l. A high-pass filtering is implicitly included in the integration that helps to emphasize the high spatial frequency components of the images, which is the basis for the name, "filtered" back-projection. The presentation of $p(l, \theta)$ in the $l - \theta$ plane is called the *sinogram*, and the reconstructed $f(x, y)$ presented in the $x - y$ plane is called the *tomogram*, which is basically the slice image of the corresponding cross-section. Hence, the basic process of tomography is to obtain a tomogram from a sinogram obtained at a fixed z point and repeat this step over different z values along the z-axis. This full process will generate the 3D image of the target object. While FBP is a popular choice for CT as mentioned above, the reconstruction can be performed by the iterative methods as well, such as the algebraic reconstruction [54] and the ordered subsets expectation maximization method [55]. Compared to FBP, the iterative methods are known to provide more accurate reconstruction with less errors, potentially with reduced projection numbers [56], but they generally require a greater computation power.

The *diffraction tomography* (DT) is similar to CT in its imaging setup, composed of a source and a detector placed on either side of the target object [57]. The 3D image of the target object is reconstructed based on the multiple images acquired for a series of different projection angles as was the case for CT. The key difference of DT from CT is that the beam from the source is assumed to be diffracted by the target object, as opposed to the CT case in which the beam presumably follows a straight-line path. With DT, the reconstruction of the 3D image of the target object is obtained through the inverse scattering method applied to the field scattered by the target object. It is noted that the reconstruction with DT is made for the distribution of the refractive index of the target object, rather than the absorption coefficient for CT. Compared to CT, DT is more useful for complex target objects with fine structures, which cause a high level of diffraction that strongly affects the projected images. Remember that diffraction is more pronounced for a longer wavelength for a given aperture size. Hence, for the cases where the dimension of the detailed structure in the target object is comparable to the wavelength of the probe beam, DT technique may be favored.

The *time-of-flight* (TOF) technique distinguishes itself from CT and DT techniques in that it is based on the reflection-mode of imaging. As such, the source and detector can be located on the same side. Also, no rotational moving part is needed in the TOF setup, a great advantage for system compactness. Pulses transmitted by the source are reflected back by the target object, and the resultant amplitude and phase change are recorded by the detector. The pulse can be ultrashort THz pulses or pulse-modulated CW THz signals. The total flight time of the signal carries information about the distance of the reflecting surface from the source and detector. Hence, a precise measurement of the temporal delay of the signal after the round trip will provide the depth information of the target surface that faces toward the source/detector. TOF is particularly useful for target objects made of multiple layers, as the reflection will occur on each interface boundary and generate depth information of each layer with a proper analysis. This will enable the reconstruction of the 3D structural image of the target object along the direction of the beam propagation [58]. It is noted that the operation of the pulse-mode radar (RAdio Distance And Ranging) and the lidar (LIght Distance And Ranging) are also based on the TOF principle. The radar operation will be described in detail in Sect. 6.4.

There have been quite a lot of THz tomography imaging results reported, earlier with pulse-based optical methods and more recently with CW-based electrical methods. Some THz CT examples based on electrical methods are reviewed here. One of the earliest THz CT images demonstrated based on electrical methods employing a semiconductor chip as either detector or source was reported in 2011 [59]. The setup is composed of a 65-nm Si CMOS detector integrated with an on-chip antenna and a multiplier-based 650-GHz signal source with parabolic

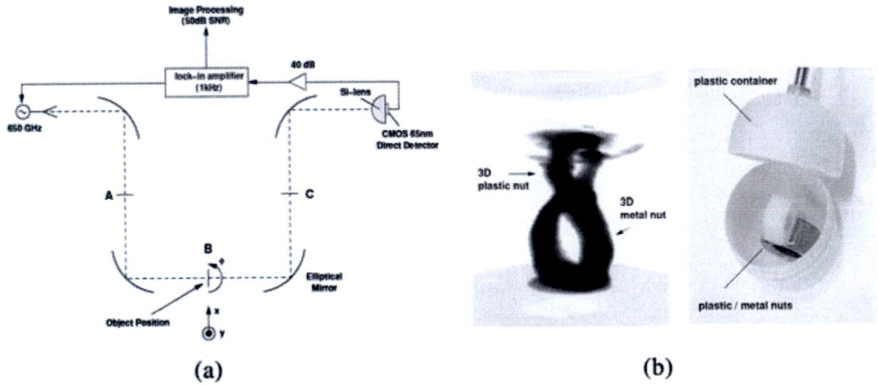

Fig. 6.23 (**a**) Imaging setup for THz tomography. (**b**) The target object and its THz CT image obtained with the given setup. (© 2011 IEEE [59])

mirrors for beam path control, as shown in Fig. 6.23a. The target object is metal and plastic nuts enclosed in a polyethylene container. The obtained 3D CT image is shown in Fig. 6.23b, which was acquired by raster-scanning the target object with rotation. It clearly shows the internal structure of the target object, an indication of the transparency of the container against THz waves. Another CT experiment by the same group working at 500 GHz was reported in [60], which includes both source and detector based on Si chips, the former in SiGe HBT and the latter in Si CMOS technologies.

In another effort, 300-GHz THz CT experiments have been carried out based on the source based on InP HBT and the detector based on SiGe HBT technologies [61, 62]. For the detector, both direct detector and heterodyne detector [25] were used and the CT images from the two types of detectors were compared. An imaging setup modified from Fig. 6.13 to allow target rotation was used to image the target object shown in Fig. 6.24a, which is composed of a metallic bolt and a washer enclosed in an acrylonitrile butadiene styrene (ABS) box. Figure 6.24b shows a set of 2D projection images (for various projection angles), from which sinograms and tomograms were obtained (for various *z*-axis positions) with FBP algorithm. From these images, the 3D image of the target object was reconstructed. The image acquisition was carried out with both direct detector and heterodyne detector for comparison. To investigate the effect of the incident CW power and the types of detector, a series of 3D images were obtained by varying the source output power level, which was repeated for both detector types. The results are shown in Fig. 6.24c. With both detectors, there exists a lower limit of the source power below which the images begin to degrade, while the limit is lower for the heterodyne detector owing to its smaller NEP (4 pW/Hz$^{1/2}$ vs. 21 pW/Hz$^{1/2}$ of direct detector). Instead, the heterodyne detector exhibited a sign of saturation at high input power level, a result of the larger responsivity (320 kV/W vs. 6 kV/W of direct detector).

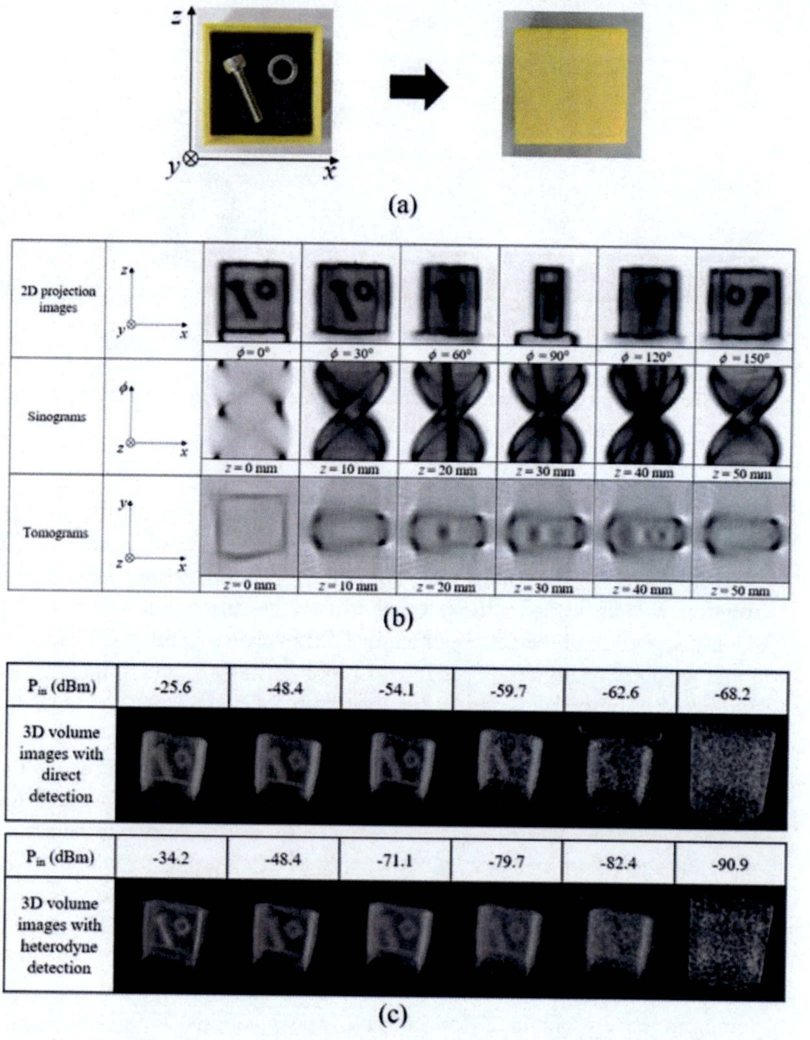

Fig. 6.24 **(a)** The target object composed of a metallic bolt and washer enclosed in a plastic box. **(b)** 2D images taken from various rotation angles of the target. Also shown are sinograms and tomograms obtained from the 2D images with FBP algorithm. **(c)** THz CT images with varied source power, with both direct detector and heterodyne detector. (© 2018 IEEE [62])

6.3 THz Communication

6.3.1 Overview

One definition of communication is the transportation of information between two physically separated locations. Modern communication systems can be configured as either a wired or a wireless scheme. The primary advantage of the latter comes from the fact that it is free of the bothersome wire connection across the link points. One downside of the wireless systems, though, has been known to be its lower data rate compared to the wired counterpart. However, we observe a rapid increase in wireless data rate around our daily lives. In fact, the pace of data rate increase for the wireless systems is so fast that some foresee that it will soon overtake that of the wired systems and eventually end up surpassing it. For this optimistic prospect to come true, the exploitation of the THz spectrum is an essential part to enable the broadband needed for information mass transportation. In this section, a brief overview of THz wireless communication systems is provided.

6.3.1.1 Opportunities and Challenges

The opportunity of using the THz band for wireless communication is rather straightforward. One effective approach to raise the data rate is to increase the bandwidth of the system, as indicated by the famous Shannon's theorem [63]:

$$C = B \log_2\left(1 + \frac{S}{N}\right), \tag{6.11}$$

where C is the channel capacity, or the maximum data rate allowed in the channel, B is the bandwidth, and S/N the signal-to-noise ratio (SNR). The most effective way to stretch the bandwidth is to increase the carrier frequency. For instance, if the same fractional bandwidth is assumed, a $10\times$ increase in carrier frequency would result in a $10\times$ increase in the bandwidth. This simple math has been the driving force behind the incessant efforts for the carrier frequency increase in the past decades, and it has now arrived at the THz regime [64–66].

Despite this obvious opportunity for THz communication, there are a great deal of challenges lying ahead that need to be addressed for its realization. Those challenges are quickly manifested with a simple link budget analysis as shown below. From the Friis transmission equation (Eq. 4.57), the signal power at the receiver, $P_{\mathrm{in,RX}}$, can be expressed as follows in dB scale (including dBm and dBi):

$$P_{\mathrm{in,RX}} = P_{\mathrm{out,TX}} + G_{\mathrm{A,TX}} + G_{\mathrm{A,RX}} - 20 \log\left(\frac{4\pi R}{\lambda}\right), \tag{6.12}$$

where $P_{\mathrm{out,TX}}$ is the signal power radiated by the transmitter, $G_{\mathrm{A,TX}}$ and $G_{\mathrm{A,RX}}$ are the antenna gains at the transmitter and receiver, respectively. The last term is the free-space path loss (FSPL) with distance R and wavelength λ. Another parameter we need to consider for the analysis is the receiver sensitivity $P_{\mathrm{sen,RX}}$, or the minimum required power received at the receiver for an SNR desired for the given application. It is given as follows in dB scale at room temperature, which can be readily derived from the definition of the noise figure (Eq. 3.10):

$$P_{\text{sen,RX}} = -174\,\text{dBm/Hz} + 10\log B + \text{NF} + \text{SNR}_{\text{min}}, \tag{6.13}$$

where B is the bandwidth given in Hz, NF is the noise figure, and SNR_{min} is the minimum required SNR at the receiver output. Note that the first term corresponds to the value of KT at room temperature expressed in dBm, which is obtained by $10\log(26\text{ mJ} \times 1.6 \times 10^{-19} / 1\text{ mJ}) = -174$ (dBm/Hz), where the relation $kT = 26$ meV at room temperature is used. Hence, the first two terms combined indicate the total noise power at the receiver input for a given bandwidth, as the noise is proportional to the bandwidth in linear scale. The sum of the first three terms are often called the noise floor.

Now, for the link to be safely established, the received power should be larger than the minimum required power at the receiver, or $P_{\text{in,RX}} > P_{\text{sen,RX}}$. That is:

$$\begin{aligned} P_{\text{out,TX}} + G_{\text{A,TX}} + G_{\text{A,RX}} - 20\log\left(\frac{4\pi R}{\lambda}\right) \\ > -174\,\text{dBm/Hz} + 10\log B + \text{NF} + \text{SNR}_{\text{min}}. \end{aligned} \tag{6.14}$$

Challenges for implementing THz communication systems are obvious from Eq. (6.14). To satisfy the inequality with an enough margin, most of all, $P_{\text{out,TX}}$ should be as large as possible and NF needs to be as small as possible. However, as discussed in Chaps. 2 and 3, the major challenges for THz sources and detectors are the limited output power and the excessive noise level, respectively, both imposing adverse effects on the relation. Besides, as indicated in Fig. 6.25, FSPL increases with frequency and the values are remarkably high at the THz band, which is another difficulty in THz link setup. What about the antenna gain terms, $G_{\text{A,TX}}$ and $G_{\text{A,RX}}$? To compensate for the large FSPL, antennas with high gains will be on high demand. High-gain antennas, however, will exhibit a narrow beam width, requiring a precise alignment between TX and RX. This is an unfavorable situation particularly for mobile applications, for which an adaptive active alignment, possibly with beam

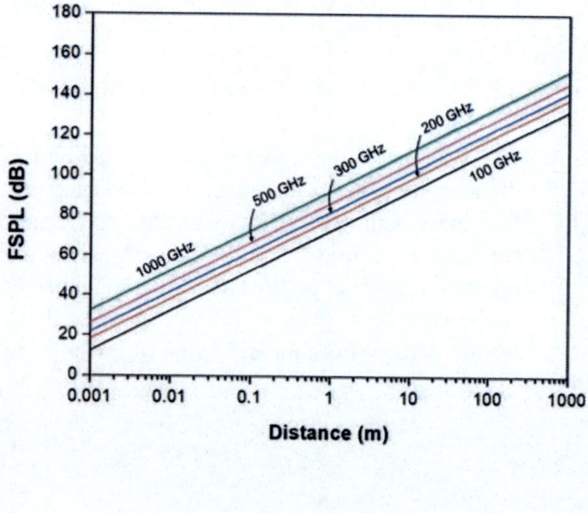

Fig. 6.25 Free space path loss (FSPL) for various frequencies

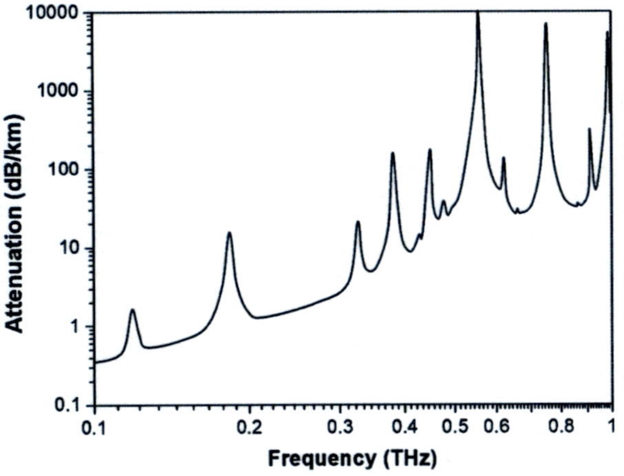

Fig. 6.26 Attenuation in the Earth's atmosphere at the standard condition (pressure of 1013 hPa, temperature of 15 °C, water-vapor density of 7.5 g/m^3)

forming, would be required to establish and retain the link. This situation will be further aggravated by the weak diffraction of THz waves, which will prevent non-line-of-sight connection. A complex relay mechanism may be needed to cope with dynamic channel conditions. This long list of challenges is something we ought to address to arrive at the reliable THz communication systems operating at a high date rate as desired.

A brief comment will be worthwhile at this point regarding the effect of the attenuation due to the absorption in the Earth's atmosphere, which will potentially affect the wireless links. Figure 6.26 presents the attenuation level in the Earth's atmosphere at the standard condition (see figure caption), which includes multiple peaks that arise from water or oxygen molecule absorption [67]. It will be useful to compare the loss due to FSPL and that of the atmospheric attenuation. For this comparison, one may be tempted to combine Figs. 6.25 and 6.26 as a single chart. However, the situation is not that straightforward. The atmospheric attenuation is given by an exponential factor driven by the distance R, or exp.$(-\alpha R)$, where α is the attenuation constant given as a function of the frequency, or $\alpha(f)$. This means that, when the attenuation level is expressed in terms of dB scale, it is linearly proportional to the distance (i.e., $\propto \alpha R$). Hence, the attenuation can be represented by attenuation (dB) per distance, as is the case of Fig. 6.26, where it is represented in dB/km. For FSPL, however, this is not true. FSPL is given as $(4\pi R/\lambda)^2$, or $(4\pi Rf/c)^2$, and thus its dependence on the distance is not linear anymore when expressed in dB scale (i.e., $\propto$ log$(4\pi Rf/c)$). Therefore, FSPL cannot be represented by dB per distance, or by dB/km. This difference would not allow a direct comparison of the atmospheric attenuation and FSPL in a single chart. More importantly, it implies that the atmospheric attenuation increases more rapidly with distance than FSPL.

If you still desire a direct comparison between the atmospheric attenuation and FSPL, one possible solution is to plot them together with a *fixed* distance. It is done in Fig. 6.27, which presents the two losses plotted together at four selected distances of interest. Also included in each plot are the cases where FSPL are compensated for

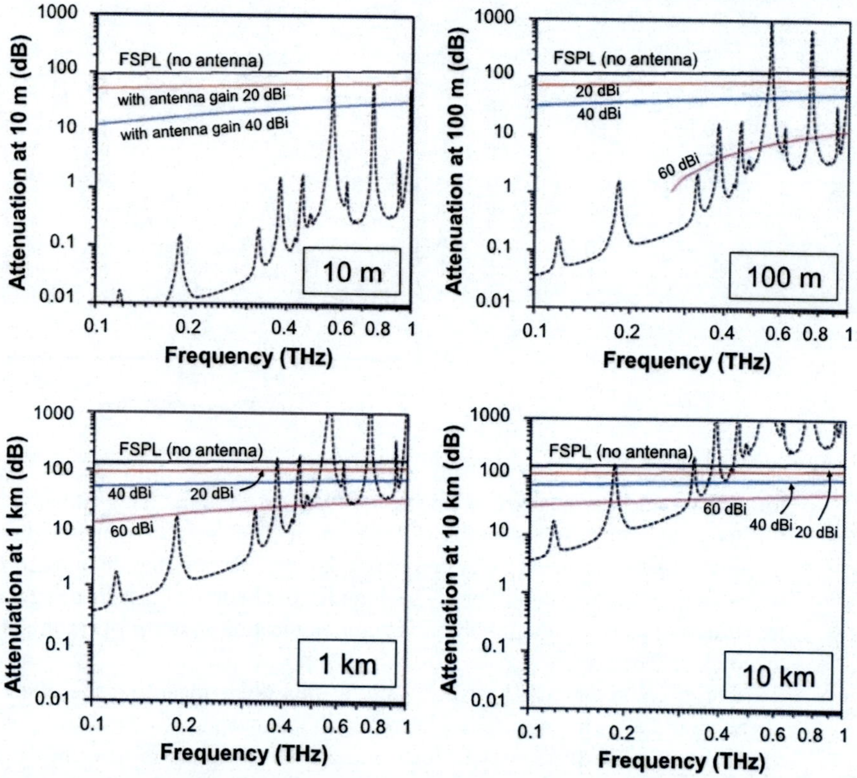

Fig. 6.27 Comparison of the atmospheric attenuation and FSPL, as well as FSPL compensated for with various antenna gains, at fixed distances of 10 m, 100 m, 1 km, and 10 km

with various antenna gains (20, 40, and 60 dBi), assuming a pair of antennas of the shown gain values are placed at TX and RX. At a distance of 10 and 100 m, it is clear FSPL is dominating, almost totally eclipsing the atmospheric attenuation, even with the cases with high-gain antennas. (For the entire frequency with the 10-m case and below ~240 GHz with the 100-m case, negative loss, or net *gain*, will be obtained in the free space with a 60-dBi antenna gain, which is not shown in the plot in the log scale. In this situation, though, the loss comparison is not quite relevant because the atmospheric attenuation is already too small.). Even with the case of 1-km distance, FSPL still prevails over the atmospheric attenuation except at a few scattered frequency intervals. The two losses become comparable beyond ~500 GHz only with very high antenna gains. At 10 km, there is a cross-over point near 400 GHz, beyond which the atmospheric attenuation is consistently larger than FSPL. However, with this level of distance, the total loss will be too large for any THz links to be useful for practical applications. Hence, it will be fair to say that the loss is dominated by FSPL for most practical THz applications. Exceptions will be very long-distance wireless links based on pencil-like beams with a very high antenna gain operating at limited frequency intervals with a very strong atmospheric molecular resonance, which is highly unlikely and most likely avoided in realistic cases. Note that the detailed atmospheric attenuation will vary with the weather condition and altitude, as is partly shown with Fig. 1.2.

6.3.1.2 Application Scenes

Despite the series of challenges mentioned above, people are marching toward real-world THz communication systems. Optimistic-minded, they expect those technical hurdles will be eventually resolved, as has been the case for many other engineering frontiers. Various usage scenarios have been proposed for future THz communication systems, some of which are introduced here.

The most rational approach for THz communication systems would be to maximize the advantage of the THz band, the wide bandwidth, while circumventing the weak point, the limited link budget. Naturally, the broadband short-range communication first appeared as the most immediate application scene of the THz communication. Much attention has been paid to the *chip-to-chip communication* application [68]. A wireless link will get rid of the physical interconnection between chips, avoiding related issues such as routing complexity and *RC* delay. THz link is well suited for this purpose as it can support the increasingly heavy inter-chip data traffic, while occupying a small chip area including on-chip antennas. The inter-chip distance will be small, in the order of centimeter or less, which will lead to FSPL around 40 dB for 300 GHz, for instance, as indicated by Fig. 6.25. This level of loss can be managed even with low-gain on-chip antennas if RX and TX of reasonable performances are employed. The idea can be extended toward more practical applications, which includes *wireless connectivity for high resolution displays*. For uncompressed data streaming, 4k (3840 × 2160 pixels) and 8k (7680 × 4320 pixels) displays will require data rate as high as ~12 Gbps and ~48 Gbps, respectively. THz link is a highly favored candidate to support this level of data rate between a set-top box and a display. The link distance may be in the order of 10 cm, with which FSPL will soar up to around 60 dB. However, unlike the chip-to-chip cases, high-gain antennas such as horns can be employed here, which would be able to provide a loss compensation of 40–50 dB as a pair. Another practical example of THz link is *kiosks for massive data download* to personal mobile devices, as was recently demonstrated by NTT [69]. For this vending machine concept for data download, the link budge will be similar to the case for the display link described above, while alignment condition will be less deterministic as the users are expected to place their mobile terminal on the kiosk body by hands. Also, the antenna gain will be limited for the mobile device side. This use case can be easily developed to *links between mobile devices and storage devices* of any types in general including PCs or servers [64].

As opposed to such a short-range communication, there is rather a bold approach as well, in which attempts are made to apply THz waves to a *long-distance point-to-point link* application [66]. The main enabler for this long-distance THz link is the high gain antenna. As briefly mentioned above, reflector antennas can provide a gain sufficiently large to compensate for the high FSPL of the long distance. One favorite

Fig. 6.28 Relation between the antenna gain and the dish diameter of dish antennas for various frequencies. e_r in Eq. (6.15) is assumed to be 1 in the curves

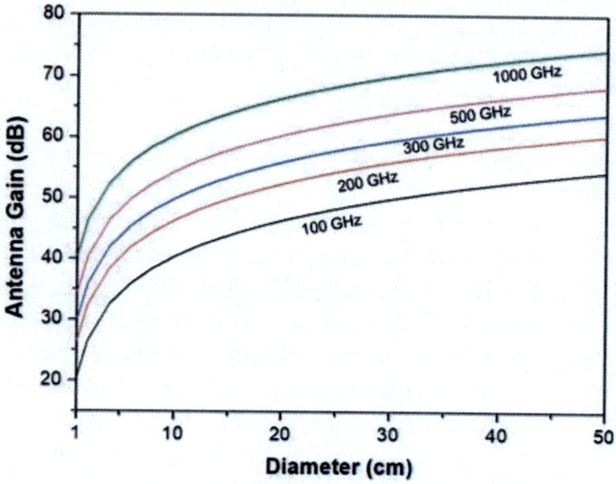

tendency for THz applications is the increasing antenna gain with increasing frequency (decreasing wavelength) when the antenna aperture size is fixed. For example, with a dish antenna with a diameter of D, the gain formula defined in Eq. (4.54) can be converted to the following form assuming that the maximum effective aperture is equal to the dish area ($A_{em} = \pi(D/2)^2$):

$$G = \frac{4\pi}{\lambda^2}A_e = \frac{4\pi}{\lambda^2}e_r A_{em} = e_r\left(\frac{\pi D}{\lambda}\right)^2. \tag{6.15}$$

This relation is plotted as a function D for various frequencies in Fig. 6.28. A good example exploiting this property was reported in [70], where a pair of Cassegrain reflector antenna, each with 55-dBi gain, was employed to cope with FSPL of 139 dB near 240 GHz for an 850-m link. Although alignment may be an issue, such a THz long-distance communication link will benefit from the wide bandwidth of the THz band to enable high data rate. In this work, 64 Gbps was reported with the given link. It can be mentioned that the 3-dB beam width is around 0.3°, or ~4.5 m at the 850-m distance, for the 55-dBi antenna, which would require a dish diameter around 20 cm near 240 GHz based on Eq. (6.15). In addition to FSPL, TX output power and RX sensitivity are also issues for long-distance applications. It is certainly true that the link budget will benefit from efforts to enhance the performance of the THz TX and RX, which would probably lead to a few-dB enhancement in both PA output power and LNA NF. However, a few dB of gain improvement will be obtained by only a moderate expansion of the antenna dish size, which will have the equal effect on the link budget. This implies that TX and RX performances may not be the critical factors in such a long-distance THz link.

6.3.1.3 Channel Modeling

As opposed to wired links that have a fairly well defined channel conditions, wireless links are based on the air channel, which is greatly affected by the surrounding environment. The channel is composed of not only the line-of-sight path but also reflection, refraction, as well as diffraction by various objects, stationary or dynamic. Movement of transmitters and receivers also need to be taken into account for mobile wireless applications. Hence, it is crucial for wireless links to have channels properly modeled for the best prediction of the signal arriving at the receiver. The model will help the reliable design of the communication systems with expected performance requirements for antenna gains and other TX/RX parameters in realistic channel environments. The channel modeling applies to both outdoor and indoor links, while the latter is considered more critical as the former case will be mostly based on the line-of-sight straight paths that are less affected by the environment.

There have been growing efforts to develop THz channel models for realistic building interiors to pave the way for the actual deployment of THz communication links [71–73]. The scattering and reflection properties of typical building interiors were investigated with rough or multilayered surface in those reports. For THz communication links, the utilized range of spectrum is extensive due to the wide bandwidth available. Hence, in addition to the angle dependence, the frequency dependence of the reflection properties will become also relevant for THz links unlike the lower frequency cases, which may be caused by interference in the multilayered surfaces [72]. Also, detailed behaviors of the reflection and the diffuse scattering on the rough surfaces will be different for THz waves than the lower frequency bands due to the small wavelengths [71]. A high molecular absorption in the atmosphere at the THz band as mentioned earlier is another additional factor to be considered for channel modeling. All these aspects suggest that channel modeling dedicated to the THz links is critically needed.

6.3.1.4 Standardization

Standardization offers compatibility between systems developed by different parties based on the consensus over the details of the system. A wide range of industrial standardization has been driven by IEEE, including one for communication. IEEE 802 family governs local area networks (LAN), metropolitan area networks (MAN), and other levels of area networks. There are many sub-standards under IEEE 802, the most widely known to people probably being IEEE 802.11, which is for Wi-Fi, ubiquitous in the modern work and home environments. Another standard in the same family is IEEE 802.15, which mainly relates personal area networks (PAN) of various types. Some notable standards in this category are IEEE 802.15.1 (Bluetooth) and IEEE 802.15.6 (body area network or BAN). THz communication is related to IEEE 802.15.3, which deals with PAN systems of high data rates. Many mm-wave engineers may be familiar with IEEE 802.15.3c, intended for a short-

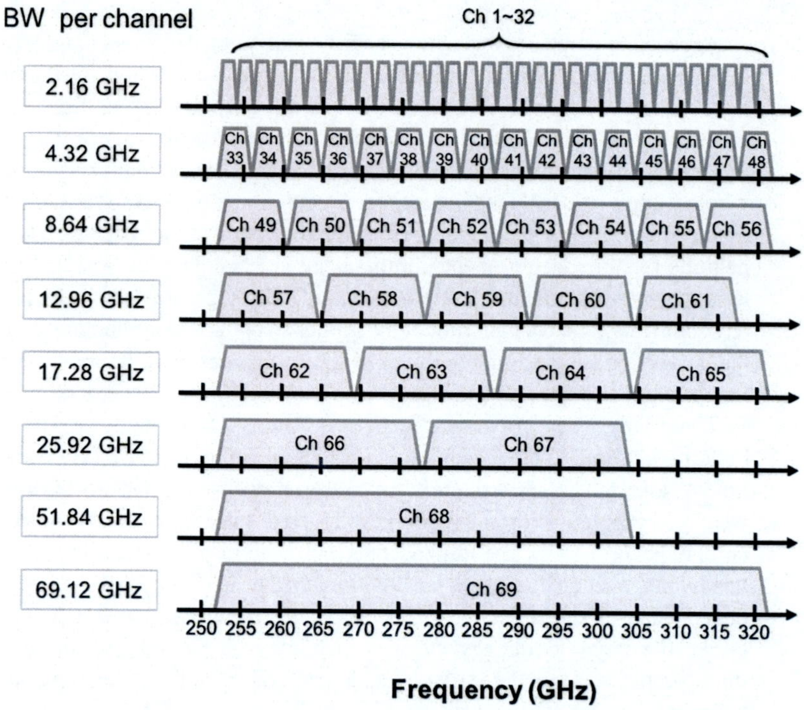

Fig. 6.29 Channelization of the 252–322 GHz frequency band by IEEE Standard 802.15.3d

range connectivity based on the band near 60 GHz for a data rate of around several Gbps. A more advanced version recently developed for a higher data rate based on a higher frequency band is IEEE 802.15.3d, which belongs to THz communication and is described below.

IEEE 802.15.3d was driven by IEEE 802.15 Task Group 3d (TG3d) formed in 2014, which had evolved from the Interest Group THz (IG THz) (2008) and Study Group 100G (SG 100G) (2013). IEEE 802.15.3d was approved in September 2017 and published in October 2018 [74]. The channel allocation by IEEE 802.15.3d is shown in Fig. 6.29. Total 69 channels are defined in the frequency range between 252 and 322 GHz, based on eight different bandwidth types ranging from 2.16 GHz to 69.12 GHz, all multiples of 2.16 GHz. For the case of the 2.16-GHz channels, for instance, total 32 channels fit into the entire band provided. This can be compared with IEEE 802.15.3c based on the 57–66 GHz band, which accommodates only four channels with the same bandwidth of 2.16 GHz. IEEE 802.15.3d considers two physical layers, single carrier (SC) and on-off-keying (OOK), with seven modulation schemes, BPSK, QPSK, 8PSK, 8APSK, 16QAM, 64QAM, and OOK (see Sect. 6.3.2 for the modulation scheme details). The major target applications suggested by IEEE 802.15.3d are kiosk downloading, intra-device communication, wireless backhauling/fronthauling, and additional wireless links in data centers, which well align with those described earlier in this section.

6.3.2 Modulation Schemes

Before moving on to the discussion of actual THz communication systems, it will be helpful to review the modulation schemes that can be adopted for THz links. Modulation, by definition, refers to the process of modifying one or multiple properties of a sinusoidal carrier waveform in accordance with an information-carrying signal. While there are rather traditional analog modulation schemes such as AM (amplitude modulation), FM (frequency modulation), or PM (phase modulation), digital modulations are almost exclusively considered for THz communication. For digital modulations, the term "modulation" is often replaced by "shift keying", since the "key" or the "tone" of the carrier signal is shifted in a discrete manner with the modulation. You may imagine the shift of keys on a piano, a good analogy with the acoustic waves. (After all, the music can be defined as the frequency- and time-domain modulations of the acoustic waves nicely interwoven, in the author's personal view.)

In the actual digital modulations, typically modulated are the *amplitude* and the *phase* (rather than the frequency) of the carrier signal. ASK (amplitude shift keying) is an example of the former, while BPSK (binary phase shift keying) and QPSK (quadrature phase shift keying) are for the latter. A hybrid modulation, which simultaneously modulates both amplitude and phase, is also a highly feasible option, further improving the spectral efficiency. QAM (quadrature amplitude modulation) belongs to this category. These various modulation schemes are briefly explained below, while a more detailed formal treatment can be found in many standard textbooks on the communication theory (for example [75]).

ASK is based on the modulation scheme that switches the amplitude between two levels. Often implemented is switching between the zero and a non-zero amplitude level. In this particular case, the modulation is called OOK (on-off keying). Hence, OOK is a special case of ASK, and, in fact, is by far the most popular variant of ASK. The signal of OOK, $x_{\mathrm{OOK}}(t)$, can be represented as:

$$x_{\mathrm{OOK}}(t) = 0 \quad (\text{if } b_n = 0) \quad \text{or} \quad A_{\mathrm{C}} \cos \omega_{\mathrm{C}} t \quad (\text{if } b_n = 1), \tag{6.16}$$

where b_n is the input state of the nth bit and ω_{C} is the carrier frequency. Another modulation scheme, PAM (pulse amplitude modulation), is also an example of ASK, which may involve more than two signals levels. With PAM, $M = 2^k$ different levels are allowed, where k is the number of bits modulated together. BPSK shares a common property with ASK, in that they are both based on two signal states, or binary levels. For BPSK, however, the two states are represented by two signals with the same amplitude but the opposite phases, 0 and 180°:

$$x_{\mathrm{BPSK}}(t) = A_{\mathrm{C}} \cos(\omega_{\mathrm{C}} t + \phi), \quad \phi = 0 \quad (\text{if } b_n = 0) \quad \text{or} \quad \pi \quad (\text{if } b_n = 1). \tag{6.17}$$

QPSK and QAM are often called the quadrature modulation, as they require both *in-phase* and *quadrature* signals. For the quadrature modulations, a stream of the

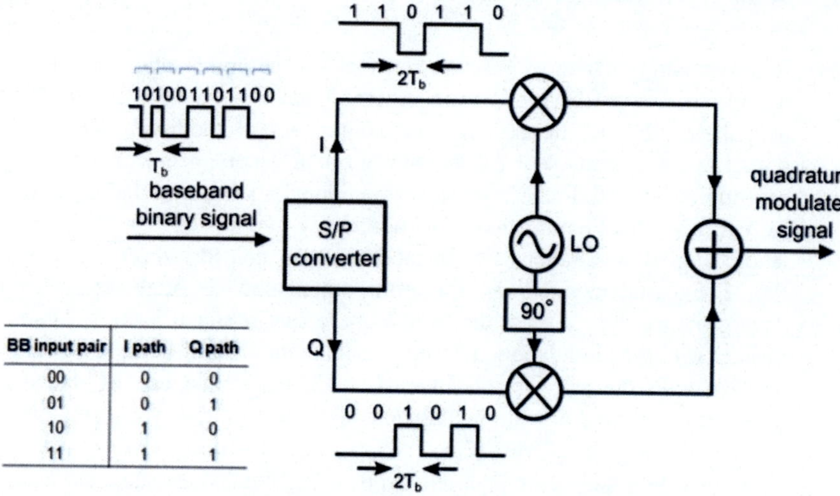

BB input pair	I path	Q path
00	0	0
01	0	1
10	1	0
11	1	1

Fig. 6.30 Schematic diagram of the quadrature modulation showing the split of the baseband binary signal. Shown in the inset table is the mapping between two consecutive baseband input bits (each bit with T_b) and corresponding I- and Q-path bits (with $2T_b$)

baseband binary signal bits is divided into two branches, in-phase (I) and quadrature (Q). The way they are split is depicted in Fig. 6.30. Every pair of two consecutive binary bits with a period of T_b are converted into two parallel binary bits with a period of $2T_b$ on two separate signal paths, I and Q. The two parallel signal streams with a relaxed rate of level transition (by a factor of two, from T_b to $2T_b$) are modulated by two orthogonal functions, sine function on one path and cosine function on the other. The modulated signals on the two paths are combined at the output and transmitted. Hence, for both QPSK and QAM, the modulated signal output can be expressed as follows:

$$x_{\text{quadrature}}(t) = \alpha_1 A_C \cos \omega_C t + \alpha_2 A_C \sin \omega_C t, \tag{6.18}$$

where α_1 and α_2 are the amplitude coefficients. It is this pair of coefficients that distinguish between QPSK and QAM, as well as between different orders of QAM. The allowed coefficient pairs for QPSK and various orders of QAM are:

$$\begin{aligned}
&\text{QPSK}: && \alpha_1 = \pm 1 \text{ and } \alpha_2 = \pm 1\\
&\text{16QAM}: && \alpha_1 = \pm 1, \pm 3 \text{ and } \alpha_2 = \pm 1, \pm 3\\
&\text{64QAM}: && \alpha_1 = \pm 1, \pm 3, \pm 5, \pm 7 \text{ and } \alpha_2 = \pm 1, \pm 3, \pm 5, \pm 7\\
&\text{256QAM}: && \alpha_1 = \pm 1, \pm 3, \pm 5, \pm 7, \pm 9, \pm 11, \pm 13, \pm 15 \text{ and}\\
& && \alpha_2 = \pm 1, \pm 3, \pm 5, \pm 7, \pm 9, \pm 11, \pm 13, \pm 15.
\end{aligned} \tag{6.19}$$

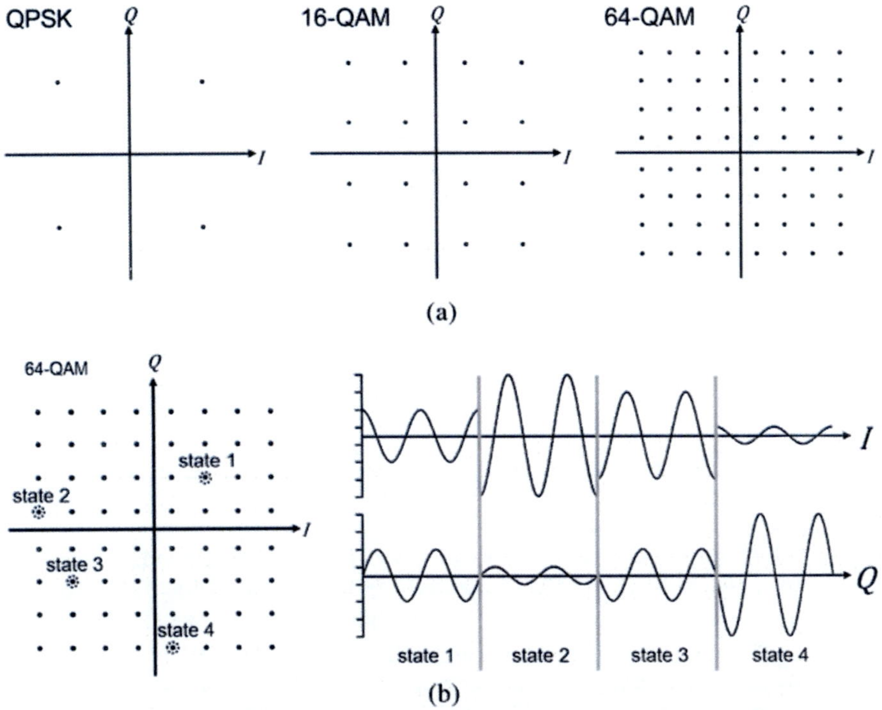

Fig. 6.31 (**a**) Constellation of various quadrature modulation schemes. (**b**) Example of waveform development due to state transitions: state 1 → 2 → 3 → 4, and corresponding waveforms along I- and Q-paths in 64-QAM system

Hence, there can be four possible distinct states for QPSK, while 16QAM, 64QAM, and 256QAM comprises 16, 64, and 256 distinct states, respectively. Of course, there are QAMs with higher orders than shown in Eq. (6.19). Also, there can be intermediate modulation orders for QAM other than those shown above, such as 32QAM or 128QAM, which omit some portion of the states. As indicated by Eq. (6.19), a higher-order modulation scheme involves a larger number of amplitude levels for a given time period, resulting in a higher information density. Equivalently, from the spectrum point of view, this implies a higher spectrum efficiency.

A widely adopted graphical representation of modulations, especially useful for quadrature modulations, is the *constellation*, which maps the location of each state of the modulation in the two-dimensional I - Q space. The name comes from the fact that the points representing the states resemble a collection of stars in the night sky that constitute a constellation. The constellations of various quadrature modulation schemes are shown in Fig. 6.31a. Each point on a constellation represents a signal state as defined in Eq. (6.19). Some of signal waveforms in I and Q paths are presented in Fig. 6.31b with 64QAM case with a few example points, which will help understand the development of waveforms due to state transitions. Every state transition in the time domain corresponds to a jump from one point to another on the

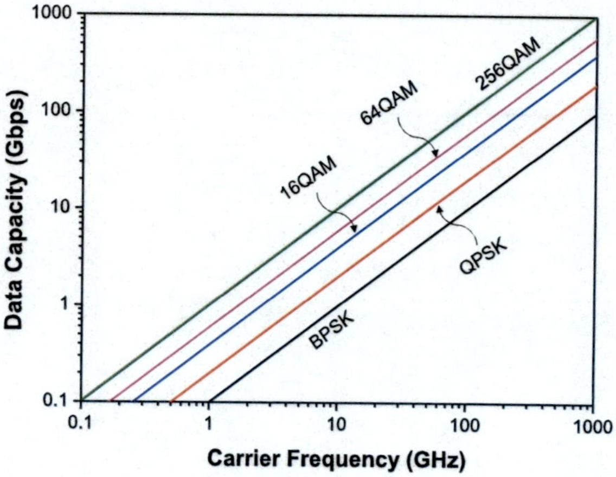

Fig. 6.32 Comparison of the data rate available from various modulation schemes. It was assumed that the data capacity is approximated by $\log_2(M) \cdot B$ with a high SNR, where M is the modulation order and B is the bandwidth (10% fractional bandwidth assumed)

constellation. As briefly mentioned above, a higher spectrum efficiency is expected for higher order QAMs than for lower order QAMs and QPSK. This is well indicated in Fig. 6.32, which compares the data rate of various modulation schemes as a function of carrier frequency. However, as can be easily seen from Fig. 6.31a, the separation between adjacent points gets smaller for higher order QAMs if a similar level of maximum amplitude level is assumed. This means that a higher SNR will be required for higher order modulations in order to maintain a tighter distribution of the actual data points around the ideal location, or, equivalently, a small error vector magnitude (EVM). A higher required SNR will lead to a tighter link budget as indicated by Eq. (6.14), which calls for improved performances of TX (high output power) and RX (low noise) as well as low FSPL (shorter distance) and higher antenna gains. Simply put from the implementation viewpoint, for low-order modulations (including OOK) to achieve a high data rate, the main challenge is to realize a system with a wide bandwidth. On the other hand, for high-order modulations, the challenge is to build a system with a high SNR.

6.3.3 OOK Modulation Systems

The OOK modulation transfers the information by selectively transmitting the carrier signal in the time domain depending on the state of the binary signal. The system that supports this function can be separately reviewed with two parts: transmitter and receiver.

On the *transmitter* side, the simplest approach to realize OOK modulation is to involve a *switch* to turn on and off the continuous wave (CW) radio-frequency (RF) carrier signal streaming out of the signal generator based on the base-band (BB) signal. The most straightforward approach is to employ a dedicated electrical switch inserted in the middle of the RF signal path, which is driven by the BB signal (see Fig. 6.31a). For this purpose, the switch needs to show a low insertion loss and a high isolation, as well as a fast switching speed to handle the data rate that may reach as high as tens of Gbps. This method typically benefits from the high data rate available with fast switching, but is affected by the inevitable loss from the switch. Alternative approaches involve switching of one of the functional circuit components that constitute the OOK transmitter core, such as oscillator or PA. The switching of the oscillator that generates the carrier signal may result in a high on-off isolation with a minimal loss, but it inherently suffers from the slow switching speed arising from the finite oscillation start-up delay. The switching of the PA may be a practical option as no extra switching loss is expected, but it typically shows a limited on-off isolation. After all, the best option needs to be selected for a given application scenario, considering loss, on-off isolation, switching speed, as well as the chip area and power dissipation [76].

Another approach to attain the OOK modulation is to employ a *mixer*, with which multiplication of the carrier signal and the BB signal is performed. Assuming the BB signal is composed of a stream of binary bits of one and zero, the multiplication will provide basically the same function as was obtained with the switching described above. The multiplication in the time domain is equivalent to the convolution in the frequency domain, leading to a frequency shift, or frequency conversion, of the BB signal upward by the carrier frequency. From the physical implementation point of view, this function can be realized with a *heterodyne* system including a mixer and an LO (local oscillator), as depicted in Fig. 6.33b. For the modulation in the transmitter, the RF carrier generated by the signal generator enters the mixer together with a binary BB signal, and the modulated signal streams out at the mixer output. There are issues surrounding the mixer when THz operation is assumed, which will be discussed shortly when heterodyne detection is described. It is noted that switching is in fact one practical way of achieving a mixing action, and thus the both approaches described above are basically based on the same principle.

In either approach for OOK modulation, the modulated signal needs to be amplified by a PA and transmitted into the air by an antenna. A major challenge for the transmitter is the availability of PA operating at the THz band, as mentioned in Sect. 2.5. Depending on the fabrication technology used for the design, amplification may not be achieved at the targeted carrier frequency for the modulation. Or, even when amplifiers are available with a desired gain, the saturation output power and the bandwidth may fall short of the system needs. In this case, as an alternative approach, a frequency multiplier ($\times N$) can be placed at the final stage, driven by a preceding PA that operates at a much lower frequency (by a factor of N), as is described in Fig. 6.33c. For this purpose, a frequency multiplier with a small conversion loss and a high saturation power will be needed. The inclusion of a frequency multiplier will also relax the frequency requirement of the signal generation by the same factor. This frequency-multiplier-last approach works fine for OOK modulation, but will not be suitable for many of the quadrature modulation schemes as will be discussed later. Another challenge is related to the radiation of the modulated signal into the air, which will be handled by an antenna, on-chip or

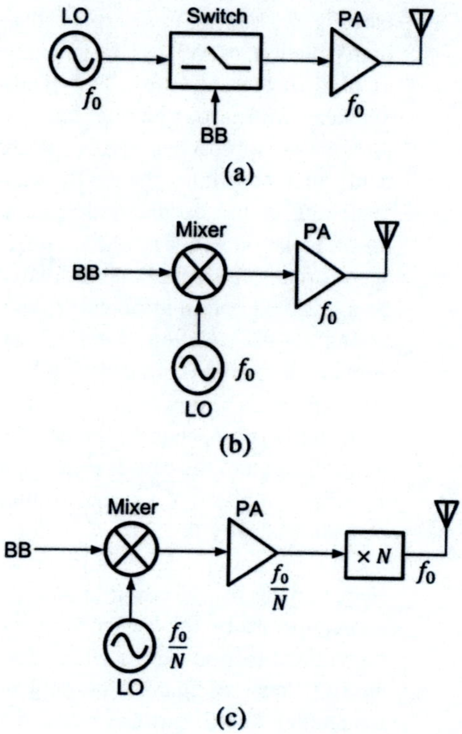

Fig. 6.33 Block diagram of various OOK transmitters: (**a**) Switch-based transmitter with a PA at the last stage, (**b**) Heterodyne transmitter with a PA at the last stage, (**c**) Heterodyne transmitter with a frequency multiplier at the last stage

off-chip. An on-chip antenna will be favored for compact implementation, but the antenna gain will be too low for most THz applications. For such a situation, an extra gain-enhancement scheme will be needed, for which a Si lens is often adopted. A large bandwidth is also a desired aspect of antennas, especially for THz communication systems that benefit from wide channel bandwidths. This need will be more pronounced for low-order modulation schemes such as OOK, as they would prefer a large channel bandwidth to compensate for the low modulation order in achieving the required data rate. Regarding the LO for signal generation, a large tuning range would be preferred if the system supports multiple channels. For this purpose, off-chip signal generation may be considered, which is followed by on-chip frequency multipliers with buffers. For a compact system, however, on-chip LO will be preferred with improved performance.

On the *receiver* side, a couple of topology options can be considered to perform demodulation. As a first option, the binary signals can be recovered (or demodulated) by the *direct detection*. In this case, the envelope of the waveform incoming through an antenna will be detected, as described in Fig. 6.34a. The envelope detector, which performs the detection, works in a similar way as the direct detector used for imaging receiver described in Sect. 6.2: it converts the arriving signal power into a voltage output. The difference is that a much higher temporal resolution is required for the detectors in communication receivers than imaging

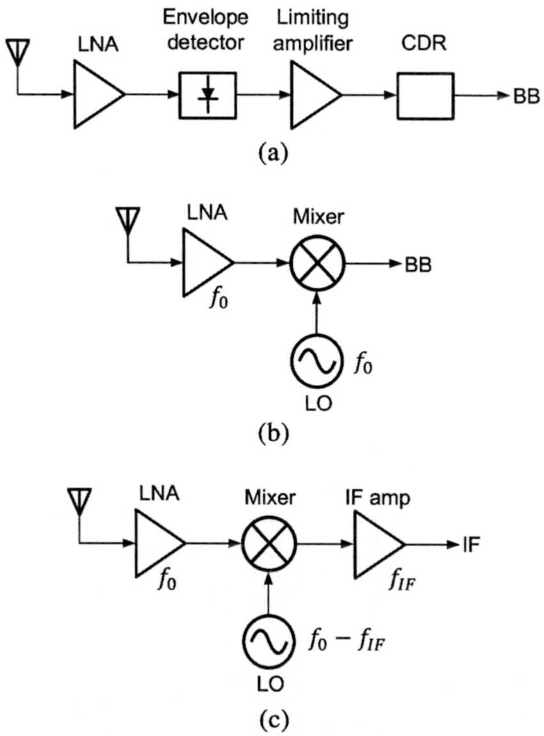

Fig. 6.34 Block diagram of various OOK receivers: (**a**) Receiver with direct detection, (**b**) Homodyne receiver, (**c**) Heterodyne receiver

receivers. As the noise figure of the envelope detector is substantially high, it may degrade the overall noise performance of the receiver chain if located at its front-end. For this reason, a low noise amplifier (LNA) is desired at the front-end of the detector, while it is not always available at the THz band for a similar challenge faced by PAs at the transmitter. The envelope detector is often followed by a limiting amplifier, which is a highly nonlinear amplifier with a high gain, so that the signal peaks (at top and bottom) of the waveform are truncated, or "limited." With the limited waveform, usually obtained with intended signal saturation, a square-shaped waveform, which is a replica of the original BB signal, can be recovered. The recovered signal streams into the BB block, optionally through an on-chip clock and data recovery (CDR) circuitry for improved data quality and clock extraction.

As an alternative approach for receivers, demodulation can be carried out by *multiplying* the incoming signal with the LO signal with a mixer. Assuming the same frequency for the LO and incoming signals, this process can be easily understood based on a basic trigonometric identity:

$$y(t) = A_C(t) \cos \omega_C t \cdot \cos \omega_C t = \frac{1}{2} A_C(t)(1 + \cos 2\omega_C t), \qquad (6.20)$$

where $A_c(t)\cos\omega_c t$ is the modulated incoming signal, $\cos\omega_c t$ is the LO signal, and $y(t)$ is the demodulated output signal. It is clear from Eq. (6.20) that the BB signal, $A_C(t)$, can be recovered with a linear factor if the high frequency component ($2\omega_c$) is filtered out with a low pass filter. This is called the *homodyne* detection, as the carrier frequency and the LO frequency are identical (*homo*geneous). This scheme is also referred to as the zero-IF detection or direct down-conversion. A typical homodyne receiver is shown in Fig. 6.34b. One caveat in Eq. (6.20), however, is that it was implicitly assumed that the LO signal has the same phase as the incoming signal. In general cases, however, there should be a finite phase difference between the received signal and the LO signal, which leads to a more generic expression of $\cos(\omega_c t + \phi)$ for LO signal where ϕ is the phase difference. This non-zero phase difference will inevitably lead to a reduced recovery signal level, the worst case occurring when $\phi = \pm\pi/2$ that leads to a complete elimination of the signal at DC. To mitigate this phase issue, a CDR or I-Q mixer may be needed.

For a slightly altered approach, the LO frequency may be set slightly off the carrier frequency. In this case, the down-conversion does not directly result in the BB signal, but in the BB signal "weakly" modulated by an intermediate frequency (IF) that equals the difference between the carrier and LO frequencies. The process will be governed by expression as follows:

$$\begin{aligned} y(t) &= A_C(t)\cos\omega_{C1}t\cdot\cos\omega_{C2}t \\ &= \frac{1}{2}A_C(t)(\cos(\omega_{C1}-\omega_{C2})t+\cos(\omega_{C1}+\omega_{C2})t), \end{aligned} \tag{6.21}$$

where ω_{C1} and ω_{C2} are the carrier frequency and the LO frequency, respectively. One can easily notice that Eq. (6.20) is a special case of Eq. (6.21) with $\omega_{C1} = \omega_{C2}$. As was the case of homodyne detection, a low-pass filtering will leave only the desired component, or the IF signal with $\omega_{IF} = |\omega_{C1} - \omega_{C2}|$. An additional mixing will be required afterwards to arrive at the BB signal This scheme is called the *heterodyne* detection (see Fig. 6.34c), as the carrier frequency and the LO frequency are different (*hetero*geneous). Occasionally, however, in a broader sense, the heterodyne detection refers to both types of the detection described above. Note that it was again assumed in Eq. (6.21) that there was no phase difference between the LO and RF signals. A finite phase difference will cause a similar issue as mentioned above, which requires the same solutions.

With both homodyne and heterodyne detection schemes based on frequency down-conversion, a much higher sensitivity can be expected when compared with the direct detection case. This is because a lower noise level is typically exhibited by mixers than envelope detectors at the frequency range. The employment of an LNA at the front-end will make the difference less pronounced, but, as mentioned earlier, the availability of LNAs is not guaranteed for THz applications due to the demanding operation frequency. Also affected by the high operation frequency in the THz receiver is the LO that needs to drive the mixer at the carrier frequency. One technique that relaxes the frequency requirement for the LO is to employ a subharmonic mixer, instead of a fundamental mixer, which will reduce the required LO frequency by a factor of the harmonic number adopted for the subharmonic mixer. It is true that mixer performances such as the conversion gain will be compromised with a subharmonic mixer [77], but it is still an attractive option when the generation of the carrier frequency is not available at the receiver side.

Yet another way to reduce the LO frequency is to insert a frequency multiplier between the LO and the mixer. This extra component will crowd the receiver circuit, but mixer performance degradation may be avoided if the frequency multiplier delivers a sufficiently large output power. The adoption of the frequency multiplier also allows off-chip LOs. An off-chip LO, often a commercial signal generator, would potentially provide a high output power along with a low phase noise. However, on-chip LOs will be still favored if a compact chip solution is desired, although an on-chip LO may require a signal-locking, on-chip or off-chip, to improve the signal integrity that is crucial for communication applications. One additional advantage to be mentioned for the heterodyne systems is that they can possibly be operated as a coherent mode, in which case the phase information of the signal can be extracted. Although OOK modulation does not necessarily require phase information, a heterodyne-based OOK receiver can be easily extended for quadrature modulation [78], for which the phase information is critical for information recovery.

There have been several notable reports on THz communication systems based on OOK (or ASK) modulation scheme [77–80]. In one of the earliest such reports [78], heterodyne TX and RX chips operating at 220 GHz were implemented based on a 50-nm GaAs mHEMT technology (see Fig. 6.35). The TX chip is composed of a mixer and an amplifier, together with a frequency doubler that up-converts the sub-harmonic LO signal (110 GHz) injected from an external signal source. The single-ended fundamental-mode mixer modulates the carrier signal with the IF signal that carries the BB data. The modulated carrier signal at 220 GHz is amplified before exiting the chip into a waveguide package, which leads to an off-chip horn antenna used for radiation. The amplifier is in fact a low-noise amplifier that is also used for the RX, which resulted in 1.4-dBm maximum TX output power with IF and LO power of 5 dBm and 10 dBm, respectively. The RX is composed of the identical set of circuits included in the TX, say, LNA, mixer, and frequency doubler. The noise figure is estimated as low as 6.5–7.5 dB at 220 GHz, which reflects the excellent low-noise performance of the HEMT technology. Based on the TX and RX enclosed in waveguide packages, OOK link experiments were conducted, with a pair of lenses placed along the air channel for beam collimation. With the link setup, 10-Gbps data transmission was successfully achieved across a 2-m distance with a bit error rate (BER) of 3.2×10^{-10} and 1.6×10^{-9} for pseudo random binary sequence (PRBS) of $2^{15} - 1$ and $2^{31} - 1$, respectively. A similar experiment was repeated at a shorter distance of 50 cm, resulting in a higher data rate of 25 Gbps. It is noteworthy that the identical chips were also used for QAM modulation experiments.

In another effort based on a 250-nm InP HBT technology, a 300-GHz heterodyne chip set was developed that contains both TX and RX [77]. In a rare approach, the

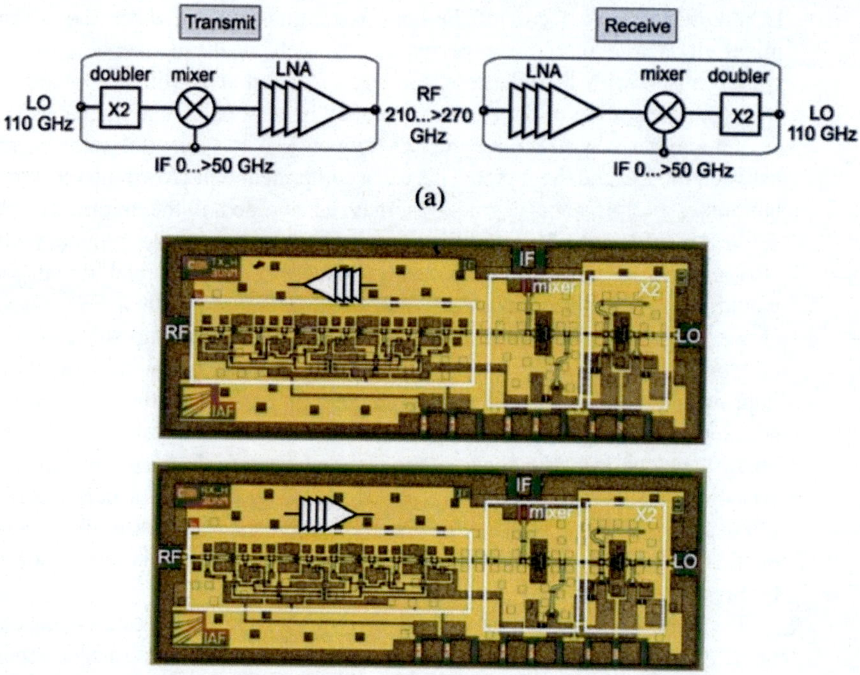

Fig. 6.35 (**a**) Block diagram of a 220-GHz transmitter and receiver chip sets based on a 50-nm GaAs mHEMT technology. (**b**) Photo of the fabricated chips. (© 2011 IEEE [78])

chip includes an on-chip local oscillator, which is shared for TX and RX to drive up-conversion and down-conversion mixers, respectively (See Fig. 6.36). With the oscillator operating at the carrier frequency of 300 GHz [29], the mixers can be realized in a fundamental mode without the need for a frequency multiplier for subharmonic mixing. No external signal generator is required, either. An RF amplifier operating at 300 GHz, a variant of the one reported in [81], enables the amplification of the carrier signal at both TX and RX. The transceiver circuit is all based on the differential configuration, which requires the inclusion of baluns at the RF and LO interfaces for connection with single-ended schemes. The TX exhibited a conversion gain up to 25 dB with a 3-dB bandwidth of 18 GHz, with an output power of −2.3 dBm after the output balun. At the RX, a conversion gain of 26 dB was attained with a 3-dB bandwidth of 20 GHz. The noise figure, which was measured with the *Y*-factor method based on a setup with liquid-nitrogen (LN), showed a range of 12.0–16.3 dB. While not demonstrated in the paper, the chipset is capable of OOK data transmission. There have been OOK transmitters and receivers reported based on Si technologies as well operating beyond 200 GHz [82, 83].

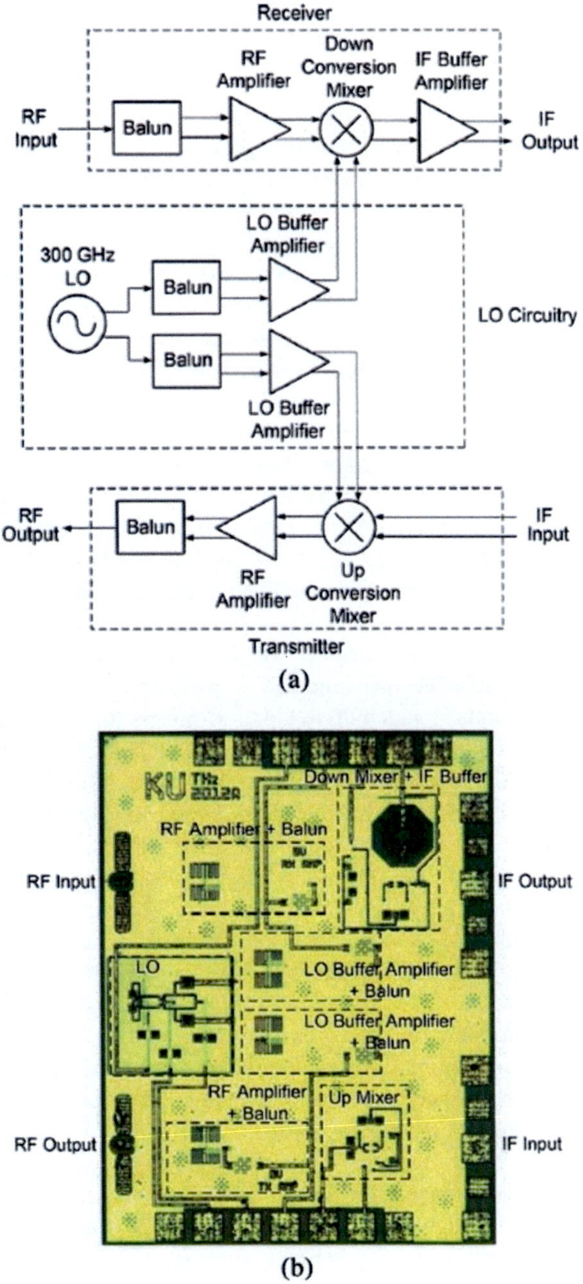

Fig. 6.36 (**a**) Block diagram of a 300-GHz transceiver based on a 250-nm InP HBT technology. (**b**) Photo of the fabricated chip. (© 2015 IEEE [77])

6.3.4 Quadrature Modulation Systems

In a sense, transmitters and receivers for quadrature modulations resemble the heterodyne systems employed for OOK modulation described above. However, there is one critical difference. For quadrature modulations, the baseband signal is split into two branches, I and Q, and the (de)modulations are carried out separately (yet simultaneously) on both I and Q channels. This mandates the inclusion of I-Q mixers, which in turn requires the generation of quadrature LO signals. A simple picture of quadrature modulations is as follows: At the transmitter side, the modulated signals on the two channels will be combined at the output and transmitted through an antenna, optionally power-boosted with a PA. Upon arrival at the receiver, the signals will be split into two branches again, optionally after an LNA, and demodulated separately with quadrature LO signals. While there are a set of different quadrature modulation schemes, ranging from QPSK to various orders of QAMs, the implemented quadrature transmitters and receivers will in principle work on any type of the quadrature schemes, although the required performance may differ. The details of this quadrature process are described below.

The block diagram of a typical quadrature *transmitter* is shown in Fig. 6.37a. The baseline LO signal is generated either on-chip or off-chip and passes through a frequency multiplier (optional) and a buffer amplifier for frequency- and power-boost. The generated LO signal at the carrier frequency is split into two quadrature signals, which drive a pair of mixers that modulate the I and Q baseband data. The modulated RF signals from each channel is combined and transmitted through an antenna after a final power boost with a PA. However, a PA at the carrier frequency may not be available if the carrier frequency is too high, as was the case for OOK transmitters. In the absence of a PA, an alternative approach is to perform the modulation at a lower LO frequency followed by a low-frequency amplification, then apply a final frequency up-conversion to reach the carrier frequency for transmission. In this case, the final frequency up-conversion can be achieved with either frequency multiplication or mixing, while the former method involves an issue and needs to be carefully approached as will be shortly discussed. Another approach to boost the power in the absence of a power amplifier is to make use of a power combining technique, in which the modulation is carried out in multiple paths (for both I and Q channels), then power-boosted with power combining over the multiple paths before transmission.

There are several aspects worthy of discussion related to the transmitter. The requirements for the LO is similar to those for OOK systems as mentioned earlier, or even tighter. Signal integrity in LO is of upmost importance, for which a PLL or any form of locking is highly desired. Also, a wide tuning capability is certainly a plus for the flexibility in the channel selection, especially when the channels allocated by the standard are spreading over a wide range. When a high-order frequency multiplication is performed to reach the carrier frequency, the suppression of the unwanted harmonics is important to avoid modulated replicas around the desired main signal [84]. For the quadrature signal generation, a quadrature coupler, such as

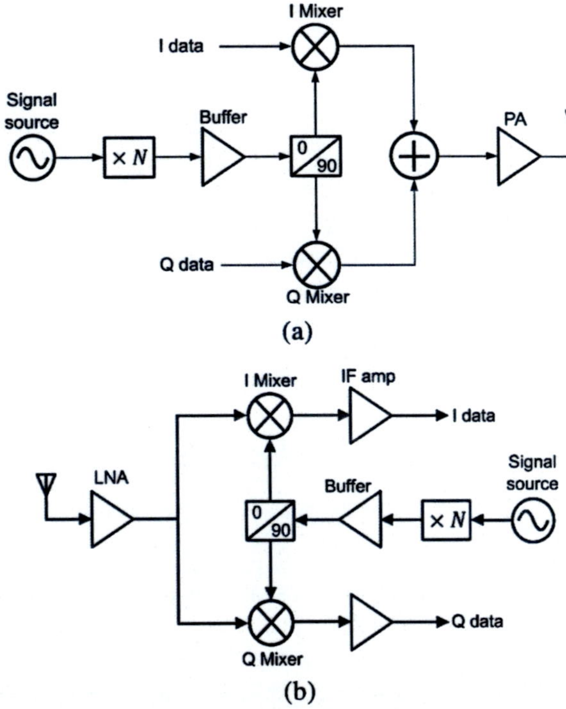

Fig. 6.37 Schematic of a typical THz quadrature system: (**a**) Transmitter, (**b**) Receiver

the branch-line coupler described in Sect. 2.5.2, can be used [85], or simply a $\lambda/4$ line section can be inserted in one of the two signal paths. Employing a quadrature VCO (QVCO) is another option [86, 87], although not popular for THz applications yet partly due to the complexity in locking for stable oscillation. As for the power combiner that merges the signals from I and Q channels, the various power combiners introduced in Sect. 2.5.2 can be considered, including rat race couplers modified for a large number of inputs [88–90] or smaller foot print [91]. The requirements for the mixers will be similar to those in OOK transmitters discussed earlier, and so will those for antennas. From the overall system point of view, maintaining balanced characteristics between the I and Q channels is critical for quadrature transmitters (and receivers), calling for extra efforts in circuit design compared to the OOK case.

As mentioned above, when PA is not available at the carrier frequency, one option is to perform modulation at a lower frequency, followed by a frequency up-conversion. Probably the simplest way for this frequency up-conversion process is to apply a frequency multiplier. However, this frequency-multiplier-last scheme works fine for OOK transmitters (see Fig. 6.33c), but not necessarily for quadrature transmitters. This scheme may corrupt the constellation of quadrature modulations with misarranged signal points. This can be easily shown with a simple math. Consider the modulated signal (before frequency up-conversion) in the form of

Eq. (6.18), which can be shown by a simpler expression without loss of generality as follows:

$$x(t) = \alpha_1 \cos \omega_0 t + \alpha_2 \sin \omega_0 t, \tag{6.22}$$

where ω_0 is used instead of ω_C to distinguish from the final carrier frequency. The condition required for this quadrature signal to maintain the same (relative) position in the constellation after frequency up-conversion is that the I signal and Q signal should scale by the same factor. In other words, the magnitude ratio of the I and Q signals should remain the same after the frequency multiplication. Now, let us consider the case where the frequency up-conversion is performed with a frequency doubler based on nonlinearity. Then, the output signal will become:

$$\begin{aligned} x^2(t) &= (\alpha_1 \cos \omega_0 t + \alpha_2 \sin \omega_0 t)^2 \\ &= \left(\frac{\alpha_1^2 - \alpha_2^2}{2}\right) \cos 2\omega_0 t + (\alpha_1 \alpha_2) \sin 2\omega_0 t + \left(\frac{\alpha_1^2 + \alpha_2^2}{2}\right). \end{aligned} \tag{6.23}$$

Hence, at the up-converted frequency of $2\omega_0$, the magnitude ratio between I and Q signals will be $2\alpha_1\alpha_2/(\alpha_1^2 - \alpha_2^2)$, altered from the original value of α_2/α_1. As a result, the constellation will lose its original configuration and get corrupted. Although a frequency doubler was assumed in the example above, it will be true for frequency triplers and other higher-order frequency multipliers. Only exceptions will be odd-ordered frequency multipliers (including triplers) applied to QPSK, in which the constellation shape will remain uncorrupted. Although the symbol points are shuffled for this case, too, each constellation point will be occupied with a single state even after the multiplication, leading to preserved modulation information.

As a way to avoid the issue of constellation corruption, one may consider applying frequency multiplication separately on I and Q channels before they are combined. In this case, however, the performance (such as conversion gain) imbalance between the two multipliers may cause asymmetry between the two channels, in addition to the increased power dissipation and area, especially when multiplication is preceded by amplification. Fortunately, we have an alternative frequency up-conversion option that works fine with quadrature modulation. *Mixing* will preserve the constellation even after the frequency up-conversion, which can be shown as follows. Let us assume the signal modulated with ω_0 is up-converted by a mixer with an LO signal at the same frequency, $\alpha_0 \cos \omega_0 t$, which will lead to:

$$\begin{aligned} x(t) \cdot \alpha_0 \cos \omega_0 t &= (\alpha_1 \cos \omega_0 t + \alpha_2 \sin \omega_0 t)\alpha_0 \cos \omega_0 t \\ &= \left(\frac{\alpha_0 \alpha_1}{2}\right) \cos 2\omega_0 t + \left(\frac{\alpha_0 \alpha_2}{2}\right) \sin 2\omega_0 t + \frac{\alpha_0 \alpha_1}{2}. \end{aligned} \tag{6.24}$$

Obviously, the ratio of α_2/α_1 between I and Q signals is preserved at the up-converted frequency of $2\omega_0$. This will be true with LO frequencies different than ω_0 as well. This indicates that the frequency up-conversion with mixing is a feasible option for quadrature modulation transmitters, when modulation at a lower frequency is desired.

The block diagram of a typical quadrature modulation *receiver* is shown in Fig. 6.37b. The received signal through the antenna is first amplified with an LNA, if available at the carrier frequency, then split into I and Q channels before reaching a pair of I-Q mixers. The LO signal, which is typically frequency- and power- boosted by a frequency multiplier and a buffer, respectively, as was the case for the transmitters, drives the mixers for the down-conversion of the carrier frequency. For the quadrature split of the LO signal, schemes similar to those used in the transmitter can be employed. As was the case for OOK receiver, the down-conversion can be made directly down to BB signal (homodyne) or down to IF (heterodyne), the latter requiring an extra down-conversion step to reach BB. The down-converted signals can optionally pass through IF amplifiers to raise the signal level.

Let us briefly turn our attention to the bandwidth. As was shown in Fig. 6.29, the channel BW may stretch well beyond 10 GHz depending on the allocated channels. A large channel BW would be favored in terms of the raised data rate, but it imposes challenges in the design of the receiver circuit components. To handle the large BW, the bandwidth of the IF amp, as well as that of the mixers, should be large enough, which is not a straightforward design task. Also, if a uniform performance across the BW is not achieved, the system performance will be significantly degraded, as it may possibly lead to a signal leak between I and Q channels. Besides, SNR at the receiver will be degraded with a large BW if noise level at the front-end is not kept low enough. All these challenges related to a large BW may boost the favor for the higher-order modulations that will need a narrower BW to achieve required data rates. The higher-order modulations, however, will suffer from their own issues, such as the power back-off needed in PA operation in the transmitter, which may reach as high as several dB, and the higher SNR required to achieve a given target bit error rate (BER) as mentioned earlier.

There have been a growing number of transmitters and receivers operating beyond 200 GHz based on quadrature modulation scheme with various process technologies. In [90], a Si CMOS-based transceiver operating around 260 GHz was reported that works on 16QAM to achieve an 80-Gbps data transmission (see Fig. 6.38). The circuit, which covers the frequency range of 250–279 GHz that contains the channels 49, 50, and 66 shown in Fig. 6.29, is designed so that it can be operated as both transmitter and receiver which share the LO function block. A 44-GHz signal externally generated is tripled by a frequency tripler, which drives the I-Q mixers for the quadrature modulation of the BB signal. The modulated signal is frequency-doubled by a square mixer to reach the carrier frequency, for which a novel frequency multiplication scheme was employed that exploits the LO signal leakage out of the mixer [88, 89]. Since a PA is not readily available with the Si CMOS technology at the carrier frequency, a multi-way power-combining over multiple signal paths was performed with a double-rat-race coupler, which resulted in a transmitter output power of −1.6 dBm with a DC power dissipation of 890 mW. When operated in the receiver mode, a positive conversion gain and a mean single-

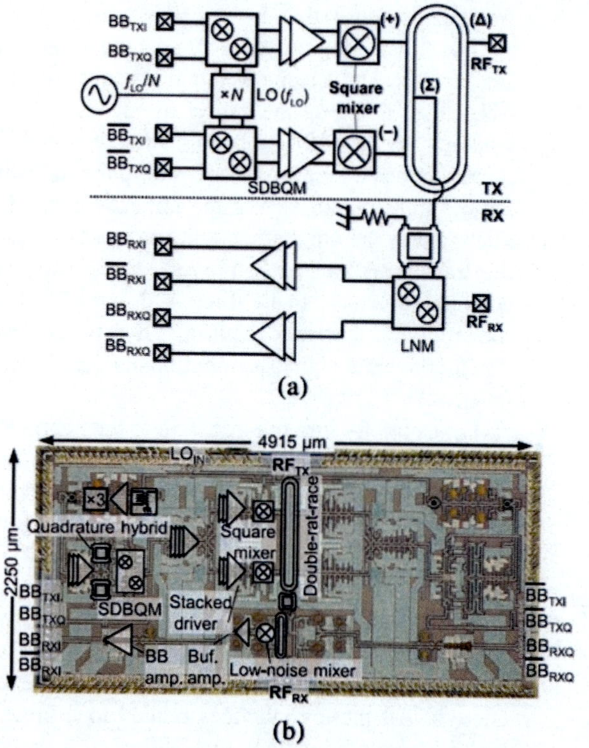

Fig. 6.38 (**a**) Block diagram of a 260-GHz quadrature transmitter and receiver chip sets based on a 40-nm Si CMOS technology. (**b**) Photo of the fabricated chips. (© 2019 IEEE [85])

side band (SSB) NF of 22.9 dB were attained in the absence of LNA. With the developed TX and RX, a data rate of 80 Gbps was obtained with a 16QAM link at a distance of 3 cm, while operations with QPSK and 32QAM were also reported.

In [92], a transmitter and a receiver operating at the 220–260 GHz frequency range was developed based on a SiGe HBT technology, which features a PA operating at the carrier frequency for both TX and RX, an indication of the superior RF performance supported by the SiGe HBT technology. In TX, an externally injected signal is converted into the 236 GHz carrier frequency with a $\times 16$ frequency multiplier followed by a buffer amplifier, which drives the I-Q mixers to modulate the BB signal after a split by a quadrature coupler. The modulated I and Q signals are combined and radiated by an on-chip ring antenna after power-boosted by a power amplifier. On the RX side, a reverse process is carried out for demodulation. The received signal is first amplified by an amplifier (the same kind as in the transmitter) then split into I and Q channels. The quadrature LO signal, generated by the same procedure as in the TX, drives the I-Q mixers to demodulate the signals on the I and Q channels. The TX showed a 3-dB RF bandwidth of 28 GHz and an output power (P_{sat}) of 8.3 dBm, while the RX exhibited a conversion gain of 24 dB, which were greatly helped by the power amplifier employed as previously mentioned. A maximum data rate of 90 Gbps was achieved over a 1-m data link with 32QAM. A similar structure was reported later on [84], in which the front-end amplifier was

Table 6.1 The conversion between data rate units

Modulation	Order	N	baud	bps
OOK/BPSK	$2 = 2^1$	1	1 Gbaud	1 Gbps
QPSK	$4 = 2^2$	2	1 Gbaud	2 Gbps
8QAM	$8 = 2^3$	3	1 Gbaud	3 Gbps
16QAM	$16 = 2^4$	4	1 Gbaud	4 Gbps
32QAM	$32 = 2^5$	5	1 Gbaud	5 Gbps
64QAM	$64 = 2^6$	6	1 Gbaud	6 Gbps
128QAM	$128 = 2^7$	7	1 Gbaud	7 Gbps
256QAM	$256 = 2^8$	8	1 Gbaud	8 Gbps

removed from the RX to avoid the imbalance between the upper-side band (USB) and lower-side band (LSB) that may arise from the asymmetric gain profile. Without the RX amplifier, in fact, a higher data rate (100 Gbps) was achieved despite the degraded receiver noise figure and conversion gain, which was helped by the symmetric response. There have been other results that have successfully demonstrated high-speed data links based on quadrature-modulation transmitters/receivers beyond 200 GHz in Si CMOS [93–95], SiGe HBTs [96], GaAs mHEMTs [70, 78], as well as in InP HBTs [91]. While the overview in this section has been focused on the electronics-based approach, similar systems that rely on carrier signals generated by optical means have also been reported exhibiting excellent data link performances [97–99].

As a final note, a brief review on the units for the data rate can be made here. The readers may have noticed in some reports that the data rate is quoted in terms of "baud" instead of (or together with) bit per second (bps). Baud rate is basically the rate of modulation (or the change rate of states), indicating the number of *symbols* transferred per second, different from bps that represents the number of *bits* transferred per second. Hence, the relation between baud and bps is simple but tricky as it depends on the modulation order. This is because the number of bits included in a single symbol varies over the different modulation orders. Here is a simple conversion rule between the two units: To obtain the data rate in bps from a given baud rate, simply multiply a factor of N to the baud rate, where 2^N is the modulation order of the modulation scheme. For example, 1 Gbaud corresponds to 2 Gbps with QPSK, for which the modulation order is 4 with $N = 2$. More examples are presented in Table 6.1 with various modulation schemes.

6.4 THz Radars

6.4.1 Overview

Radar (RAdio Detection And Ranging) is a system that determines the range (distance) and velocity of a target object based on the detection of the reflected radio waves from the object. The modern development of radars has allowed imaging of the target objects by radar operation as well. The history of radars can be traced back to the early twentieth century when they attempted to detect the presence of a target object based on the reflection of radio waves, but the major development of radar systems was made in the time period around World War II, which clearly indicates the early needs driven by military purposes. This legacy continues till today, as the main application area of radars is still heavily linked to military applications. However, the commercial application of radars is growing fast. The weather forecast of today is heavily dependent on radars, and the airport traffic control radars are a critical part in modern airport operation. Major airplanes and ships are equipped with radars, and speed guns based on radars are measuring car speeds and ball speeds on the roads and ballparks. More importantly, the market for automotive radars is growing rapidly, as indicated by recent mid-size cars rolling out of the factories with radars installed, which used to happen only for limited luxury cars several years ago. There are growing efforts to expand the application envelope of radars into various commercial sectors including industrial, medical, and security-related areas. Along with this trend, also increasing is the interest in radars operating in extended frequencies, including THz radars.

There are many ways to categorize radars depending on various aspects of radar operation. Most of all, they can be categorized into pulse radars and CW radars, depending on the waveform of the radar signals. Pulse radars can be divided into radars with short pulses and radars with pulse compression, while CW radars can be grouped into unmodulated and modulated radars. They can be further sub-divided based on the details of the waveform and modulation. We can define other types of radars based on different features as well. There can be military and commercial radars, as well as ground-based, ship-based, air-borne, and space-borne radars. They can be grouped by operation frequency and the extent of the detection ranges, too. There are radars with special features, such as imaging radars and synthetic aperture radars (SARs). In this section, the discussion will be made in terms of pulse radars and CW radars, which will be covered in the following two subsections.

Before embarking on the subsections for a separate treatment of pulse and CW radars, it would be helpful to review the common basic principles of radars. A key equation that governs the radar operation is the radar range equation. A brief derivation to arrive at the equation will provide a useful snapshot over the radar operation. Let us first consider a configuration depicted in Fig. 6.39. Electromagnetic

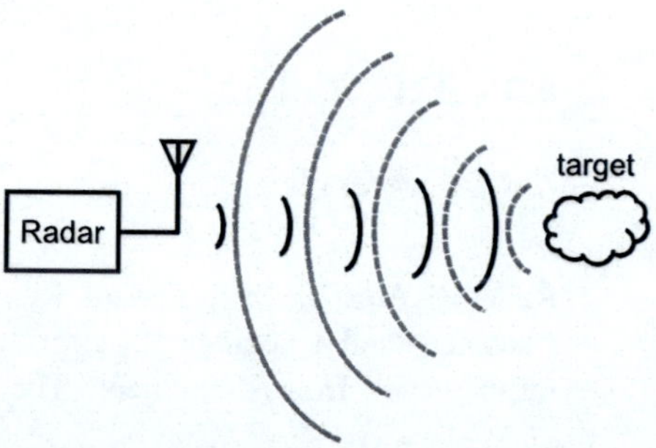

Fig. 6.39 Conceptual sketch describing the transmitted signal from the radar system and reflected echo signal from the target object

waves emitted from the antenna of a radar arrive at the target and are reflected back to the antenna and then detected. The incident power density at the target can be obtained as Eq. (4.55), which is repeated here for convenience:

$$S_{\text{target}} = \frac{G_t P_t}{4\pi R^2}, \tag{6.25}$$

where G_t is the transmitter antenna gain at the radar, P_t is the power emitted from the antenna, and R is the range or distance between the antenna and the target. The target will have a finite dimension, and the intercepted power by the target will be obviously given by the incident power density multiplied by the effective target area, or:

$$P_{\text{target}} = \sigma S_{\text{target}} = \sigma \frac{G_t P_t}{4\pi R^2}, \tag{6.26}$$

where σ is called the radar cross section (RCS), which can be roughly understood as the effective surface area of the target, but is better defined as a measure of how effectively the target reflects the waves back to the radar in that direction. The incident power will be reflected and scattered by the target surface upon its arrival. In fact, this scattering event can be considered as a re-radiation of the reflected power by the target. According to this view, the illuminated target serves as an antenna, which is assumed to be isotropic. Hence, the scattered power will propagate toward all the directions, only a portion of which will travel back to the radar. The power density arriving at the receiver antenna of the radar is given by:

$$S_{\text{receiver}} = \frac{P_{\text{target}}}{4\pi R^2} = \frac{\sigma G_t P_t}{\left(4\pi R^2\right)^2}. \tag{6.27}$$

Hence, the power received by the antenna can be expressed in terms of the effective aperture A_e such that:

$$P_{\text{receiver}} = A_e S_{\text{receiver}} = A_e \frac{\sigma G_t P_t}{\left(4\pi R^2\right)^2} = \frac{\lambda^2 \sigma G_r G_t}{(4\pi)^3 R^4} P_t, \tag{6.28}$$

where the relation for the effective antenna aperture of $A_e = G_r \lambda^2 / 4\pi$ is used (Eq. 4.54). Here, G_r is the gain of the receiver antenna at the radar and λ is, of course, the wavelength of the radar signal. Equation (6.28) is a key relation between the transmitted and received powers at the radar after the round trip to the target. It shows that the returned power is inversely proportional to R^4, rapidly decreasing with the increasing distance. This can be compared to the point-to-point FSPL that is inversely proportional to R^2.

Now, let us obtain the output SNR of the radar system. Note that the power obtained in Eq. (6.28) is the *input* power that enters the radar system. From the definition of the noise factor F as given in Eq. (3.10), the output SNR can be expressed as:

$$\mathrm{SNR_{out}} = \frac{1}{F}\mathrm{SNR_{in}} = \frac{1}{F}\frac{S_{\mathrm{in}}}{N_{\mathrm{in}}}. \tag{6.29}$$

Here, S_{in} is the input signal power and simply given by P_{receiver} obtained in Eq. (6.28). N_{in} is the input noise power and can be expressed as follows:

$$N_{\mathrm{in}} = kT_{\mathrm{o}}B, \tag{6.30}$$

where k is Boltzmann constant, T_{o} is the reference temperature (290 K), and B is the system bandwidth. Hence, SNR at the radar output ($\mathrm{SNR_{out}}$) is obtained by substituting Eqs. (6.28) and (6.30) into Eq. (6.29), leading to:

$$\mathrm{SNR_{out}} = \frac{\lambda^2 \sigma G_{\mathrm{r}} G_{\mathrm{t}} P_{\mathrm{t}}}{(4\pi)^3 kT_0 BFR^4}. \tag{6.31}$$

If needed, the loss factor L can be included in denominator of Eq. (6.31), which is omitted here. Now, the maximum detectable range is given as the distance that corresponds to the minimum output SNR required for the radar systems [100], which becomes,

$$R_{\max} = \left(\frac{\lambda^2 \sigma G_{\mathrm{r}} G_{\mathrm{t}} P_{\mathrm{t}}}{(4\pi)^3 kT_0 BF \cdot \mathrm{SNR_{out,\,min}}}\right)^{1/4}. \tag{6.32}$$

This relation is widely referred to as the radar range equation. A typical number for $\mathrm{SNR_{out,min}}$ in radar systems is 13 dB [101], while it can be as large as 20 dB [102].

Let us examine the maximum range given in Eq. (6.32) from the THz application point of view. One parameter that is directly affected by the operation frequency is, of course, the wavelength λ. The maximum detectible range decreases with decreasing wavelength, or increasing frequency, which is not a favored trend for THz application. Other parameters affected by the frequency is the transmission power P_{t} at the transmitter and the noise factor F at the receiver. Both of them generally degrade with increasing frequency, which makes the range of THz radars further limited. In fact, for these reasons, the operation range of typical radars has been mainly determined by the operation frequency. Starting from the long-range radars (hundreds of km) based on L-band (1–2 GHz) and the medium-range radars (tens of km) based on S-band (2–4 GHz), the operation range generally reduces with increasing frequency. From this viewpoint, we can expect that the THz radars will best fit for the very short-range applications, as was the case for THz communication. On a promising side, though, the range and image resolutions will be improved with increasing frequency, in a similar way for THz imaging, which would be the main motivation for THz radars. The issue of resolution in radars will be discussed later in this section.

The topic of radar comprises not only the hardware but also lots of signal processing issues. However, the discussion below will be mostly confined to the hardware side, with a main focus on its application based on the THz band.

6.4.2 Pulse Radars

In pulse radars, a train of *pulses* are transmitted from the antenna and the reflected signals (called "echo") are received, from which the range and the velocity of the target is determined. In typical cases for pulse radars, the same antenna is used for both emission and detection, time-controlled by a duplexer, which is a main feature of the monostatic radars. Let us first obtain the *range* of the target that can be obtained with pulse radars. We can start with the simplest case, where a single pulse (with a pulse width of τ) is transmitted and then received after reflection as is shown in Fig. 6.40a. Then, the range is given by:

$$R = \frac{c\Delta t}{2}, \tag{6.33}$$

where Δt is the time delay of the pulse and c is the speed of light. Hence, this is basically a time-of-flight (ToF) technique as mentioned earlier in this chapter. Now consider a more practical case with a train of multiple pulses transmitted and received, as described by Fig. 6.40b. A couple of parameters can be defined for this multiple-pulse situation. The pulse repetition interval (PRI), denoted as T in the figure, indicates the time interval between the pulses. The pulse repetition frequency (PRF), or f_r, is simply the inverse of PRI, representing the frequency of the pulse train. The readers, however, should be aware that a pulse is basically a switched CW, which has its own frequency that needs to be distinguished from PRF. In the given

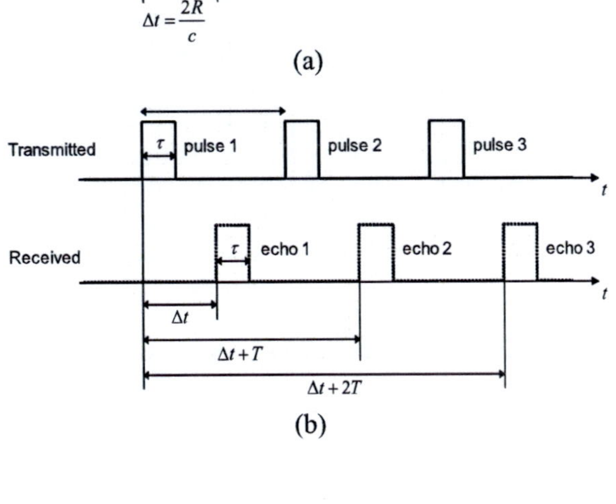

Fig. 6.40 Transmitted and received pulses of a pulse radar: (**a**) Single pulse, (**b**) A train of multiple pulses

configuration with a train of pulses, the range appears to be determined by Eq. (6.33) again. However, this will be true only when the echo 1 is the response of pulse 1. There is a possibility that echo 2 is the response of pulse 1 for a longer range, in which case R is given as $c(\Delta t + T)/2$. Likewise, there is a chance that echo 3 is the response of pulse 1, leading to $R = c(\Delta t + 2\,T)/2$. Hence, there exists a range ambiguity of $cT/2$, which can be avoided by allowing enough time between adjacent pulses or increasing PRI.

Next, the *range resolution* is discussed. For this, consider the case given in Fig. 6.41, where two targets are located at different locations, R_1 and R_2, along the direction of signal transmission. If the separation between the two targets is not large enough (see Fig. 6.41a), the two echo signals from target 1 and target 2 will partially overlap and cannot be separated, depending on the pulse width. For the two targets to be resolved (see Fig. 6.41b), the two echo signals need to be separated, which is satisfied by the following condition:

$$\frac{2R_2}{c} - \frac{2R_1}{c} = \frac{2}{c}(R_2 - R_1) > \tau. \tag{6.34}$$

If we define the range resolution ΔR as the minimum allowed separation between target 1 and 2, then from Eq. (6.34) we have:

$$\Delta R = \frac{c\tau}{2} = \frac{c}{2B}, \tag{6.35}$$

where B is the radar bandwidth which is defined as the inverse of the pulse width τ (Note that B is not the RF bandwidth of the radar receiver included in the radar range equation, Eq. (6.32)). Hence, to improve the resolution, τ needs to be small. However, a small τ will lead to a reduced received power, degrading SNR of the radar system. Therefore, there arises a trade-off between the resolution and SNR. Fortunately, there is a technical solution that resolves this trade-off: modulating the frequency or the phase of the pulses. Officially known as the *pulse compression*, this technique will allow improved resolution together with a high SNR when combined with the matched filtering. The details of the pulse compression and the matched filtering can be found in standard radar textbooks (e.g., [100]) and will not be discussed here further. Just one typical example of the pulse modulation will be introduced with the help of Fig. 6.42. Called the linear frequency modulation (LFM), the frequency of the CW wave within one pulse cycle is linearly increased. With this frequency sweep from f_1 to f_2, the pulse width will be effectively reduced by a factor of $(f_2 - f_1)\tau$ after matched filtering, improving the range resolution while maintaining the receiver signal power. Apparently, a larger reduction factor can be obtained with higher frequencies (because $f_2 - f_1$ will be larger), which is an attractive aspect for THz radars.

Another parameter to be acquired by pulse radars is the target *velocity*. The principle behind the velocity acquisition is the Doppler effect, with which the readers should be familiar: When the wave source is moving toward the observer, the

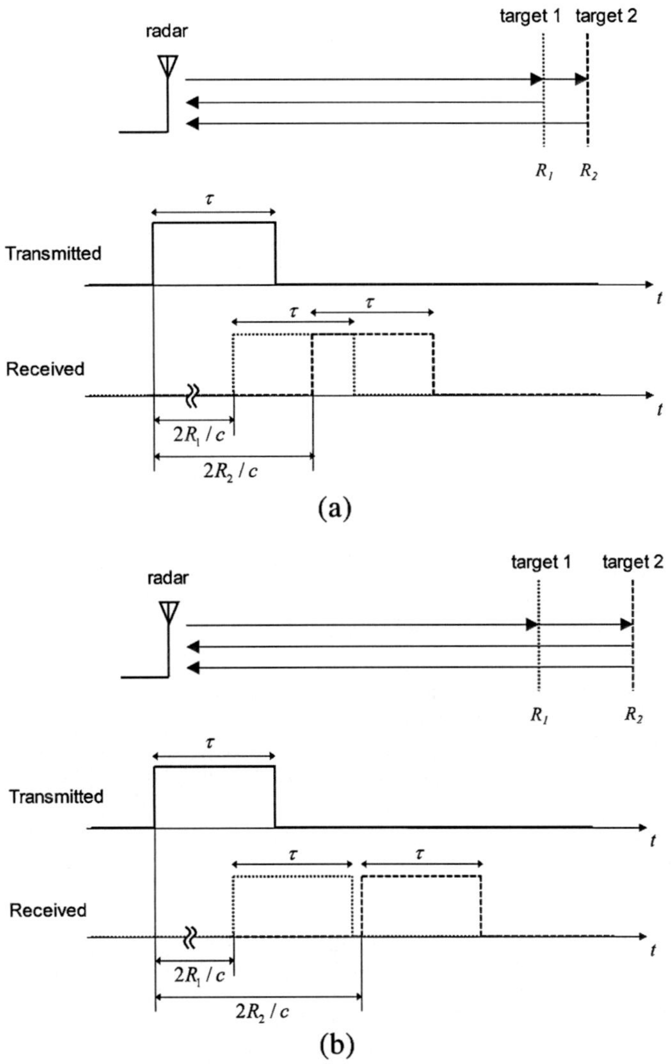

Fig. 6.41 Transmitted and received pulses of a pulse radar with two targets: (**a**) Small separation between two targets, (**b**) Large separation between two targets

frequency of the wave increases, while the frequency decreases when the wave source is moving away from the observer. It applies to the reflected waves as well, a relevant situation for radars: When the wave reflector (target) is moving toward or away from the wave source, the frequency increases or decreases, respectively. Hence, the main equations for the Doppler effect apply to reflected waves as well. From the radar perspective, when the target is moving toward the radar with a velocity of v (in a rigorous sense, it is speed, not velocity, as it is considered as a

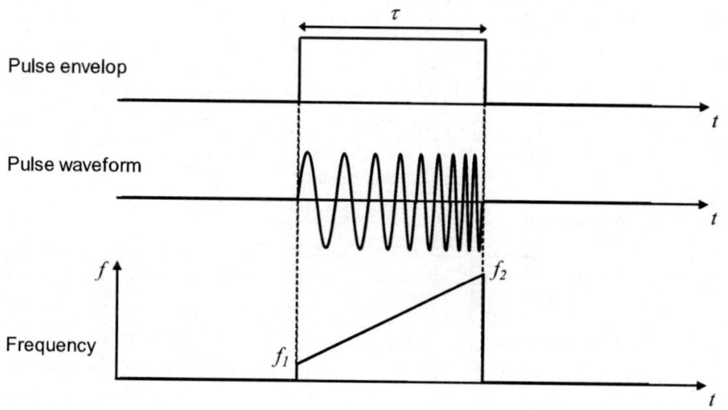

Fig. 6.42 Pulse envelop, pulse waveform, and the frequency profile of the linear frequency modulation (LFM)

scalar parameter in the range analysis; however, in the discussion throughout this subsection, we will maintain the term "velocity" to follow the convention), the new frequency f' is related to the original frequency f with:

$$f' = \frac{c+v}{c-v} f: \quad \text{target moving toward radar,} \tag{6.36}$$

where c is the speed of light. Likewise, when the target is moving away from the radar, the relation becomes:

$$f' = \frac{c-v}{c+v} f: \quad \text{target moving away from radar.} \tag{6.37}$$

Now let us express the relations in terms of the pulse width instead of the frequency. If the frequency of the CW wave inside a pulse increases (decreases), the pulse width will decrease (increase) by the same factor, when we assume that the number of wave cycles inside a pulse remains unchanged, which is certainly true. In other words, $f'/f = \tau/\tau'$, where τ' is the new pulse width. Hence, the following relations hold for pulse radars:

$$\tau' = \frac{c-v}{c+v} \tau: \quad \text{target moving toward radar,} \tag{6.38}$$

$$\tau' = \frac{c+v}{c-v} \tau: \quad \text{target moving away from radar.} \tag{6.39}$$

Therefore, by detecting the ratio of τ'/τ, the target velocity can be obtained. Here, it was assumed that the target is moving along the radial direction. When the target is moving in the direction off the radial direction by an angle θ, v in Eqs. (6.38) and (6.39) needs to be replaced by $v \cos \theta$, or the radial component of v.

From the implementation point of view, the challenges for pulse radars comes from the need for the fast and high-isolation switches, which are increasingly difficult for higher frequencies, including the THz band. The heavy load on the baseband circuit with the large bandwidth also leads to design challenges for the baseband design. The limited transmitter output power and receiver noise figure in high-frequency operation will reduce the maximum detectable range as mentioned above, which is in fact a general issue that affects not only radars but also any high-frequency transmitters and receivers.

Despite the challenges, there have been recent notable reports on pulse radars operating beyond 100 GHz. In [102], a 160-GHz pulse radar implemented in 65-nm Si CMOS technology was reported. The radar employs a 2×2 phased-array transceiver architecture to increase the maximum detectable range with improved directivity. The output signal was generated with a 40-GHz PLL, which was subsequently followed by phase shifters, $\times 4$ frequency multiplier, 160-GHz injection locked oscillator, and a PA. The generated signal was merged with a pulse signal inside the PA, leading to a pulse output. With a pulse width of 100 ps, a range resolution of 1.5 cm was expected with a bandwidth of 20 GHz, while the pulse repetition frequency of 100 MHz led to a range ambiguity of 1.5 m. With the measurement, the transmitter exhibited an EIRP of 18.8 dBm for continuous wave output, while the receiver showed a gain of 42.5 dB and noise figure of 22.5 dB with a 7-GHz bandwidth, with a total DC power dissipation of 2.2 W.

Based on a similar but slightly different approach than pulse radars are the *phase radars* that determine the target range by detecting the delay of the phase (instead of pulse edges) in CW waves. This is the basis for the 144-GHz Si CMOS phase radar reported in [103], which was used to acquire 3D images for raster-scanned targets with a depth resolution of 0.76 cm. To deal with the range ambiguity, which is in the order of the wavelength (much smaller than the case of regular pulse radars) and thus a more serious issue with higher frequencies for phase radars, a successive-approximation technique was employed in this work.

6.4.3 CW Radars

In CW radars, a sinusoidal *CW signal* is transmitted from the antenna and the signal reflected from the target is received. As for the antenna configuration of CW radars, the transmitter and receiver antennas are usually separated as they continuously transmit and receive signals, differently from pulse radars. Hence, it is technically a bistatic configuration as the two antennas are separate, but it is often considered monostatic especially when the two antennas are integrated on a single chip. Or, sometimes, it is considered quasi-monostatic since the distance between the two antennas are in the order of a wavelength [104]. Still, there are cases with FMCW radars employing a single antenna for both transmission and reception of the signal, in which case directional couplers (or circulators) or controlling antenna polarization are utilized to control the signals on two paths [105, 106].

For the basic *unmodulated* CW radars, which is based on a single-frequency monotonic signal, target velocity can be determined with a frequency shift from the Doppler effect, but the range information cannot be obtained because there is no apparent timing mark that provides a clue for the time delay. A good example of this type of radar is the radar gun for speed check, for which range information is not needed. While the function is limited, these unmodulated CW radars benefit from the low cost, as components for frequency chirping function are not required. For these radars, the velocity is determined by solving Eq. (6.36) for v:

$$v = c\frac{f' - f}{f' + f} \approx c\frac{f_D}{2f}, \tag{6.40}$$

where an approximation of $f' + f \approx 2f$ was made with $c \gg v$. f_D is the Doppler frequency shift, which is the difference between the transmitted and received frequencies. From circuit implementation point of view, f_D can be easily determined with a simple mixing operation between the transmitted and received signals. Hence, unmodulated CW radars such as radar guns can be implemented with a relatively simple system configuration.

In most radar applications, however, both range and velocity need to be determined. This can be achieved with *modulated* CW radars. The most popular modulation scheme is the linear frequency modulation (LFM), which is the basis for FMCW (Frequency-Modulated CW) radars. The remaining part of this subsection will focus on FMCW radars.

Let us first obtain the target *range* with FMCW radars. For simplicity, we will first consider a case with a stationary target. The frequency profile of the FMCW signal for this case is shown in Fig. 6.43a as a function of time. While the triangular profile is assumed in Fig. 6.43a, other profiles can also be adopted such as saw-tooth shape that consists of ramp-up cycles only. For the profile presented by Fig. 6.43a, the frequency of FMCW signal linearly ramps up and down with a rate of f_m, the modulation frequency, which is usually in the order of kHz or tens of kHz range. Beware that the profile shown in the figure is not for the signal magnitude but for the frequency of the signal, as indicated by the vertical axis representing the frequency. The received signal is a replica of the transmitted signal but with a finite time delay. In addition to the modulation frequency, there are a couple of other key parameters that define the range of the target. The beat frequency, f_b, is the frequency difference between the transmitted and received signals at a given time point. The peak frequency deviation, Δf, is the modulation range of the frequency, which determines the range resolution of FMCW radars. It is also often referred to as the bandwidth B, but the readers should not confuse this modulation bandwidth with the bandwidth in the pulse radars (defined as $1/\tau$) or the system RF bandwidth (they are all related, though, because a large system RF bandwidth is needed to support a large signal bandwidth). Now, from the slope of the frequency profile in Fig. 6.43a, it is clear that:

$$\frac{f_b}{\Delta t} = \frac{\Delta f}{\frac{1}{2f_m}}. \tag{6.41}$$

By substituting the relation $\Delta t = 2R/c$ into Eq. (6.41), the range can be expressed in terms of the three frequency parameters and the speed of light c, so that:

$$R = \frac{cf_b}{4f_m \Delta f}. \tag{6.42}$$

Now let us consider the range resolution of FMCW radars. For this, assume that two targets are located at ranges R_1 and R_2 ($R_2 > R_1$), which lead to beat frequencies of f_{b1} and f_{b2}, respectively, at the radar receiver. For R_1 and R_2 to be resolved in the spatial domain, f_{b1} and f_{b2} need to be resolved in the frequency domain. If the two echo signals from the two targets are arriving at the radar receiver simultaneously, it takes a certain time interval to resolve the two signal profiles centered at f_{b1} and f_{b2} in the frequency domain when they are obtained by the Fourier transform [107]. The rule of thumb is that a frequency resolution of $1/T$ can be obtained by an observation time interval of T. For FMCW radars, the maximum allowed observation time is the maximum linear stretch of the frequency modulation profile in the time domain, or the full ramp-up (or ramp-down) time, which corresponds to $1/2f_m$. This means that the best frequency resolution available in the FMCW modulation profile given in Fig. 6.43 is $2f_m$. Hence, for the two beating signals at f_{b1} and f_{b2} to be resolved, it follows that:

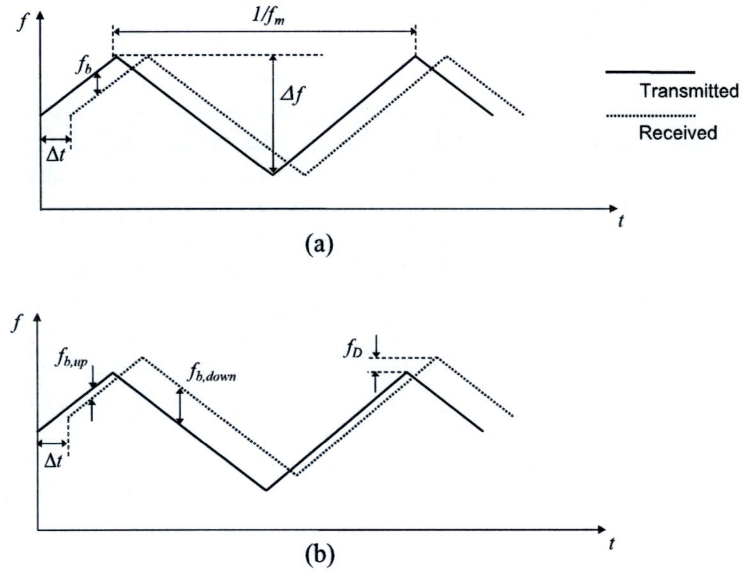

Fig. 6.43 Transmitted and received signal of a FMCW radar: (**a**) Stationary target, (**b**) Target is moving toward the radar

$$\Delta f_{\mathrm{b}} = f_{\mathrm{b2}} - f_{\mathrm{b1}} > 2f_{\mathrm{m}}. \tag{6.43}$$

With the relation between f_b and R given in Eq. (6.42), Eq. (6.43) becomes:

$$\Delta f_{\mathrm{b}} = \frac{4f_{\mathrm{m}}\Delta f}{c}(R_2 - R_1) = \frac{4f_{\mathrm{m}}\Delta f}{c}\Delta R > 2f_{\mathrm{m}}, \tag{6.44}$$

which leads to the condition for R_1 and R_2 to be resolved:

$$R_2 - R_1 = \Delta R > \frac{c}{2\Delta f}. \tag{6.45}$$

Hence, the desired range resolution ΔR is now obtained as:

$$\Delta R = \frac{c}{2\Delta f} = \frac{c}{2B}. \tag{6.46}$$

The readers will notice that this expression is identical to the range resolution expression for pulse radars as shown in Eq. (6.35). In fact, this expression is valid for any radar signals regardless the modulation type [108], although the exact definition of B varies over modulation and radar types (for instance, $B = \Delta f$ for FMCW, $B = 1/\tau$ for pulse radars, as mentioned above). Equation (6.46) clearly indicates that a high carrier frequency is desired for enhanced resolution, since a large Δf can be more conveniently supported by a higher carrier frequency with a wide RF bandwidth available. Note that higher carrier frequency will be favored for cross-range (normal to the radial direction) resolution as well because it is limited by the diffraction, which reduces with increasing frequency. This frequency dependence of the resolutions that favor the high frequencies is the major opportunity for THz radars. While Δf is typically in the order of hundreds of MHz for automobile radars operating near 77 GHz, it can be extended to be as large as tens of GHz for radars working in the THz band as we will see shortly. In fact, for sub-cm range resolution, Δf larger than 15 GHz is needed according to Eq. (6.46), which can be reasonably accomplished with operation frequencies above 100 GHz.

Now, let us consider a more general case, where the target is moving toward the radar with a velocity v. In this case, the frequency profile of the received signal is shifted up in the frequency axis due to the Doppler effect, in addition to the shift in the time frame due to the time delay, which is depicted in Fig. 6.43b. In this case, the beat frequency is split into two cases, $f_{\mathrm{b,up}}$ and $f_{\mathrm{b,down}}$, which correspond to the beat frequency for ramp-up and -down cycles, respectively. For the ramp-up cycle, the beat frequency is reduced by the Doppler frequency f_{D}, while the beat frequency is increased by f_{D} during the ramp-down cycle, with reference to the case with stationary targets. One can easily show that the expression for the range in Eq. (6.42) is modified as:

$$R = \frac{c}{4f_{\mathrm{m}}\Delta f}\frac{f_{\mathrm{b,down}} + f_{\mathrm{b,up}}}{2} = \frac{c}{8f_{\mathrm{m}}\Delta f}\left(f_{\mathrm{b,down}} + f_{\mathrm{b,up}}\right), \tag{6.47}$$

which indicates that the averaged beat frequency, or $(f_{\mathrm{b,down}} + f_{\mathrm{b,up}})/2$, replaces f_{b} of the stationary target. What about the target velocity v? As mentioned above, the two beat frequencies are shifted by f_{D} in the opposite directions due to the Doppler effect, so that:

$$f_{\mathrm{b,down}} - f_{\mathrm{b,up}} = 2f_{\mathrm{D}}. \tag{6.48}$$

By combining Eqs. (6.48) and (6.40), we obtain:

$$v = \frac{c}{4f}\left(f_{\mathrm{b,down}} - f_{\mathrm{b,up}}\right) = \frac{\lambda}{4}\left(f_{\mathrm{b,down}} - f_{\mathrm{b,up}}\right), \tag{6.49}$$

where f and λ are the frequency and the wavelength of the transmitted signal. From a rigorous point of view, f and λ are not constants and periodically change due to the modulation, but the values at the center frequency can be taken in this case. It is noted again that v obtained with Eq. (6.49) is the radial component of the velocity, or the radial velocity. As mentioned earlier, frequency difference can be readily obtained by a mixer, and thus $f_{\mathrm{b,down}}$ and $f_{\mathrm{b,up}}$ can be extracted by mixing the transmitted and received signals during the ramp-up and -down cycles, respectively, of the frequency chirping. Equation (6.49) is valid for targets moving away from the radars as well, in which case v will assume negative values.

A typical implementation example of FMCW is shown in Fig. 6.44. The linearly modulated signal, which can be achieved by a PLL that employs a direct digital synthesizer (DDS) as a varied-frequency reference signal, is amplified by a PA and transmitted by an antenna. There are other ways to generate modulated signals, such as a fractional-N PLL in which the divide modulus of the frequency divider is swept [109], a fixed-frequency PLL whose output is mixed with a varied-frequency VCO [110], or a direct frequency multiplication of a DDS output [108]. On the receiver

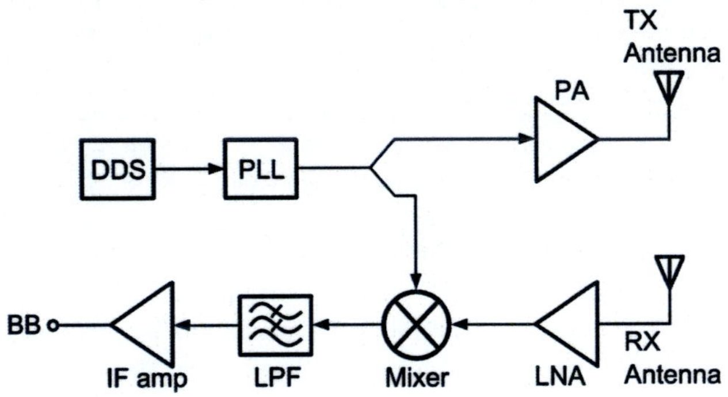

Fig. 6.44 Schematic of a typical FMCW radar

channel shown in Fig. 6.44, the received signal from the antenna enters a mixer, optionally after an LNA, and then mixed with a portion of the modulated signal that is taken from the transmitter channel by a power divider or coupler. Optionally, a quadrature topology based on I-Q mixers can be adopted for the mixing, which will provide additional phase information of the echo signal. It will also help to reduce noise at the receiver as the double-side band (DSB) noise will be effective instead of the single-side band (SSB) noise, leading to an NF reduction up to 3 dB [111]. The IF signal generated by the mixer is in fact the beat signal between the transmitted and received signals and is used to obtain the range and velocity of the target based on the relations provided above. The mixer output will also include the up-converted signal, which will be filtered out with a low-pass filter (LPF) that is placed next to the mixer. The beat signal amplified by an IF amplifier is accepted by the baseband block for various signal processing steps to acquire the desired information and present the data in the proper formats. Note that the structure of unmodulated CW radars will be similar, except for the absence of the modulating components as well as the less heavy BB block.

As mentioned earlier, FMCW radars benefit from the high operation frequency for improved range resolution, for which THz radars are an attractive option. Yet, it is still challenging to have a large voltage tuning at the THz band, which may partially limit the efforts to attain a wide modulation bandwidth (Δf). At the same time, the components in the system, including the amplifiers, mixers, as well as on-chip antennas, if employed, should have RF bandwidths large enough to cover the modulation bandwidth. Another issue to be considered for FMCW radars, especially when monolithically implemented on a single chip, is the signal leakage from the transmitter to the receiver channels. As the leaked signal will have the same frequency with the LO frequency for the receiver, it will cause a DC component in the IF signal, which needs to be filtered out with a DC block capacitance. The linearity in the frequency modulation is also important because any nonlinearity will cause a spread in the demodulated frequency, leading to degradation in the range resolution. One relief is that the nonlinearity can be compensated in the baseband signal-processing steps.

Recently, a growing number of THz radars beyond 100 GHz are being reported, majority of them being FMCW radars [105, 106, 108, 110, 112–114]. One recent example is briefly introduced below, which reports a 220-GHz FMCW radar based on a 55-nm SiGe BiCMOS technology [113]. The basic block diagram of the radar and the chip photo are shown in Fig. 6.45a, b, respectively. For modulated ramp signal generation at the desired frequency band, a 220-GHz push-push VCO with a wide tuning range was employed, which was driven by a ramp generator that sweeps the tuning voltage of the VCO with a rate $f_{\mathrm{m}} = 20$ kHz. The generated signal is transmitted by an on-chip slot antenna. For the receiver chain, the received signal detected by an on-chip folded-dipole antenna is mixed with an LO signal coupled from the VCO in the transmitter chain, then filtered and amplified. With a modulation bandwidth of 62.4 GHz around the center frequency of 221 GHz, the radar exhibited a range resolution of 2.7 mm, close to the theoretical value of 2.4 mm as expected by Eq. (6.46). The lateral cross-range resolution was 2 mm. The transmitter

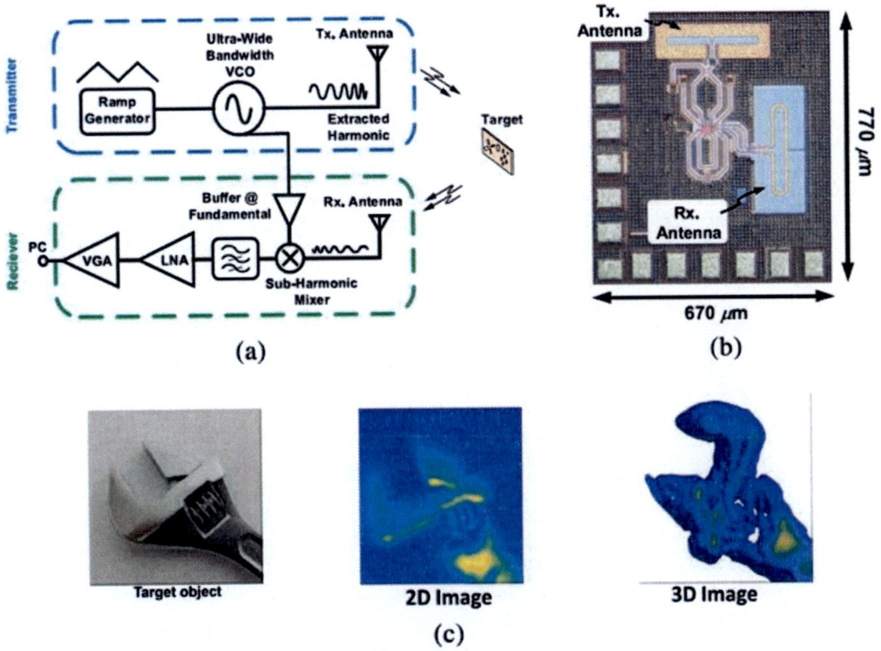

Fig. 6.45 A 220-GHz SiGe BiCMOS FMCW imaging radar: (**a**) Block diagram, (**b**) Photo of the fabricated chip, (**c**) Example images acquired by the radar. (© 2018 IEEE [113])

exhibited an output power of −4.6 dBm and EIRP of 14 dBm with the help of a Si lens. Total DC power dissipation was 87 mW. With the fabricated radar, 3D images were successfully acquired based on the inverse synthetic aperture radar (ISAR) technique, one example shown in Fig. 6.45c.

References

1. E. Bründermann, H.-W. Hübers, M.F. Kimmitt, *Terahertz Techniques* (Springer, Berlin, 2012)
2. R. Wood, The echelette grating for the infra-red. Philos. Mag. J. Sci. **20**(118), 770–778 (1910)
3. P. Connes, Early history of Fourier transform spectroscopy. Infrared Phys. **24**(2–3), 69–93 (1984)
4. D. Auston, K. Cheung, P. Smith, Picosecond photoconducting Hertzian dipoles. Appl. Phys. Lett. **45**(3), 284–286 (1984)
5. C. Fattinger, D. Grischkowsky, Point source terahertz optics. Appl. Phys. Lett. **53**(16), 1480–1482 (1988)
6. J. Neu, C.A. Schmuttenmaer, Tutorial: an introduction to terahertz time domain spectroscopy (THz-TDS). J. Appl. Phys. **124**(23), 231101 (2018)
7. P. Han, M. Tani, M. Usami, S. Kono, R. Kersting, X.-C. Zhang, A direct comparison between terahertz time-domain spectroscopy and far-infrared Fourier transform spectroscopy. J. Appl. Phys. **89**(4), 2357–2359 (2001)

8. D. Grischkowsky, S. Keiding, M. Van Exter, C. Fattinger, Far-infrared time-domain spectroscopy with terahertz beams of dielectrics and semiconductors. J. Opt. Soc. Am. B. **7**(10), 2006–2015 (1990)
9. M. Van Exter, D. Grischkowsky, Carrier dynamics of electrons and holes in moderately doped silicon. Phys. Rev. B **41**(17), 12140 (1990)
10. S. Howells, L. Schlie, Transient terahertz reflection spectroscopy of undoped InSb from 0.1 to 1.1 THz. Appl. Phys. Lett. **69**(4), 550–552 (1996)
11. J. Chen, Y. Chen, H. Zhao, G.J. Bastiaans, X.-C. Zhang, Absorption coefficients of selected explosives and related compounds in the range of 0.1–2.8 THz. Opt. Express **15**(19), 12060–12067 (2007)
12. A. Tang, B. Drouin, Y. Kim, G. Virbila, M.F. Chang, 95–105 GHz 352 mW all-silicon cavity-coupled pulsed echo rotational spectroscopy system in 65 nm CMOS. IEEE Trans. Terahertz Sci. Technol. **7**(3), 244–249 (2017). https://doi.org/10.1109/TTHZ.2017.2692041
13. T.M. Korter, D.F. Plusquellic, Continuous-wave terahertz spectroscopy of biotin: vibrational anharmonicity in the far-infrared. Chem. Phys. Lett. **385**(1–2), 45–51 (2004)
14. N. Kim et al., Tunable continuous-wave terahertz generation/detection with compact 1.55 μm detuned dual-mode laser diode and InGaAs based photomixer. Opt. Express **19**(16), 15397–15403 (2011)
15. N. Karpowicz, H. Zhong, J. Xu, K.-I. Lin, J.-S. Hwang, X.C. Zhang, Comparison between pulsed terahertz time-domain imaging and continuous wave terahertz imaging. Semicond. Sci. Technol. **20**(7), S293–S299 (2005). https://doi.org/10.1088/0268-1242/20/7/021
16. A.I. Meshkov, F.C. De Lucia, Broadband absolute absorption measurements of atmospheric continua with millimeter wave cavity ringdown spectroscopy. Rev. Sci. Instru. **76**(8), 083103 (2005)
17. K. Sakai (ed.), *Terahertz Optoelectronics* (Springer, Berlin, 2004)
18. C. Wang, B. Perkins, Z. Wang, R. Han, Molecular detection for unconcentrated gas with ppm sensitivity using 220-to-320-GHz dual-frequency-comb spectrometer in CMOS. IEEE Trans. Biomed. Circuits Syst. **12**(3), 709–721 (2018)
19. M. Phillips, J. Herrera, S. Krishnan, M. Zain, J. Greenberg, R.N. Cataneo, Variation in volatile organic compounds in the breath of normal humans. J. Chromatogr. B. Biomed. Sci. Appl. **729** (1), 75–88 (1999). https://doi.org/10.1016/S0378-4347(99)00127-9
20. B. Buszewski, M. Kęsy, T. Ligor, A. Amann, Human exhaled air analytics: biomarkers of diseases. Biomed. Chromatogr. **21**(6), 553–566 (2007)
21. Q. Zhong, W. Choi, C. Miller, R. Henderson, K.K. O, A 210-to-305GHz CMOS receiver for rotational spectroscopy, in *2016 IEEE International Solid-State Circuits Conference (ISSCC)*, (IEEE, 2016), pp. 426–427
22. K. Schmalz, N. Rothbart, P.F.-X. Neumaier, J. Borngräber, H.-W. Hübers, D. Kissinger, Gas spectroscopy system for breath analysis at mm-wave/THz using SiGe BiCMOS circuits. IEEE Trans. Microw. Theory Tech. **65**(5), 1807–1818 (2017)
23. U.R. Pfeiffer et al., A CMOS focal-plane array for terahertz imaging, in *International Conerence on Infrared, Millimeter and Terahertz Waves*, (2008), pp. 1–3
24. E. Ojefors, U.R. Pfeiffer, A. Lisauskas, H.G. Roskos, A 0.65 THz focal-plane array in a quarter-micron CMOS process technology. IEEE J. Solid-State Circuits **44**(7), 1968–1976 (2009)
25. D. Yoon, J. Kim, J. Yun, M. Kaynak, B. Tillack, J. Rieh, 300-GHz direct and heterodyne active imagers based on 0.13-μm SiGe HBT technology. IEEE Trans. Terahertz Sci. Technol. **7**(5), 536–545 (2017). https://doi.org/10.1109/TTHZ.2017.2715419
26. V. Wallace et al., Terahertz pulsed imaging of basal cell carcinoma ex vivo and in vivo. Br. J. Dermatol. **151**(2), 424–432 (2004)
27. Z. Taylor et al., Reflective terahertz imaging of porcine skin burns. Opt. Lett. **33**(11), 1258–1260 (2008)

28. J. Yun et al., Terahertz reflection-mode biological imaging based on InP HBT source and detector. IEEE Trans. Terahertz Sci. Technol. **7**(3), 274–283 (2017). https://doi.org/10.1109/TTHZ.2017.2673549
29. J. Yun, D. Yoon, H. Kim, J.-S. Rieh, 300-GHz InP HBT oscillators based on common-base cross-coupled topology. IEEE Trans. Microw. Theory Tech. **62**(12), 3053–3064 (2014). https://doi.org/10.1109/TMTT.2014.2364608
30. N. Skou, D. Le Vine, *Microwave Radiometer Systems: Design and Analysis*, 2nd edn. (Artech House, 2006)
31. M. Tiuri, Radio astronomy receivers. IEEE Trans. Mil. Electron. **8**(3), 264–272 (1964)
32. J.W. May, G.M. Rebeiz, Design and characterization of W-band SiGe RFICs for passive millimeter-wave imaging. IEEE Trans. Microw. Theory Tech. **58**(5), 1420–1430 (2010)
33. J.W. Kooi et al., Submillimeter InP MMIC low-noise amplifier gain stability characterization. IEEE Trans. Terahertz Sci. Technol. **7**(3), 335–346 (2017)
34. V. Ossenkopf, The stability of spectroscopic instruments: a unified Allan variance computation scheme. Astron. Astrophys. **479**(3), 915–926 (2008)
35. R.H. Dicke, The measurement of thermal radiation at microwave frequencies. Rev. Sci. Instru. **17**(7), 268–275 (1946)
36. A. Tomkins, P. Garcia, S.P. Voinigescu, A passive W-band imaging receiver in 65-nm bulk CMOS. IEEE J. Solid-State Circuits **45**(10), 1981–1991 (2010)
37. L. Zhou, C.-C. Wang, Z. Chen, P. Heydari, A W-band CMOS receiver chipset for millimeter-wave radiometer systems. IEEE J. Solid-State Circuits **46**(2), 378–391 (2011)
38. L. Gilreath, V. Jain, P. Heydari, Design and analysis of a W-band SiGe direct-detection-based passive imaging receiver. IEEE J. Solid-State Circuits **46**(10), 2240–2252 (2011)
39. Q.J. Gu, Z. Xu, A. Tang, M.-C.F. Chang, A D-band passive imager in 65 nm CMOS. IEEE Microw. Wirel. Compon. Lett. **22**(5), 263–265 (2012)
40. Z. Chen, C.-C. Wang, H.-C. Yao, P. Heydari, A BiCMOS W-band 2×2 focal-plane array with on-chip antenna. IEEE J. Solid-State Circuits **47**(10), 2355–2371 (2012)
41. E. Dacquay et al., D-band total power radiometer performance optimization in an SiGe HBT technology. IEEE Trans. Microw. Theory Tech. **60**(3), 813–826 (2012)
42. T. Kanar, G.M. Rebeiz, A low-power 136-GHz SiGe total power radiometer with NETD of 0.25 K. IEEE Trans. Microw. Theory Tech. **64**(3), 906–914 (2016)
43. A. Tang, Y. Kim, Q.J. Gu, 25.4 A 0.43 K-noise-equivalent-ΔT 100GHz dicke-free radiometer with 100% time efficiency in 65nm CMOS, in *2016 IEEE International Solid-State Circuits Conference (ISSCC)*, (IEEE, 2016), pp. 430–431
44. A. Tang et al., A W-band 65nm CMOS/InP-hybrid radiometer & passive imager, in *2016 IEEE MTT-S International Microwave Symposium (IMS)*, (IEEE, 2016), pp. 1–3
45. P. Hillger et al., A 128-pixel 0.56 THz sensing array for real-time near-field imaging in 0.13 µm SiGe BiCMOS, in *2018 IEEE International Solid-State Circuits Conference-(ISSCC)*, (IEEE, 2018), pp. 418–420
46. J.F. Federici et al., THz imaging and sensing for security applications—explosives, weapons and drugs. Semicond. Sci. Technol. **20**(7), S266 (2005)
47. D.L. Donoho, Compressed sensing. IEEE Trans. Inform. Theory **52**(4), 1289–1306 (2006). https://doi.org/10.1109/TIT.2006.871582
48. H. Sherry et al., A 1kpixel CMOS camera chip for 25fps real-time terahertz imaging applications, in *2012 IEEE International Solid-State Circuits Conference*, (IEEE, 2012), pp. 252–254
49. R. Al Hadi et al., A 1 k-pixel video camera for 0.7–1.1 terahertz imaging applications in 65-nm CMOS. IEEE J. Solid-State Circuits **47**(12), 2999–3012 (2012)
50. K. Song, Development of THz Imaging Arrays Based on Si CMOS Technology, Ph.D. Dissertation, Korea University, 2020
51. J.P. Guillet et al., Review of terahertz tomography techniques. J. Infrared Millim. Terahertz Waves **35**(4), 382–411 (2014)
52. B. Ferguson, S. Wang, D. Gray, D. Abbot, X.-C. Zhang, T-ray computed tomography. Opt. Lett. **27**(15), 1312–1314 (2002)

53. S. Wang, B. Ferguson, D. Abbott, X.-C. Zhang, T-ray imaging and tomography. J. Biol. Phys. **29**(2–3), 247–256 (2003)
54. R. Gordon, A tutorial on ART (algebraic reconstruction techniques). IEEE Trans. Nucl. Sci. **21** (3), 78–93 (1974)
55. L.A. Shepp, Y. Vardi, Maximum likelihood reconstruction for emission tomography. IEEE Trans. Med. Imaging **1**(2), 113–122 (1982)
56. B. Recur et al., Investigation on reconstruction methods applied to 3D terahertz computed tomography. Opt. Express **19**(6), 5105–5117 (2011)
57. B. Ferguson, S. Wang, D. Gray, D. Abbott, X. Zhang, T-ray diffraction tomography, in *The Thirteenth International Conference on Ultrafast Phenomena*, vol. 72, (Citeseer, 2002), pp. 450–451
58. J. Takayanagi et al., High-resolution time-of-flight terahertz tomography using a femtosecond fiber laser. Opt. Express **17**(9), 7533–7539 (2009)
59. U. Pfeiffer, Sub-millimeter wave active imaging with silicon integrated circuits, in *2011 International Conference on Infrared, Millimeter, and Terahertz Waves*, (IEEE, 2011), pp. 1–4
60. P. Hillger, A. Schlüter, R. Jain, S. Malz, J. Grzyb, U. Pfeiffer, Low-cost 0.5 THz computed tomography based on silicon components, in *2017 42nd International Conference on Infrared, Millimeter, and Terahertz Waves (IRMMW-THz)*, (IEEE, 2017), pp. 1–2
61. D. Yoon et al., 3-D THz tomography with an InP HBT signal source and a SiGe HBT imaging receiver operating near 300 GHz, in *40th International Conference on Infrared, Millimeter, and Terahertz Waves (IRMMW-THz)*, (2015), pp. 1–2. https://doi.org/10.1109/IRMMW-THz.2015.7327606
62. J. Kim et al., Three-dimensional terahertz tomography with transistor-based signal source and detector circuits operating near 300 GHz. IEEE Trans. Terahertz Sci. Technol. **8**(5), 482–491 (2018). https://doi.org/10.1109/TTHZ.2018.2851542
63. C.E. Shannon, Communication in the presence of noise. Proc. IRE **37**(1), 10–21 (1949)
64. T. Kleine-Ostmann, T. Nagatsuma, A review on terahertz communications research. J. Infrared Millim. Terahertz Waves **32**(2), 143–171 (2011)
65. H.-J. Song, T. Nagatsuma, Present and future of terahertz communications. IEEE Trans. Terahertz Sci. Technol. **1**(1), 256–263 (2011)
66. S. Koenig et al., Wireless sub-THz communication system with high data rate. Nat. Photonics **7**(12), 977 (2013)
67. ITU, *RECOMMENDATION ITU-R P.676–9—Attenuation by atmospheric gases*, 2012
68. J.-D. Park, S. Kang, S.V. Thyagarajan, E. Alon, A.M. Niknejad, A 260 GHz fully integrated CMOS transceiver for wireless chip-to-chip communication, in *2012 Symposium on VLSI Circuits (VLSIC)*, (IEEE, 2012), pp. 48–49
69. H.-J. Song et al., Demonstration of 20-Gbps wireless data transmission at 300 GHz for KIOSK instant data downloading applications with InP MMICs, in *2016 IEEE MTT-S International Microwave Symposium (IMS)*, (IEEE, 2016), pp. 1–4
70. I. Kallfass et al., 64 Gbit/s transmission over 850 m fixed wireless link at 240 GHz carrier frequency. J. Infrared Millim. Terahertz Waves **36**(2), 221–233 (2015)
71. R. Piesiewicz, C. Jansen, D. Mittleman, T. Kleine-Ostmann, M. Koch, T. Kurner, Scattering analysis for the modeling of THz communication systems. IEEE Trans. Antennas Propag. **55** (11), 3002–3009 (2007)
72. C. Jansen, R. Piesiewicz, D. Mittleman, T. Kurner, M. Koch, The impact of reflections from stratified building materials on the wave propagation in future indoor terahertz communication systems. IEEE Trans. Antennas Propag. **56**(5), 1413–1419 (2008)
73. C. Han, A.O. Bicen, I.F. Akyildiz, Multi-ray channel modeling and wideband characterization for wireless communications in the terahertz band. IEEE Trans. Wirel. Commun. **14**(5), 2402–2412 (2014)

74. IEEE, IEEE standard for high data rate wireless multi-media networks—Amendment 2: 100 Gb/s wireless switched point-to-point physical layer, in *IEEE Std 802.15.3d-2017*, (2017), pp. 1–55. https://doi.org/10.1109/IEEESTD.2017.8066476
75. J.G. Proakis, M. Salehi, *Digital Communications*, 5th edn. (McGraw-Hill, New York, 2008)
76. C.W. Byeon, C.S. Park, A high-efficiency 60-GHz CMOS transmitter for short-range wireless communications. IEEE Microw. Wirel. Compon. Lett. **27**(8), 751–753 (2017)
77. S. Kim et al., 300 GHz integrated heterodyne receiver and transmitter with on-chip fundamental local oscillator and mixers. IEEE Trans. Terahertz Sci. Technol. **5**(1), 92–101 (2015). https://doi.org/10.1109/TTHZ.2014.2364454
78. I. Kallfass et al., All active MMIC-based wireless communication at 220 GHz. IEEE Trans. Terahertz Sci. Technol. **1**(2), 477–487 (2011)
79. M. Abbasi et al., Single-chip 220-GHz active heterodyne receiver and transmitter MMICs with on-chip integrated antenna. IEEE Trans. Microw. Theory Tech. **59**(2), 466–478 (2010)
80. H.-J. Song, J.-Y. Kim, K. Ajito, M. Yaita, N. Kukutsu, Fully integrated ASK receiver MMIC for terahertz communications at 300 GHz. IEEE Trans. Terahertz Sci. Technol. **3**(4), 445–452 (2013)
81. H.J. Park, J.-S. Rieh, M. Kim, J.B. Hacker, 300 GHz six-stage differential-mode amplifier, in *IEEE MTT-S International Microwave Symposium Digest*, (2010), pp. 49–52
82. Z. Wang, P.-Y. Chiang, P. Nazari, C.-C. Wang, Z. Chen, P. Heydari, A CMOS 210-GHz fundamental transceiver with OOK modulation. IEEE J. Solid-State Circuits **49**(3), 564–580 (2014)
83. A. Tang, M.-C.F. Chang, A 294 GHz 0.47 mW caterpillar amplifier based transmitter in 65nm CMOS for THz data-links, in *2013 IEEE 13th Topical Meeting on Silicon Monolithic Integrated Circuits in RF Systems*, (IEEE, 2013), pp. 27–29
84. P. Rodríguez-Vázquez, J. Grzyb, B. Heinemann, U.R. Pfeiffer, A 16-QAM 100-Gb/s 1-M wireless link with an EVM of 17% at 230 GHz in an SiGe technology. IEEE Microw. Wirel. Compon. Lett. **29**(4), 297–299 (2019)
85. S. Lee et al., An 80Gb/s 300GHz-band single-chip CMOS transceiver, in *2019 IEEE International Solid-State Circuits Conference—(ISSCC)*, (2019), pp. 170–172. https://doi.org/10.1109/ISSCC.2019.8662314
86. J.-Y. Kim, H.-J. Song, K. Ajito, M. Yaita, N. Kukutsu, A 325 GHz quadrature voltage controlled oscillator with superharmonic-coupling. IEEE Microw. Wirel. Compon. Lett. **23** (8), 430–432 (2013)
87. K. Song, J. Kim, H. Son, J. Yoo, M. Cho, J. Rieh, 300-GHz InP HBT quadrature VCO with integrated mixer. IEEE Trans. Terahertz Sci. Technol. **10**(3), 2020. https://doi.org/10.1109/TTHZ.2020.2972539
88. K. Katayama et al., A 300GHz 40nm CMOS transmitter with 32-QAM 17.5 Gb/s/ch capability over 6 channels, in *2016 IEEE International Solid-State Circuits Conference (ISSCC)*, (IEEE, 2016), pp. 342–343
89. K. Takano et al., A 105Gb/s 300GHz CMOS transmitter, in *2017 IEEE International Solid-State Circuits Conference (ISSCC)*, (IEEE, 2017), pp. 308–309
90. S. Lee et al., An 80-Gb/s 300-GHz-band single-chip CMOS transceiver. IEEE J. Solid-State Circuits **54**(12), 3577–3588 (2019)
91. H.-J. Song, J.-Y. Kim, K. Ajito, N. Kukutsu, M. Yaita, 50-Gb/s direct conversion QPSK modulator and demodulator MMICs for terahertz communications at 300 GHz. IEEE Trans. Microw. Theory Tech. **62**(3), 600–609 (2014)
92. P. Rodríguez-Vázquez, J. Grzyb, B. Heinemann, U.R. Pfeiffer, Performance evaluation of a 32-QAM 1-meter wireless link operating at 220–260 GHz with a data-rate of 90 Gbps, in *2018 Asia-Pacific Microwave Conference (APMC)*, (IEEE, 2018), pp. 723–725
93. Q. Zhong, Z. Chen, N. Sharma, S. Kshattry, W. Choi, K. Kenneth, 300-GHz CMOS QPSK transmitter for 30-Gbps dielectric waveguide communication, in *2018 IEEE Custom Integrated Circuits Conference (CICC)*, (IEEE, 2018), pp. 1–4

94. S. Kang, S.V. Thyagarajan, A.M. Niknejad, A 240 GHz fully integrated wideband QPSK transmitter in 65 nm CMOS. IEEE J. Solid-State Circuits **50**(10), 2256–2267 (2015)
95. S.V. Thyagarajan, S. Kang, A.M. Niknejad, A 240 GHz fully integrated wideband QPSK receiver in 65 nm CMOS. IEEE J. Solid-State Circuits **50**(10), 2268–2280 (2015)
96. D. Fritsche, P. Stärke, C. Carta, F. Ellinger, A low-power SiGe BiCMOS 190-GHz transceiver chipset with demonstrated data rates up to 50 Gbit/s using on-chip antennas. IEEE Trans. Microw. Theory Tech. **65**(9), 3312–3323 (2017)
97. G. Ducournau et al., 32 Gbit/s QPSK transmission at 385 GHz using coherent fibre-optic technologies and THz double heterodyne detection. Electron. Lett. **51**(12), 915–917 (2015)
98. X. Yu et al., 60 Gbit/s 400 GHz wireless transmission, in *2015 International Conference on Photonics in Switching*, (2015), pp. 4–6. https://doi.org/10.1109/PS.2015.7328934
99. T. Nagatsuma, G. Ducournau, C.C. Renaud, Advances in terahertz communications accelerated by photonics. Nat. Photonics **10**(6), 371 (2016)
100. B.R. Mahafza, *Radar Systems Analysis and Design Using MATLAB* (Chapman and Hall/CRC, New York, 2016)
101. M.C. Budge Jr., S.R. German, *Basic Radar Analysis* (Artech House, 2015)
102. B.P. Ginsburg, S.M. Ramaswamy, V. Rentala, E. Seok, S. Sankaran, B. Haroun, A 160 GHz pulsed radar transceiver in 65 nm CMOS. IEEE J. Solid-State Circuits **49**(4), 984–995 (2014). https://doi.org/10.1109/JSSC.2014.2298033
103. A. Tang et al., A 144GHz 0.76 cm-resolution sub-carrier SAR phase radar for 3D imaging in 65nm CMOS, in *2012 IEEE International Solid-State Circuits Conference*, (IEEE, 2012), pp. 264–266
104. M. Jahn, A. Stelzer, A 120 GHz FMCW radar frontend demonstrator based on a SiGe chipset. Int. J. Microw. Wirel. Technol. **4**(3), 309–315 (2012)
105. S. Thomas, C. Bredendiek, N. Pohl, A SiGe-based 240-GHz FMCW radar system for high-resolution measurements. IEEE Trans. Microw. Theory Tech. **67**(11), 4599–4609 (2019)
106. J. Grzyb, K. Statnikov, N. Sarmah, B. Heinemann, U.R. Pfeiffer, A 210–270-GHz circularly polarized FMCW radar with a single-lens-coupled SiGe HBT chip. IEEE Trans. Terahertz Sci. Technol. **6**(6), 771–783 (2016)
107. S. Rao, Introduction to mmWave sensing: FMCW radars, in *Texas Instruments (TI) mmWave Training Series*, (2017)
108. T. Bryllert, V. Drakinskiy, K.B. Cooper, J. Stake, Integrated 200–240-GHz FMCW radar transceiver module. IEEE Trans. Microw. Theory Tech. **61**(10), 3808–3815 (2013)
109. J. Lee, Y.-A. Li, M.-H. Hung, S.-J. Huang, A fully-integrated 77-GHz FMCW radar transceiver in 65-nm CMOS technology. IEEE J. Solid-State Circuits **45**(12), 2746–2756 (2010)
110. M. Hitzler et al., Ultracompact 160-GHz FMCW radar MMIC with fully integrated offset synthesizer. IEEE Trans. Microw. Theory Tech. **65**(5), 1682–1691 (2017)
111. K. Ramasubramanian, T. Instruments, Using a complex-baseband architecture in FMCW radar systems. *Texas Instruments*, 2017
112. H.J. Ng, M. Kucharski, W. Ahmad, D. Kissinger, Multi-purpose fully differential 61-and 122-GHz radar transceivers for scalable MIMO sensor platforms. IEEE J. Solid-State Circuits **52**(9), 2242–2255 (2017)
113. A. Mostajeran et al., A high-resolution 220-GHz ultra-wideband fully integrated ISAR imaging system. IEEE Trans. Microw. Theory Tech. **67**(1), 429–442 (2018)
114. A. Mostajeran, A. Cathelin, E. Afshari, A 170-GHz fully integrated single-chip FMCW imaging radar with 3-D imaging capability. IEEE J. Solid-State Circuits **52**(10), 2721–2734 (2017)

Index

© Springer Nature Switzerland AG 2021
J.-S. Rieh, *Introduction to Terahertz Electronics*,
https://doi.org/10.1007/978-3-030-51842-4

Q

R

S

T